U.S. Marine Corps Wilderness Medicine Survival Course

Plus

Marine Pistol Marksmanship

Department of Defense

Pentagon
Publishing

UNITED STATES MARINE CORPS
Mountain Warfare Training Center
Bridgeport, California 93517-5001

WILDERNESS MEDICINE COURSE STUDENT HANDBOOK

TABLE OF CONTENTS

TOC

TABLE OF CONTENTS (CONT.)

UNITED STATES MARINE CORPS
Mountain Warfare Training Center
Bridgeport, California 93517-5001

FMST.07.18
04/02/02

STUDENT HANDOUT

MOUNTAIN SAFETY (SUMMER)

TERMINAL LEARNING OBJECTIVE: Given a unit in a wilderness environment and necessary equipment and supplies, apply the principles of mountain safety to prevent death or injury per the reference. (FMST.07.18)

ENABLING LEARNING OBJECTIVE: Without the aid of references and given the acronym "BE SAFE MARINE", list in writing the 12 principles of mountain safety, in accordance with the references. (FMST.07.18a)

OUTLINE

1. **PLANNING AND PREPARATION:** (FMST.07.18a) As in any military operation, planning and preparation constitute the keys to success. The following principles will help the leader conduct a safe and efficient operation in any type of mountainous environment. We find this principle in the acronym "BE SAFE MARINE". Remember the key, think about what each letter means and apply this in any type of environment.

B - Be aware of the group's ability.

E - Evaluate terrain and weather constantly.

S - Stay as a group.

A - Appreciate time requirements.

F - Find shelter before storms if required.

E - Eat plenty and drink lots of liquids.

M - Maintain proper clothing and equipment.

1

A - Ask locals about conditions.

R - Remember to keep calm and think.

I - Insist on emergency rations and kits.

N - Never forget accident procedures.

E - Energy is saved when warm and dry.

a. Be Aware of the Group's Ability. It is essential that the leader evaluates the individual abilities of his men and uses this as the basis for his planning. In his evaluation, the leader must include the group's overall physical conditioning, and the consideration of change in climate and how long the unit has had to acclimatize.

(1) Mental attitude of your group. Is morale high? How much tactical training has the group had in a particular type of terrain?

(2) Technical aspect of your group. Have they been on skis, snowshoes, etc.?

(3) Individual skills. At this point, you must choose who is most proficient at the individual skills that will be required for your mission, navigation techniques, security, call for fire, rope installations, track plans, bivouac site selection, skijoring, etc.

b. Evaluate Terrain and Weather Constantly.

(1) Terrain. During the planning stages of your mission, the leader must absorb as much information as possible on the surrounding terrain and key terrain features involved in your area of operation. Considerations to any obstacles must be clearly planned for. Will you need such things as fixed ropes, rope bridges, climbing gear, etc?

(a) Stress careful movement in particularly dangerous areas, such as loose rock and steep terrain.

(b) Always know your position. Knowing where you are on your planned route is important.

(2) Weather. Mountain weather can be severe and variable. Drastic weather changes can occur in the space of a few hours with the onset of violent storms, reduced visibility, and extreme changes. In addition to obtaining current weather data, the leader must plan for the unexpected "worst case". During an operation he must diagnose weather signs continually to be able to foresee possible weather changes.

(a) Constantly evaluate the conditions. Under certain conditions it may be advisable to reevaluate your capabilities. Pushing ahead with a closed mind could spell disaster for the mission and the unit.

c. Stay as a Group. Individuals acting on their own are at a great disadvantage in this environment.

(1) Give the unit adequate rest halts based upon the terrain and elevation, physical abilities of the unit, combat load and mission requirements.

(2) Remember to use the buddy system in your group.

(3) Maintain a steady pace so that it will allow accomplishment of the mission when all members of the unit reach the objective area.

d. Appreciate Time Requirements. Efficient use of available time is vital. The leader must make an accurate estimate of the time required for his operation based on terrain, weather, unit size, abilities, and on the enemy situation. This estimate must take into account the possibility of unexpected emergencies and allow sufficient leeway to make unplanned bivouacs in severe conditions.

(1) Time-Distance Formula (TDF). This formula is designed to be a guideline and should not be considered as the exact amount of time required for your movement. Furthermore, this formula is for use in ideal conditions:

3 km/mph + 1 hour for every 300 meters ascent; and/or + 1 hour for every 800 meters descent.

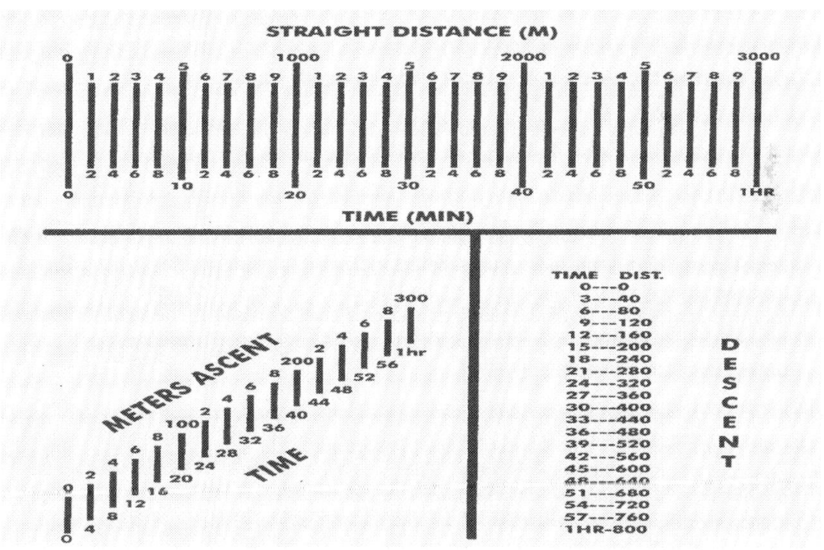

mountains, the loss of body fluids is much less noticeable. High water intake, at least 6 quarts per day when in bivouac, 8 quarts per day when active, will help to prevent dehydration.

g. Maintain Proper Clothing and Equipment

(1) Clothing

(a) Our clothing has to perform an important function in our mission; therefore, when choosing our clothing we have to take into account some essential requirements.

1. Protection against wind and rain.

2. Layered and easily adjustable.

3. Lightweight and durable.

(b) To help us remember how to maintain and wear our clothing, we use the acronym, "COLD".

C - Keep clothing Clean.

O - Avoid Overheating.

L - Wear clothing loose and in Layers.

D - Keep clothing Dry.

NOTE: In the mountains a man should never be separated from his gear. Here are some basic and essential items that should be considered during your planning stage.

(2) Required equipment:

(a) A daypack should always be carried per squad and one Marine should carry a combat load.

(b) Map and compass. Every individual in a leadership position and his assistant should carry a map and compass. The maps should be weatherproofed and extra maps should be distributed throughout the unit.

(c) Repair kit. This kit should include those items necessary to do emergency repairs on your equipment.

(d) Survival Kit. Always carried on your person. The contents of a survival kit will be covered in another period of instruction dealing specifically with survival kits.

h. Ask Locals About Conditions. An often-overlooked source of information is the indigenous population of an area. Local weather patterns, rockslide areas, watering points, and normal routes can all be obtained by careful questioning. The leader must obtain current information of the actual conditions along his intended route. Of particular importance are recent precipitation and enemy sightings.

i. Remember to Keep Calm and Think.

(1) Emergency situation. Having recognized that you are lost and that an emergency situation exists, the following principles should be followed:

(a) Keep calm and do not panic. At this point you must make every effort to conserve body heat and energy.

(b) Think. When an individual is cold, tired, hungry or frightened he must force himself to organize his thoughts into a logical sequence.

(c) The group must try to help itself by either finding the way back to safety or by preparing shelters and procuring food.

(d) Above all else, the group must act as a tight-knit unit. In emergency situations, individual dissension can cause a total loss of control and unit strength.

(2) If the decision is reached that the group should seek its way back to safety, several possibilities exist. In most situations, the safest approach will be to retrace the route to the last known point and continue from there. The other course of action is to get a group consensus on the present location and send out a small search party to locate a known point. This party must ensure that they mark their trail adequately to return to the group. If all attempts at finding a way back to known terrain fails, a definite survival situation exists and actions discussed later in this section must be instituted.

j. Insist on Emergency Rations and Kits. Just like what was covered in the SUMMER MOUNTAIN WARFIGHTING LOAD REQUIREMENTS class, survival rations and a survival kit should always be carried.

k. Never Forget Accident/Emergency Procedures.

(1) Causes of accidents. The general procedures used to handle accidents differ little in this environment, but several distinct points should be kept in mind. The most frequent causes of accidents are as follows:

(a) Overestimation of physical and technical abilities.

(b) Carelessness.

(c) General lack of observation of one's surroundings.

(d) Lack of knowledge and experience by leaders.

(e) The failure to act as a group.

(f) Underestimation of time requirements to move through mountainous terrain and underestimation of the terrain itself.

(2) Preventive measures. The only truly effective preventive measures for the above lie in the education and experience of leaders at all levels. Too often, leaders sit by watching during training and as a result have no concept of the requirements involved in the mountainous environment. Only by active involvement can a leader gain the knowledge and experience needed to effectively lead in this environment.

(3) General procedures for handling an accident. These require only a good dose of common sense as outlined below.

(a) Perform basic first aid.

(b) Protect the patient from the elements to include insulation on top and bottom.

(c) Evacuate if necessary.

(d) Send for help if required. If possible, never send a man for help alone.

(e) Send the following information regarding the accident:

1. Time of accident.
2. Nature and location of accident.
3. Number injured.
4. Best approach route to accident scene.

(4) If one man of a two-man team is injured, the injured man must be given all available aid prior to going for help. If the injured man is unconscious, he should be placed in all available clothing and sleeping gear and anchored if on steep terrain. A note explaining the circumstances, and reassuring him, should be left in a conspicuous spot. This note must also contain the following information:

(a) When you expect to return.

(b) Where you went.

(c) What you did before you left (medication, etc.).

(5) International distress signal (whistle):

(a) Six short blasts in 1 minute from person requesting help.

(b) The return signal is three blasts in 1 minute from the respondent.

(6) Other methods if help is required:

(a) Red pyrotechnics.

(b) SOS, (... --- . . .).

(c) "Mayday" by voice communications.

l. Energy is saved when warm and Dry. With the previous 11 principles in mind this one should fall right into place. Save your heat and energy by following these steps:

(1) Dress properly.

(2) Eat properly.

(3) Drink properly.

(4) Ensure shelter meets criteria.

(5) Produce external heat (fires, stove, extra clothing, etc.) to save body heat and energy for future use.

(6) Don't lose body heat by getting wet.

UNITED STATES MARINE CORPS
Mountain Warfare Training Center
Bridgeport, California 93517-5001

FMST.07.01
03/08/02

STUDENT HANDOUT

NUTRITION

TERMINAL LEARNING OBJECTIVE: Given a unit in a mountainous environment, and the necessary equipment and supplies, apply the principles of nutrition in a mountainous environment to prevent death or injury per the reference. (FMST.07.01)

ENABLING LEARNING OBJECTIVES:

(1) Without the aid of references and from a given list, choose the correct caloric intake needed per day in a mountainous environment, in accordance with the references. (FMST.07.01a)

(2) Without the aid of references and from a given list, choose which type of diet increases your tolerance to cold, in accordance with the references. (FMST.07.01b)

(3) Without the aid of references and from a given list, choose one of the preventive measures of excessive water loss in a mountainous terrain, in accordance with the references. (FMST.07.01c)

OUTLINE.

1. **STARVATION FACTORS.** Factors that combine to result in "Near Starvation" during military operations in any mountainous environment:

a. Difficulty in carrying sufficient food.

b. Lack of time for preparing meals and beverages.

c. Loss of appetite and low food palatability resulting from hypoxia/hypothermia.

d. Fat intolerance resulting from altered metabolism.

2

2. **PHYSIOLOGICAL ASPECT TO COLD STRESS.**

a. Man's adaptation to cold is essentially behavioral. Shelters and clothing are utilized to provide a micro-environment, which will allow survival. Body composition adjustments such as increases in body weight and fat are also examples of the body's adaptation to the cold. The winter months lend themselves to greater periods of inactivity, leading to these changes.

b. In the Falkland War, constipation was a major concern. Dehydration was found to be the culprit in the majority of cases, but diet had a significant role as well. MRE's are low in fiber, containing less than 10 grams. Man needs at least 31 grams of fiber per day to ease passage of stool. Constipation can be a debilitating condition, affecting overall unit readiness.

(1) Low fiber diets can lead to diverticulosis, diverticulitis and Irritable Bowel Syndrome (IBS). IBS flare-ups tend to occur during upsetting situations and emotional stress. Educate your troops on the importance of fiber, especially in a cold weather environment.

(2) Fiber is supplied by foods derived from plants. The best sources are:

(a) Whole grain cereals.

(b) Whole wheat breads.

(c) Fresh and dried fruits.

NOTE: Vegetables also help prevent constipation, but the fiber from whole grain cereals and whole wheat bread is more effective than vegetables.

3. **CALORIC REQUIREMENTS.** (FMST.07.01a) Recommended caloric intake (Naval Medical Command Instruction 10110.1) for troops operating in a cold weather environment is 4,500 calories per day.

a. Energy Expenditure. Military operations in a cold weather environment usually involve strenuous activity. Walking on snow expends roughly twice the energy that walking on hard ground does. If troops are unfamiliar with personal over-snow mobility equipment, i.e. snowshoes and skis, awkward movement and thus wasted energy will result.

(1) The heavy clothing necessary for functioning in the cold increases energy expenditures by 5-15%. You will have the opportunity to experience this effect in the upcoming field evolution.

(2) Properly clothed, experienced individuals should have negligible heat loss from radiation, conduction, and convection. However, losses associated with respiration and evaporation can be significant.

(3) In intense cold, particularly with high winds, heat loss to the environment may be inevitable.

4. **NUTRITION IN A HIGH ALTITUDE ENVIRONMENT.** A diet high in carbohydrates and high in fat is the most beneficial, while a high protein diet is generally the least desirable.

a. The marked weight loss seen in individuals at high altitude is largely due to increased metabolic demands and anorexia secondary to hypoxia. Total caloric intake commonly does not meet energy demands.

b. Altered metabolic processes that have been reported at high altitude include delayed gastric emptying, intestinal malabsorption and decreased gastric enzyme secretion.

5. **BEST DIET FOR COLD WEATHER**. (FMST.07.01b)

a. High carbohydrate and high fat foods should be included rather than high protein foods.

b. Small meals (snacks) spaced at time intervals of two hours or less should be consumed rather than the usual three meals-a-day.

c. During a typical eight-hour day of cold exposure, three high-fat (unsaturated fat) meals providing 60% of the day's calories, coupled with two normally scheduled meals (breakfast and dinner) should provide good cold tolerance.

6. **BASIC CONSTITUENTS OF FOOD**. The three basic constituents of food are carbohydrates, fats and proteins.

a. Carbohydrates. Also known as the quick energy foods that provide energy to produce heat with the byproducts of carbon dioxide and water when oxidized by the body.

(1) Carbohydrates are present in food mainly as sugars and starches and are broken down during digestion into simple sugars that are converted into glucose.

(2) Carbohydrates are stored as glycogen in the liver and muscles and can be broken down quickly into glucose to provide a rapid source of energy. However, these stores are not large and are markedly depleted by fasting for as little as 24 hours.

(3) Eating a high carbohydrate diet for several days will double one's glycogen stores and can increase endurance by up to three times compared to the ordinary balanced diet.

(4) The main carbohydrate food sources are: CARBOHYDRATE SOURCES

(a) Fruits.

(b) Vegetables.

(c) Cereals.

(d) Sugar.

b. Fats: In the body, fat serves as the main storage form of energy. One gram of fat has the equivalent of 9.3 Cal/gm as opposed to 4.1 Cal/gm for carbohydrates and proteins. They also produce energy, heat, CO_2 and H_2O.

(1) In cases of starvation, body fat tends to be broken down into acidic compounds. If these compounds accumulate in the blood faster than they can be burned, they cause the body tissues and blood to become excessively acidic (acidosis).

(2) During brief physical activity, energy is derived equally from fat and carbohydrates, but as the duration of the activity lengthens, the percent of energy supplied by fat increases as carbohydrate energy is quickly depleted.

(a) Fat sources:

1. Butter.
2. Lard.
3. Cooking oil.
4. Mayonnaise.
5. Ice cream.

FAT SOURCES

(4) Fat is also found in lesser amounts in:

(a) Dairy products. (d) Nuts.

(b) Meat. (e) Vegetables.

(c) Eggs. (f) Cereals.

c. Proteins: Protein is a reparative food of complicated molecules composed of chains of amino acids.

(1) The protein structure is made up of 25 different amino acids. Eleven of these are important for growth while eight are essential for normal tissue maintenance.

(2) Common protein food sources are:

PROTEIN SOURCES

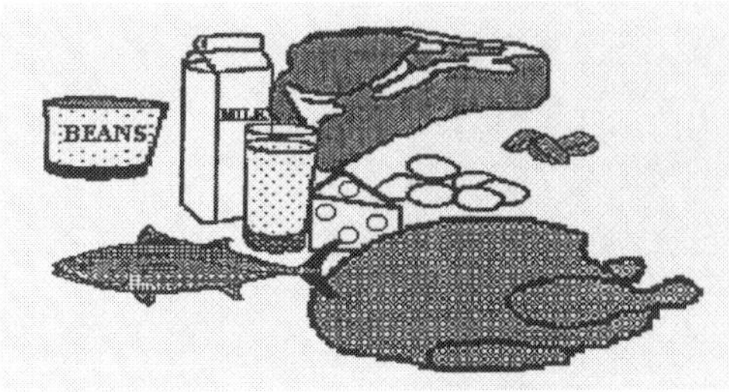

(a) Eggs.

(b) Dairy products.

(c) Meat.

(d) Poultry.

(e) Fish.

(f) Legumes (peas and beans, nuts and cereals with animal products, especially eggs, being better sources of complete protein).

(3) A pure protein diet may cause fatalities in 3-8 weeks from "rabbit starvation". This refers to those who have tried to live on relatively fat-free rabbit meat only. Fat must be a part of a normal diet.

7. **PERCENTAGE OF FATS, PROTEINS AND CARBOHYDRATES IN A SUMMER MOUNTAINOUS ENVIRONMENT**.

a. Requirements:

(1) 20% Protein.

(2) 35% Fats.

(3) 45% Carbohydrates.

b. When troops move to higher altitudes and/or are doing strenuous activity, carbohydrates should definitely be increased to at least 70 - 80% of energy intake. (Experienced mountaineers find carbohydrates more palatable at altitude).

8. **VITAMINS.** Vitamins are essential to the metabolic functioning of the body and are even more important in a cold weather environment because of an increase in metabolism and stress. Because the body cannot make vitamins, they are supplied in the food we consume. A daily multivitamin tablet supplement is recommended in a cold weather environment, but mega doses (more than 10 times the RDA) may be harmful.

a. Water-soluble vitamins are not stored in the body and need to be consumed daily. They are not stable to cooking and oxidation.

(1) B complex, Niacin, Pantothenic acid, Biotin, Choline, Folic Acid and C. (Note: Vitamin C facilitates absorption of iron from food.)

b. Fat-soluble vitamins can be stored in the body and are more stable to cooking and oxidation.

(1) A, D, E and K.

9. **FOODS THAT MAKE YOU FEEL WARM**.

a. Hot Beverages. Increases peripheral vasodilatation.

b. Adequate Dietary Iron. Mild iron deficiency increases chilling, thus decreasing cold tolerance.

c. Fat in the Diet. Fat slows gastric emptying and anecdotal evidence suggests less chilling and more restful sleep at night.

d. Carbohydrates. Keeps energy levels up. "Tired" is associated with "Cold".

e. Spices. Red pepper (Tabasco sauce) and monosodium glutamate increase peripheral vasodilatation, while capsicum and capsaicin, which are found in Cayenne, produces the same result when applied externally to the feet.

f. Water. Dehydration leads to lethargy, which tends to reduce food intake resulting in less heat production.

10. **INCREASING COLD TOLERANCE**.

a. Frequent Feedings. A high carbohydrate diet is superior to a high protein diet in increasing tolerance to cold.

(1) Diets deficient in both calories and protein reduce physical work capacity in the cold. The provision of both components is required to maintain work capacity.

11. **HIGH FAT DIETS AND SUSCEPTIBILITY TO FROSTBITE**.

a. High fat diets consumed at frequent (2 hour) intervals, maintain body temperatures better than high carbohydrate diets, but not if normal (4 hour) intervals are observed between meals.

b. Fat enriched diets lead to a propensity toward increased arterial thrombosis and increased platelet aggregation, which translates into increased potential for thrombosis.

c. Frostbite injuries are associated with disruption of the microcirculation, progressive micro vascular thrombosis, alteration of platelet function, and fibrin deposition in affected areas.

d. Therefore, diets high in (saturated) fat may predispose to frostbite.

CAUTION! A high-fat diet is important in a high-altitude, cold-weather environment. The benefits associated with the high-energy source called Fat far outweigh the risks of arteriosclerosis from the diet itself. A high-fat diet is appropriate in this specific environment during the time frame of cold, high-altitude exposure. In garrison, at sea level, a high fat diet should be avoided in order to maintain a healthy lifestyle.

12. **IRON DEFICIENCY AND THERMOREGULATION**.

a. Iron deficiency results in 9% less heat production.

b. Consuming only 1/3 of the RDA for iron results in a 29% greater heat loss during cold exposure. "The furnace is going, but the windows are open" is a phrase that helps describe this effect.

c. Iron absorption is increased by consuming iron rich foods:

(1) Lean meat.

(2) Poultry.

(3) Beets.

(4) Beans.

(5) Green leafy vegetables.

(6) Fish.

(7) Vitamin C.

d. Bottom line, IRON improves the body's response to cold.

13. **WATER REQUIREMENTS**.

a. On average, water makes up about 60% of the one's entire body weight. The average sedentary person excretes about 2.7 quarts of water each day. Of this total, 1.3 quarts are lost in the urine, 1.1 quarts through the skin and lungs and 10 ounces in the stool.

b. At high altitude, and during strenuous exercise, the amount of water lost through the skin and lungs due to the cold, dry air increases greatly.

c. Cold weather decreases the sense of thirst, which may lead to a state of chronic, mild dehydration. At temperatures below freezing and at elevations above the snow line, the lack of liquid water and the time and effort required to melt snow compound the problem.

d. Prevention of excessive water loss: (FMST.07.01c)

(1) Force fluids. Drink 6-8 quarts of water daily.

(2) Observe for dark urine and other symptoms of dehydration. (Headache, nausea/vomiting, dizziness, constipation, etc.)

(3) Limit caffeine to 1-2 cups per day.

(4) Do not rely on thirst as an indicator of hydration status.

NOTE: Eating snow or ice will cause extreme inflammation of the mucous membranes of the mouth after two days. Warm canteens of water placed in your sleeping bag will give you water to drink in the morning. They will initially act as hot water bottles. (Place them upright, as they tend to leak ...keep bag dry.)

e. Health Problems Associated with Excessive Water Loss:

(1) Dehydration.

(2) Constipation.

(3) Hypothermia.

(4) Frostbite.

(5) Heat Exhaustion.

f. The importance of hydration is essential while in a cold weather environment. If these guidelines are followed then complications arising from poor hydration should be prevented.

14. **SPECIAL PURPOSE MILITARY RATIONS**.

RATION	APPLICATION
Ration, Cold Weather, RCW.	Cold weather operations where weight and freezing are a problem.
MRE Supplement Packet.	Cold weather calorie supplements for the MRE.
Ration, Light Weight, RLW.	Long range patrols, reconnaissance, and assault.
Nutrition Sustainment Module, NSM.	Calorie dense food, bar ration for mission lasting 3-5 days where soldier must carry his own food.

Family of Operational Rations, FOR.	Modular rations that allow an optimum mix of meals for tactical operations under all climatic conditions.

15. **FIELD STRIPPING OF MRE'S.**

a. Tips on how to conserve space and weight:

(1) Remove the entire contents from the plastic wrapper.

(2) Remove the cardboard, and all materials that you will not use.

(3) Be sure to place the toilet paper and other materials from the accessory pack either back in your pocket or into the brown packaging.

(4) Carefully repack the stripped meal back into the wrapper and seal with tape.

(5) Your meal will be half its original size and two-thirds its original weight.

UNITED STATES MARINE CORPS
Mountain Warfare Training Center
Bridgeport, California 93517-5001

FMST.07.09
03/31/02

WILDERNESS PATIENT ASSESSMENT

TERMINAL LEARNING OBJECTIVE: Given a casualty in a cold weather environment and the necessary equipment and supplies, perform a patient assessment in a mountainous environment to determine the nature and extent of injury per the reference. (FMST.07.09)

ENABLING LEARNING OBJECTIVES:

(1) Without the aid of references, from a given list, choose when CPR may be legally discontinued, in accordance with the references. (FMST.07.09a)

(2) Without the aid of references and from a given list, choose which situations CPR may not be started in a remote wilderness environment, in accordance with the references. (FMST.07.09b)

(3) Without the aid of references, in the proper sequence, correctly perform a patient assessment in the proper sequence in accordance with the references. (FMST.07.09c)

OUTLINE

1. **INTRODUCTION.**

   a. An organized prioritized assessment of a wilderness trauma victim is essential to proper care. A modified Advanced Trauma Life Support (ATLS) assessment system will be introduced in this chapter.

2. **TRIAGE.**

   a. Two situations of triage may exist in a wilderness environment.

(1) The number of patients and the severity of their injuries do not exceed the ability of the care-giver and his/her equipment to render care. In this situation, treat life-threatening problems first.

3

(2) The number of patients and the severity of their injuries exceed the capacity of the care-giver and his/her equipment to render care. In this situation, those with the greatest chance of survival are treated first

3. **CPR: URBAN vs. WILDERNESS SPECIFICS.**

a. In a remote wilderness situation where advanced life support assistance may be hours or days away, CPR probably has no chance of success. In addition, administering CPR under wilderness conditions may put group members in serious danger, due to physical hazards and the effects of exhaustion.

b. Based on the above, several suggestions regarding CPR in remote wilderness areas can be made. Even though the following suggestions seem reasonable, they should be viewed as recommendations, without legal force, and not doctrine. At the present time, a rescuer who begins CPR is legally obligated to continue unless or until one of the following conditions is fulfilled:

(1) CPR may be discontinued legally when: (FMST.07.09a)

(a) The patient revives.

(b) ACLS is started by higher authority.

(c) Relieved by another rescuer.

(d) Too exhausted to continue.

(e) Pronounced dead by proper authority.

(2) In a remote wilderness environment, CPR may not be started under one of the following situations: (FMST.07.09b)

(a) The patient is in cardiac arrest caused by trauma.

(b) The patient is a drowning victim who has been immersed for over an hour.

(c) The patient is in cardiac arrest and advanced life support is more than an hour away, especially if the patient must be carried out.

(d) The patient's cardiac arrest was unwitnessed and the time of onset is unknown.

(e) The patient appears to be dead, based on rigor mortis, lethal injuries, or a body temperature below 60°F.

(f) Giving CPR would be hazardous to rescuers.

(3) After 30 minutes of CPR with no signs of life, further CPR is probably useless and may reasonably be discontinued. Administering CPR to a victim who is being evacuated by litter or sled is very difficult, if not impossible. Unless an ambulance or helicopter can be brought in rapidly, chances of survival are slim.

(4) Exceptions include patients in cardiac arrest caused by hypothermia; patients with another illness or injury complicated by hypothermia (avalanche burial and near drowning in cold water), and patients in cardiac or respiratory arrest caused by lightning injury. In these cases, the outlook is probably more favorable and CPR should be aggressively administered.

4. **<u>PATIENT ASSESSMENT</u>** (FMST.07.09c). There are different considerations to think about in a wilderness setting. The environment plays a major role in treatment, care and even the extraction of the patient. Not only do we have to be mindful of the enemy situation, by keeping tactical. We also have to take into consideration the elements. Snow, rain ice, and the terrain can and will pose a great danger to the patient and the rescuers. We must protect the patient from the elements. We can do this by placing an insulating barrier between the patient and the ground. Also exposing one body part at a time and recovering as you move on will help to fight hypothermia.

A) SCENE SIZE-UP.

1) Determine scene safety:

a. Rescuers: Your safety is priority. You are no good to the team as a casualty.

b. Patient: Protect the patient from further injuries, enemy, and the elements.

c. Bystanders: They can cause many accidents. Limit the personnel on scene.

d. Tactical: Have the Marines maintain security and find cover for the rescuer and the patient.

e. Fire: Forest fires can pose a formidable threat.

f. Electrical: Make sure to shut off electricity first. Rushing in injures many rescuers.

g. Special Rescues: Swift water and high angle rescues are very dangerous and only trained professionals should attempt the rescue.

2) Determine the mechanism of injury / nature of illness:

a. Obtain SAMPLE and Mechanism of Injury (MOI) history from patient, witnesses, or other caregivers.

S = Signs and/or Symptoms.
A = Allergies.
M = Medications.
P = Past Illnesses.
L = Last Meal.
E = Events of Injury.

MOI = What, when, where, and how injury occurred.

b. Environment: Survey the area to get clues. (i.e. falling rocks, next to a cliff, broken branches, etc…)

c. Bystanders: Ask them what they saw.

3) Determine number of patients:

a. Survey area for number of patients.

b. Look for the hidden patients. Patients could be buried in rock, sand or snow. They also could be spread out in bushes and trees. One of the most over looked places is in the bystanders. Be sure to have people search and identify all patients.

4) Request additional help / Organize help if needed:

a. Medical assistance: Due to the lack of medical personnel and equipment, you might need to request additional help.

b. Technical assistance: When the scope of care is over your ability or higher than the casevac corpsman can provide, you should request higher trained personnel on the casevac bird.

c. Organize help at the scene: Use the people around you to help out.

5) Consider stabilization of the spine:

a. If indicated from: Falls of 10 feet or greater, high speed collisions, unwitnessed loss of consciousness, etc…

b. Designate an assistant: Use the bystanders / marines. Tell them exactly what you want them to do and show them how to do it.

c. Improvise control: If you are alone you can use the patient's boots, sand bags, rocks, etc as forms of temporary control.

B) INITIAL ASSESSMENT.

1) Make general impressions of the patient: What can you obtain from the patient's condition as you walk up? Is the patient talking, or unconscious?

2) Determine responsiveness / level of consciousness.

a) Use AVPU.

b) Assess pupils for size, equality, and reactivity.

3) Determine chief complaint / Apparent life threats.

4) Assess Airway and breathing.

a) Assessment of airway.

- Perform chin lift or jaw thrust maneuver, with C-Spine precautions.
- Clear the airway of foreign bodies.
- Insert an oropharyngeal or nasopharyngeal airway.
- Use of other airways as needed, and as equipment and training allow.

b) Assessment of breathing.

- Expose the neck and chest, assure immobilization of the head and neck.
- Determine the rate and depth of respirations.
- Auscultate the chest bilaterally. (Ear to chest wall, if necessary)
- Check for symmetry. (Rise and fall of the chest)

c) Management.

- Ventilate. (Face-mask, bag-valve-mask, if available)
- Alleviate tension pneumothorax.
- Seal open pneumothorax.
- Administer high concentration oxygen, if available.

5) Assess Circulation.

a) Assess and control major bleeding.

- Identify and control major bleeding.
- Pulse: Quality, rate, and regularity.
- Pulse Pressure: Check radial, then carotid.
- Skin color: cyanosis, pallor and temperature.

b) Management.

- Insert two large bore IV's with saline locks. IV fluid replacement, if indicated.
- Place in shock position.
- Prevent hypothermia.

6) Identify priority patients and make transportation decisions.

a) Triage patients at this point.

b) Make transportation decisions. In this type of environment it is possible that a casevac vehicle might not gain access to your position. At this point you will have to organize how the patient will be transported out. Also, you might have to set up relay points and rewarming stations.

C) DETAILED PHYSICAL EXAM: This is the head to toe assessment. We will find and treat all secondary injuries at this time. When performing this exam in the wilderness, you will need to expose one body part at a time and cover it back up as you go. When assessing the patient, you are looking for deformities, contusions, abrasions, penetrations/punctures, bruising, tenderness, lacerations, and swelling (DCAP-BTLS).

1) Head.

a. Assessment.

- Inspect and palpate entire head and face for lacerations, contusions, and fractures.
- Re-evaluate pupils.
- Re-evaluate level of consciousness.
- Assess eyes for retinal hemorrhage, optic disc bulging, visual acuity disturbances, and contact lenses.
- Evaluate cranial nerve function.
- Inspect ears and nose for CSF leakage, hemorrhage
- Inspect mouth, tongue, and teeth.

b. Management.

- Maintain airway, continue ventilation, and oxygenate as indicated.
- Control hemorrhage.
- Prevent secondary brain injury.
- Remove Contact Lenses.

2) Assess the neck.

a. Assessment.

- Inspect for signs of blunt trauma, penetrating injuries, tracheal deviation, and use of accessory muscles during respiration, and jugular vein distention.
- Palpate for tenderness, deformity, swelling, or subcutaneous emphysema.

b. Management.

- Maintain open airway.
- Maintain immobilization and protection of cervical spine.
- Mark trachea.

3) Assess the chest.

3

a) Assessment.

- Inspect chest wall for signs of blunt trauma, penetrating injuries, and use of accessory muscles during respiration.

- Auscultate the chest wall and posterior bases for breath sounds.
- Auscultate for heart sounds.
- Palpate the entire chest wall for evidence of blunt trauma, penetrating injuries, flail chest, tenderness, crepitus, or subcutaneous emphysema.
- Percuss for hyperresonance or dullness.

c. Management.

- Needle thoracenthesis, if tension pneumothorax is suspected and proper training and equipment are present.
- Three-sided dressing for open chest wounds.
- Pericardiocentesis, if cardiac tamponade is suspected and proper training and equipment are present.

4) Abdomen.

a. Assessment.

- Inspect abdomen for signs of blunt trauma, penetrating injuries, and internal bleeding.
- Auscultate for the presence / absence of bowel sounds.
- Palpate the abdomen to elicit subtle tenderness, rebound, or guarding.

- Check pelvis for stability.

b b. Management.

c

- Correctly dress wounds.
- Keep patient NPO until definitive care is delivered.

5) Perineum / Rectum / Penis / Vagina.

d a) Assessment.

e

(1) Perineum.

- Contusions / hematomas.
- Lacerations.

(2) Rectum.

- Blood. Sphincter tone.
- High-riding prostate.

(3) Penis.

- Urethral bleeding.

(4) Vagina.

- Bleeding.
- Lacerations.

6) Musculoskeletal.

a) Assessment.

- Inspect extremities for evidence of blunt / penetrating injury. (deformities, altered range of motion).
- Palpate extremities for tenderness, crepitus, and abnormal movements.
- Palpate all pulses.
- Assess pelvis for stability. (Fractures) - Palpate spine for tenderness, crepitus, step-off's, depressions, and abnormal movements.

b) Management.

- Splint all suspected fractures in the position of function or comfort.

- Reduce dislocations as soon as possible, depending on training.
- Maintain immobilization and alignment of cervical, thoracic, and lumbar spine.

7) Reassess the vitals signs.

a) Assessment.

- Assess the vitals, pulse rate, breathing rate, pulse pressure.
- Re-evaluate the pupils and level of consciousness (GCS).
- Evaluate extremities for motor and sensory deficits.

b) Management.

- Maintain adequate immobilization of entire patient, as indicated.

5. **PATIENT PACKAGING AND EVACUATION.**

a. See the Mountain Casualty Evacuation chapter.

UNITED STATES MARINE CORPS
Mountain Warfare Training Center
Bridgeport, California 93517-5001

FMST.07.13
03/30/02

STUDENT HANDOUT

HIGH-ALTITUDE ILLNESSES

TERMINAL LEARNING OBJECTIVE: Given a simulated casualty, in a high altitude wilderness environment, treat high altitude health problems, in accordance with the references. (FMST.07.13)

ENABLING LEARNING OBJECTIVES:

(1) Without the aid of references, select from a given list the definition of acute mountain sickness, in accordance with the references. (FMST.07.13a)

(2) Without the aid of references, select from a given list the four main mechanisms of acute mountain sickness, in accordance with the references. (FMST.07.13b)

(3) Without the aid of references, select from a given list the most common and prominent symptom of acute mountain sickness, in accordance with the references. (FMST.07.13c)

(4) Without the aid of references, select from a given list the field management of acute mountain sickness, in accordance with the references. (FMST.07.13d)

(5) Without the aid of references, select from a given list the seven high altitude health preventive measures, in accordance with the references. (FMST.07.13e)

(6) Without the aid of references, select from a given list the correct definition of high altitude cerebral edema, in accordance with the references. (FMST.07.13f)

(7) Without the aid of references, select from a given list the pathophysiology of high altitude cerebral edema, in accordance with the references. (WMC.6.4g)

4

(8) Without the aid of references, select from a given list five of the symptoms of high altitude cerebral edema, in accordance with the references. (FMST.07.13h)

(9) Without the aid of references, select from a given list the field management of high altitude cerebral edema, in accordance with the references. (FMST.07.13i)

(10) Without the aid of references, select from a given list the correct definition of high altitude pulmonary edema, in accordance with the references. (FMST.07.13j)

(11) Without the aid of references, select from a given list three of the nine symptoms of high altitude pulmonary edema, in accordance with the references. (FMST.07.13k)

(12) Without the aid of references, select from a given list the field management of high altitude pulmonary edema, in accordance with the references. (FMST.07.13l)

(13) Without the aid of references, select from a given list choose the most important treatment for both high altitude cerebral edema and high altitude pulmonary edema, in accordance with the references. (FMST.07.13m)

OUTLINE

1. **ACCLIMATIZATION**.

a. Successful acclimatization depends on three factors:

(1) Degree of hypoxic stress (i.e. altitude)

(2) Rate of onset of hypoxic stress (i.e. ascent rate)

(3) Individual Physiology (i.e. genetic differences between individuals)

b. Physiologic Changes:

(1) Respiratory - Hypoxic Ventilatory Response (HVR)

- Hypobaric hypoxia
- Triggers carotid body
- Central respiratory center stimulated
- Ventilatory rate increases

HVR can be measured:

- "High" leads to successful acclimatization and superior performance.
- "Low" tends to be found in victims of altitude illnesses.
- HVR can be influenced by depressants (ETOH) and stimulants. (caffeine), and progesterone.
- **May** be negatively influenced by physical conditioning.

(2) Respiratory – Fast response: minutes to hours

- HVR leads to increased Ventilatory rate.
- Increased ventilations "blows off' C02."
- Low C02 levels yield a respiratory alkalosis and rising pH.
- Elevation of pH feeds back to depress the central respiratory center.
- No further immediate rise in ventilatory rate (a brake is placed on any further increase in respiration.)

(3) Renal – Slower response: kidneys respond over the next 24 - 48 hours:

- Kidneys respond to alkalosis by inducing a bicarbonate diuresis.
- PH slowly declines - brake on central respiratory center eases.
- Respiratory rate gradually rises over 4 - 7 days to new baseline.
- Final respiratory rate is altitude dependent.

(4) Cardiovascular - changes are a result of increased sympathetic tone (due to hypoxia).

- Increased: Blood Pressure, venous tone and Cardiac output (increased heart rate and contractility.)
- Stroke Volume declines in first 1 - 3 days due to bicarbonate diuresis.
- Left ventricular pressures are normal, usually; therefore, wedge pressures are usually normal.

(5) Pulmonary Circulation - critical in understanding the pathophysiology of HAPE.

- Ascent results in global Hypoxic Pulmonary Vasoconstriction.
- Elevated Pulmonary Artery Pressure (PAP) occurs.
- Exertion at altitude worsens this Pulmonary Hypertension.
- Damage to pulmonary vascular endothelium leading to leaky capillaries.

(6) Cerebral Circulation - critical in understanding pathophysiology of HACE

- Hypoxia -----------Vasodilatation
- Hypocapnia ----------Vasoconstriction
- Hypoxia is the dominant influence and vasodilatation results.
- On average, Cerebral Blood Flow (CBF) increases 20-40% on ascent to 12.000ft.

(7) Hemopoietic Response

- Erythropoietin levels increase within hours of ascent.

- New immature Red Blood Cells (RBC) seen within days.
- New mature RBC's present within 4-10 days
- RBC mass increase seen over weeks to months (altitude dependent).
- Early changes due to hemoconcentration.
- Polycythemia is a potential danger – this can lead to a condition known as CMS, or Chronic Mountain sickness.

(8) Muscle Tissue

- Intramuscular capillary and mitochondrial density increases after three weeks exposure to altitude.

(9) Sleep Pattern

- Stage 1 increase, Stage 2 unaffected
- Stage 3 & 4 decreases
- Rapid Eye Movement (REM) decreases
- Arousals increase
- Periodic Breathing (Cheyne-Stokes).
    Increased respiratory rate leads to respiratory alkalosis
    Elevated pH depresses central respiratory center
    Apnea
    02 saturation declines, C02 levels rise
    Carotid body stimulated
    Ventilatory rate rises

(10)Summary: Physiologic changes seen at altitude result from the body's responses to hypo baric hypoxia. It has been speculated that man would be limited to an 8000-foot elevation limit without supplemental oxygen if not for the acclimatization process. As Messner and Haebler demonstrated on Mt. Everest, the process of acclimatization enables man to explore the entire planet.

2. **ACUTE MOUNTAIN SICKNESS (AMS)**.

a. Definition - AMS is an acute, self-limiting illness, which results when an unacclimatized individual ascends rapidly to high altitude. (FMST.07.13a)

(1) It is rare below 8,000 feet but will occur in approximately 42% of persons rapidly exposed to altitudes greater than 10,000 feet in elevation.

(2) Aerobic fitness is no predictor for risk of AMS with exposure to altitude. There is some evidence that prior aerobic fitness leads to increased AMS incidence, most likely due to excessively rapid ascent or over-exertion at altitude. Elite athletes may also be at increased risk due to depressed cardiovascular and Ventilatory responses to exertion.

(3) Prior ascents to altitude without symptoms of AMS are no guarantee against having symptoms of AMS with future ascents. However, those with past AMS are at an increased risk of AMS with repeat exposure to high altitude. Generally, there is no correlation between the severity of the illness and increasing altitudes. The incidence is slightly increased in younger persons.

(4) There is still much we do not know about altitude-induced physiologic changes. Current explanations in the literature center around genetic differences in hypoxic ventilatory response and vascular endothelium release of inflammatory mediators and Nitric Oxide.

b. Pathophysiology of AMS: The physiologic changes associated with the symptom complex known as AMS are linked to: (FMST.07.13b).

(1) Poor ventilatory response: An individual's Hypoxic Ventilatory Response (HVR) dictates how well he/she will respond to the hypobaric hypoxia of altitude. Those with a brisk HVR will do well, whereas those with a blunted response tend to suffer the symptoms of AMS. A poor ventilatory response leads to hypoxia and hypercapnia (elevated C02 levels). These two combine to augment cerebral blood flow by inducing cerebral vasodilatation.

(2) Fluid retention: The increased sympathetic tone associated with hypobaric hypoxia explains many of the changes seen in the renal system at altitude. Low Renal Blood Flow, increased anti-diuretic hormones, low Glomerular Filtration Rates, and decreased Urine Output all contribute to a Net Fluid Gain. This retained fluid redistributes within the body's fluid spaces. As a result, intravacular and extravascular fluid shifts occur, especially in the brain, lung, and peripheral interstitial tissues.

(3) Increased Intra-cranial Pressure: Increased cerebral blood flow coupled with CNS inflammation resulting in increased vascular permeability leads to increased intra-cranial pressure and associated symptoms.

(4) Altered Pulmonary Mechanics. On ascent to altitude, there is an increase in extravascular lung water. This is thought to be mediated, in part, by a phenomenon known as hypoxic pulmonary vasoconstriction. Blood vessels in the lungs constrict when they are exposed to low oxygen levels, this global vasoconstriction leads to increased right-sided heart pressures. The result of increased right-sided heart pressures causes fluid to leak from the vascular space into the alveoli, which results in impaired gas exchange. This has been demonstrated via Pulmonary Function Testing bronchoalveolar lavage, right sided-heart catheterization and arterial blood gas analysis showing decreased Vital Capacity, decreased Peak Expiratory Flows, pulmonary exudates rich in proteins and Increased Alveolar-Arterial oxygen Gradients. One study showed a 54% increase in estimated lung mass. It is not uncommon to hear crackles in all lung fields when auscultating asymptomatic individuals. The presence of intra-alveolar fluid inhibits the diffusion of oxygen

across the alveolar membrane, thereby reducing oxygenation across the pulmonary capillary membrane.

Signs and Symptoms of AMS. Symptoms tend to occur within the first 1- 3 days of exposure to altitude. Symptoms should resolve spontaneously within 15 - 24 hours, if ascent is arrested. The three most prominent symptoms are Headache, Nausea and Vomiting. (FMST.07.13c) Although nonspecific, headache must be present to make a diagnosis of AMS. 95% of AMS victims will complain of one or more of these three. Other common complaints include dizziness, anorexia, drowsiness, malaise, weakness, and insomnia. It is possible that findings of dyspnea on exertion and mental status changes along with ataxic gait may accompany the classic AMS signs and symptoms. These are clear indications for descent, as these are the hallmark signs of progression to HACE and the development of HAPE. Again, headache is the most common and prominent symptom of AMS. The Lake Louise Consensus criteria was created to give clinicians formal guidance in diagnosing AMS. The criteria require the presence of headache plus one or more symptoms from the above list. (i.e. nausea, vomiting, anorexia, insomnia, dizziness, fatigue, and dyspnea on exertion.)

c. Field Management (FMST.07.13d)

(1) Stop any further ascent. Light duty. No tobacco.

(2) Symptomatic Treatment. Mild analgesics such as Aspirin or Tylenol for the headache, Compazine or Phenergan for nausea/vomiting.

(3) Other Meds: Acetazolamide (Diamox), Dexamethasone (Decadron), and oxygen. The effects of Diamox and Decadron are additive when used together and are a very effective means for treating AMS.

(4) Hyperventilate: It is unrealistic to try to consciously control one's respiratory rate over any appreciable period of time. It is nearly impossible to maintain conscious hyperventilation while functioning in the field.

(5) Descend. If none of these methods help, then individuals should descend 1,000-3,000 feet. This will usually result in marked relief of symptoms. Most cases, however, improve in 1-2 days with symptomatic treatment.

(6) ALL PATIENTS WITH AMS MUST BE EVALUATED FOR HAPE AND HACE.

HOW?

- Field sobriety test.
- Mini-mental status exam.
- Auscultation of lung fields.

d. Preventive Measures: (FMST.07.13e)

(1) The best approach to high-altitude travel is staged, gradual ascent, when moving at altitudes greater than 3000 meters. Gradual ascent means no faster than 3000 ft/day up to 14,000 and no greater than 1,000 ft/day over 14,000 ft.

(2) Avoid any alcohol and/or sedatives during the first two nights.

(3) Maintain adequate fluid intake.

(4) Work high - Sleep low.

(5) High Carbohydrate Diet (approximately 70% of diet).

(6) Avoid overexertion, but remember that mild exercise assists acclimatization.

(7) No smoking.

(8) Pharmacological options: Prevention of AMS is possible with Acetazolamide (Diamox). It dramatically lowers the incidence of AMS when taken prophylactically. The usual dose is 125 mg PO BID starting 24-48 hours before ascent and should be continued for 3-4 days after the start of the mission. It is also effective when used as a treatment especially when combined with Decadron. In individuals with contraindications or allergies to Diamox use Decadron (dosed PO/IM/IV 4 mg Q6).

3. **HIGH ALTITUDE CEREBRAL EDEMA (HACE)**.

a. Definition. HACE is a high-altitude illness characterized by swelling of the brain. (FMST.07.13f)

(1) HACE can occur as low as 8,000 feet, but typically occurs at more than 12,000 feet.

(2) HACE, considered by most, to be a progression of AMS, occurs as a failure to heed the warnings of AMS signs and symptoms and continue ascent.

b. Pathophysiology. Increased cerebral blood flow along with the fluid shifts mentioned above lead to vasogenic edema (leaky blood vessels) inside the confined space of the skull. It is also thought that inflammation plays a key role in the development of HACE and occurs as follows. Hypoxia triggers the release of inflammatory mediators (likely Transforming Growth Factor Beta, TGF-β and Vascular Endothelial Growth Factor, VEGF.) These mediators cause increased vascular permeability that result in fluid accumulation within the brain – edema and increased intracranial pressure. Once this pressure nears arterial pressure, cerebral blood flow will be impaired. Eventually, intra-cranial pressure will rise further with a concomitant increase in the risk of brain herniation leading to marked alterations in mental status, ataxia and even cranial nerve palsies. Increased CSF production further compounds the problem. (FMST.07.13g)

c. Differential Diagnosis: Other illnesses are possible in individuals ascending to altitude. Your differential should include the following when neurologic deficits are involved:

(1) Cerebrovascular spasm and Transient Ischemic Attacks.

(2) Cerebral vascular thromboembolic events.

(3) Intra-cranial hemorrhage (aneurysm, AVM).

(3) Hypothermia.

(4) Meningitis

(5) Encephalitis

(6) "Hangover"

d. Signs and Symptoms:

(1) Early signs and symptoms are those found in AMS. (i.e. H/A, NV)

(2) When the above warning signs go unheeded the true signs and symptoms of HACE can quickly follow: (FMST.07.13h)

(a) Ataxia (loss of muscle coordination leading to difficulty maintaining balance), especially prominent in the victim's gait.

(b) Mental Status Changes, poor judgement, personality change which may progress to stupor, coma and death relatively rapidly without proper treatment.

(c) Lassitude, confusion, hallucinations, convulsions, behavioral aberrations.

(d) Paralysis of one or more extremities (see Differential Diagnosis).

(3) The most important impediment to early recognition of HACE is its insidious onset. Early signs and symptoms frequently go unrecognized or are ignored by patients, as well as their companions, who also may be suffering to some degree from the effects of altitude.

e. Field Treatment (FMST.07.13i)

(1) Early recognition is the KEY. Look for Ataxia and Mental Status Changes.

(2) Treatment should be immediate since fatalities can occur within a few hours in severe cases.

(3) Once diagnosed, the patient should be placed in the most comfortable position possible, descent should be directed immediately and administration of high concentration O2 should be given, in route, if available.

(4) Decadron (early), 10mg PO/IM/IV followed by 4 mg Q6. Diamox can also be used 125-250 mg BID.

(5) Consider diuretics – use carefully.

(6) If loss of consciousness, think airway management and ABC's.

(7) Casevac to an advanced medical facility ASAP!

(8) A Gamow Bag can be a lifesaver if decent is not immediately possible. Perform 4-6 compressions per day each treatment should last 40-60 minutes.

NOTE: The Gamow Bag is a one-man portable hyperbaric chamber. It weighs 14.5 lbs., is operated by a foot-pump, and is made of nylon with an airtight zipper. With the HAPE or HACE victim inside, the bag is pressurized with the foot-pump to an internal pressure of 2 psi. This is the equivalent to a 5000 - 6000-foot descent in regards to the partial pressure of oxygen inside the bag. Air is circulated within the bag via the foot-pump. This can be a life-saving resource in the event a victim is caught at altitude and cannot descend due to weather or other uncontrollable variables.

f. Prevention

(1) Preventive measures are the same as those discussed for AMS.

(2) Decadron can be used as a prophylactic medication along with Diamox, in cases of unavoidable rapid ascent to extreme altitude. However one must continue this regimen, while at altitude, or face rapid onset of symptoms once the medication is stopped.

4. **<u>HIGH ALTITUDE PULMONARY EDEMA (HAPE)</u>**.

a. Definition. HAPE is a high-altitude illness that is characterized by filling of the lungs with edema fluid. (FMST.07.13j)

(1) HAPE rarely occurs below 12,000 feet.

(2) HAPE victims in the past were commonly described, after autopsy, as victims of "pneumonia" and "Inflammation of the lungs". Since 1960, medical science has had the technology to investigate the complexities of the physiological changes associated with hypobaric hypoxia.

(3) Incidence: Occurs in 1-2 % of people brought rapidly to 12,000ft.

(4) Commonly occurs within 2-4 days of ascent.

(5) Risk for developing HAPE remain the same as above:

(a) Rate of ascent
(b) Sleeping altitude
(c) Level of exertion
(d) Gender (M>F)
(f) Individual physiology

b. Pathophysiology of HAPE:

(1) Hypoxic Pulmonary Vasoconstriction

(a) Normal part of the acclimatization process (elevated PAP).
(b) Adaptive at sea level (i.e. lobar pneumonia).
(c) Maladaptive at altitude (global hypoxia).
(d) Tends to be uneven, leading to high pressure/high flow area.

(2) Overperfusion

(a) Localized areas exposed to high pressure and high flow.
(b) Shearing forces and stress failure at the microvascular level.
(c) Endothelial damage to capillaries leads to inflammatory response and increased vascular permeability.
(d) Elevated hydrostatic pressure combined with increased vascular permeability leads to leakage of protein rich fluids into the alveoli and decreased oxygenation

(3) Nitric Oxide deficiency

Individuals prone to HAPE may have a deficiency in the amount of Nitric Oxide produced in the vascular endothelium resulting in more profound hypoxic pulmonary vasoconstriction.

c. Pathologic Findings of HAPE: Postmortem studies of victims of HAPE have yielded some interesting results:

(1) Grossly, the lungs are congested and swollen, as seen with other causes of pulmonary edema. Average lung weight on autopsy is 2-4 times normal. Histologically, one sees pulmonary edema with a protein rich exudate filling the alveoli. It is also common to find the alveolar spaces filled with hyaline membranes.

(2) The left heart is completely normal in appearance. However, the right ventricle, right atrium, and pulmonary arteries are distended and dilated.

(3) Greater than 50% of HAPE victims are found to have evidence of HACE.

d. Signs and Symptoms. These tend to occur within 2-4 days of arrival at altitude. Usually the symptoms of AMS are present before or occur with the symptoms of HAPE. (FMST.07.13k)

(1) Early signs:

(a) Dry cough, frequently occurring at night.

(b) Dyspnea on Exertion (DOE), especially with ambulation uphill.

(c) Mild chest pain - usually perceived as an ache beneath the sternum.

(d) Decreased work performance and increased recovery time between events.

(e) Peripheral or Central Cyanosis.

(2) Later signs:

(a) Dyspnea at Rest.

(b) Productive cough which yields large amounts of pink, frothy sputum.

(c) Rapid pulse and respiratory rates.

(d) Audible crackles on auscultation.

(e) **Mental status changes, Ataxia**, Loss of Consciousness.

(f) CXR findings:

- Prominent pulmonary arteries.
- Normal heart size.
- Patchy infiltrates.

<u>ECHO findings:</u>

-Non-cardiogenic pulmonary edema -increased PAP
-Tricuspid regurgitation.
-Normal left ventricle
-Prominent pulmonary arteries -enlarged right atrium
-Enlarged right ventricle.

NOTE: Crackles heard on lung exam, as an isolated finding is NOT an indication for descent or grounds for the diagnosis of HAPE. It is common (up to 15% of climbers on Mt. Rainier) to find asymptomatic rales in climbers at altitudes of 10,000-12,000ft. Concern is heightened when crackles occur in constellation with other signs and symptoms.

e. Field Treatment (FMST.07.13l)

(1) The most important emergency care measure is immediate descent to a lower altitude, since fatalities can occur within 6-12 hours in severe cases.

(a) A descent of at least 2,000-3,000 feet can be a definitive treatment if the condition is caught early enough.

(2) The patient should be placed in the most comfortable position (usually sitting) and given **high-flow O2 if available**. It is important to minimize activity, as exertion increases pulmonary artery pressures. Remember oxygen is Lasix for the lungs.

(3) Medication regimens include:

(a) Nifedipine (10mg PO initially then 30mg SR Q12) – calcium channel blocker - pulmonary artery/arteriole dilator. SR=sustained relief.

(b) Diamox and Decadron may be appropriate regimens, depending on your medication availability.

(c) Morphine (5mg IV initially decreased PAP) as a last resort.

(4) Oxygen is always appropriate, however, it is not always available.

(5) Gamow Bag can be a lifesaver if decent is not immediately possible.

f. Prevention Measures. Preventive measures are the same as those discussed for AMS.

(1) Acetazolamide (Diamox) has been proven to be successful in the prevention of HAPE. 125 - 250 mg PO bid helps prevent HAPE in individuals with recurrent episodes.

(2) Nifedipine (30 mg SRq12h) - also effective in individuals with recurrent HAPE.

(3) It should be noted that a history of HAPE is NOT a contraindication to the individual participating in high-altitude activities. Each case should be evaluated individually.

NOTE: Remember decent is the most important treatment for both HACE and HAPE. (FMST.07.13m)

5. **SYSTEMIC EDEMA AT HIGH ALTITUDES**.

Occasionally, certain individuals, especially women, will develop edema of the face, hands, and feet at altitude. This edema usually tends to occur in the absence of other symptoms. In susceptible persons, repeat episodes are common. The edema goes away upon return to lower altitude. There is no need for descent in these cases because the patient is usually able to tolerate the edema and continues to function well. However, they should be examined carefully to rule out the presence of HAPE or HACE. Diuretics are usually not warranted. It is also probably reasonable to restrict salt intake. The only indication for descent is if the edema is causing functional disability, i.e. inability to don one's boots, or facial edema severe enough to restrict vision.

6. **SICKLE CRISIS**.

Persons with sickle cell trait are at risk for developing sickling attacks when exposed to hypoxic stress, as might occur at high altitudes. However, sickling can occur at lower elevations. Individuals with Mediterranean family ties should be considered at risk for having sickle cell trait. The trait is more likely in black individuals as well. Any patient complaining of Left Upper Quadrant pain, SOB, and/or arthralgia at altitude should have sickling crisis included in the differential diagnosis regardless of race. It has been known to occur in white females as well. It has been recommended by some that individuals with sickle cell trait should be restricted to an altitude of 8,000 feet and below. However, more liberal recommendations exist, suggesting that the individual and their symptoms dictate ultimate altitude.

7. **ALTITUDE THROAT**.

At higher altitudes, mouth breathing is often increased during exertion. Inhalation of this cold, dry air can result in drying of the mucous membranes in the throat and upper airway. This condition responds to humidified air, salt-water gargles or lozenges. If exudate or edema is present, throat culture and/or antibiotics may be warranted.

8. **HIGH ALTITUDE FLATUS EXPULSION HAFE).**

Ascent to altitude can result in rapid expansion of intestinal gas, due to the decreased pressure of altitude. This produces abdominal discomfort and the passage of colonic gas. This condition is not associated with serious consequences; however, it may create tension amongst tent-team members. Treatment consists of oral administration of anti-flatulence such as simethicone, Mylanta II, or Gaviscon. Know and avoid your gas producing foods!

9. **HIGH ALTITUDE RETINAL HEMORRHAGE.**

Up to 60% of individuals ascending to altitudes greater than 18,000 feet will develop asymptomatic retinal hemorrhages. On fundoscopic exam, there will be hyperemia of the disc and dilation and tortuosity of the vessels. "Flame" hemorrhages and "cotton wool spots" can be seen on exam. The only indication for descent in an individual with retinal hemorrhages is loss of visual acuity. This usually results when a large hemorrhage involves the macular region of the retina. Remember that Carbon Monoxide poisoning can present with retinal hemorrhage as well.

10. **THROMBOEMBOLIC EVENTS.**

Thromboembolic disease occurs with increased frequency at higher altitudes. It is believed that the common circumstances of volume depletion due to dehydration and Polycythemia due to altitude exposure predispose individuals to coagulopathic events. In addition, extended periods of inactivity due to “going to ground” during storms contribute to this predisposition. There does not seem to be any direct influence of hypobaric hypoxia on the body's normal coagulation system.

11. **ACETAZOLAMIDE.**

Acetazolamide (Diamox) is a Carbonic Anhydrate Inhibitor, which is especially active in the brain, the lung, and the kidney.

a. Kidney:

(1) Uptake of bicarbonate ion is inhibited.

(2) Bicarbonate diuresis is induced.

(3) Metabolic acidosis ensues within one hour of medication.

(4) Sequence mimics and accelerates the natural acclimatization process.

(5) Enhanced HVR is the end result.

(6) Diuresis addresses problems associated with fluid retention.

b. Brain:

(1) Direct effects at the level of the Central Respiratory Center.

(2) Diminishes periodic breathing during sleep.

(3) Fewer episodes of apnea and extreme hypoxemia.

(4) CSF production is inhibited and CSF absorption is enhanced.

c. Side Effects:

(1) Polyuria.

(2) Flattens taste of carbonated beverages.

(3) Peripheral paraesthesias, myopia, impotence, nausea, and drowsiness.

(4) Contraindicated for Sulfa Allergic and G6PD-deficient individuals (reversible when medication is discontinued.)

UNITED STATES MARINE CORPS
Mountain Warfare Training Center
Bridgeport, California 93517-5001

FMST.07.35
03/13/02

STUDENT HANDOUT

HEAT ILLNESSES

TERMINAL LEARNING OBJECTIVE: Given a casualty in a wilderness environment, and necessary equipment and supplies, manage common heat related injuries to prevent death or further injury per the reference (FMST.07.35)

ENABLING LEARNING OBJECTIVES:

(1) Without the aid of references, given a list of heat illnesses match the definition to the correct condition, in accordance with the references. (FMST.07.35a)

(2) Without the aid of references, select from a given list the correct field treatment for heat cramps, in accordance with the references. (FMST.07.35b)

(3) Without the aid of references, select from a given list the correct field treatment for heat exhaustion, in accordance with the references. (FMST.07.35c)

(4) Without the aid of references, select from a given list the correct field treatment for heat stroke, in accordance with the references. (FMST.07.35d)

1. **CAUSES OF HEAT ILLNESSES**. Heat illness is caused by one of or a combination of the following factors:

a. Excessive environmental heat and humidity.

b. Faulty heat dissipation mechanisms within the body.

c. Stressed heat dissipation mechanisms.

d. Prolonged strenuous physical activity with inadequate fluid and electrolyte replacement.

2. **THERMOREGULATION**. The body is constantly adjusting itself in order to maintain its optimal temperature. This is called thermoregulation. A basic physics principle applies, in that heat is transferred from a higher to a lower temperature. What this means is that when the environmental temperature is higher than the skin temperature, the body gains heat, but when the environmental temperature is lower, the body loses heat. As a result, elevated environmental temperature adds to the heat burden of the body and can interfere with heat dissipation.

a. Heat Transfer Mechanisms. Keeping in mind the above principle we will now discuss the four methods of heat transfer.

1) Conduction: This is heat exchange between two surfaces in direct contact. An example would be lying on a hot or cold surface.

2) Convection: This is heat exchange from a surface to a gas or liquid, usually air or water. The amount of heat transferred depends on many variables, such as density, flow rate and surface area exposed.

3) Radiation: This is heat transfer between the body and its surroundings. All matter either absorbs heat or emits thermal radiation. The sun emits heat and the body absorbs it. A interesting fact is that highly pigmented skin while protective against UV radiation absorbs about 20% more heat than lightly pigmented skin.

4) Evaporation: This is the heat flow from the body surface to the environment by sweating.

b. If the body is unable to maintain thermal equilibrium by radiation, convection, and conduction and core temperature rises, sweating must occur to permit heat loss by vaporization of water. Since gastric emptying of water is about 1L/hr and sweat rates can exceed 1L/hr, a healthy diet along with constant replenishment of water lost is imperative to prevent illness.

3. **PREDISPOSING FACTORS**.

a. An underlying illness such as infection.

b. The temperature and humidity of previous days effect the likelihood of becoming a heat casualty.

c. Most heat illnesses occur during the morning hours.

d. Thirst is not an accurate indication of hydration; therefore you must drink even when you don't feel thirsty.

e. People usually do not voluntarily drink as much water as they lose.

f. All persons in the heat should be considered dehydrated, unless they have been forced to drink more than they desired.

4. **PREVENTIVE MEASURES**.

a. Drink 6-8 qts every day to replace fluids lost by sweating or from other bodily functions.

b. Consumption of a solute rich meal or a fluid replacement beverage about 2 hours before exercise can potentially counteract the sweat/salt losses.

5. **EARLY SIGNS OF HEAT RELATED ILLENSSES**. Changes in cognitive functions appear to occur frequently prior to the physical symptoms of heat illnesses. It is therefore imperative that medical personnel and leaders watch for confusion and impaired ability to function or work in the members of their group.

6. **SPECIFIC HEAT RELATED ILLNESSES**.

a. Heat Induced Syncope. (FMST.07.35a) This event is caused by the gravitational displacement of blood into the legs of an upright and stationary person combined with hemodynamic displacement of blood into peripheral circulation to support heat transfer at the body surface.

1) Prodromal symptoms.

(A) Restlessness

(B) Sighing

(C) Dysphoria

(D) Nausea

(E) Yawning

2) Treatment.

(A) The fainting is usually brief and responds to horizontal positioning.

(B) Rest in a cool, shaded area.

(C) Oral fluids.

(D) Persons should be cautioned against standing in hot environments and advised to flex leg muscles repeatedly while standing still to enhance venous return and warned to sit or lie down if warning signs reappear.

b. Heat Cramps. These are painful spasms of muscles, caused by sweating resulting in excessive loss of body salt. Generally, the spasms involve voluntary skeletal muscle subjected to prior exertion. This can occur even if a person is replacing water loss, but not salt losses.

1) Precipitating Factors: Cramps are preceded by several hours of sustained effort, combining heavy sweating in hot surroundings, with the ingestion of large volumes of water.

2) Symptoms: Painful muscle cramps in the arms, legs, and abdomen.

3) Treatment: (FMST.07.35b)

(A) Rest in a cool, shaded area.

(B) Gentle massage of affected muscles (may relieve spasms) or place the affected muscle in a stretch.

(C) Generally, heat cramps respond quickly to salt solutions.

1.) 1/4 to 1/2 teaspoon of salt in a quart of water (sip).

2.) In severe cases IV isotonic saline 0.9% NaCL.

NOTE: Salt tablets are not recommended due to the gastric irritation they cause.

c. Heat Exhaustion. This occurs when body salt losses and dehydration from sweating are so severe that a person can no longer maintain adequate blood pressure. It tends to develop over several days, that present ample opportunity exists for electrolyte and water imbalances to occur. Some 20% of cases are found to also have some form of gastroenteritis. Additionally, 60% of heat exhaustion casualties will also have symptoms of heat cramps.

1) Problems in recognition.

(A) Early signs may go unrecognized or be mistaken for malingering.

(B) Headache and visual disturbances may mimic prodromal of migraine headache.

(C) Dizziness and syncope may lead to an evaluation for arrythmia.

(D) Chills, ^ temperature, and sweating may lead to an evaluation for infection.

(E) Altered mental status or ataxia may lead to an evaluation for Meningitis or high altitude medical problem.

2. Symptoms:

(A) Headache

(B) Fatigue

(C) Piloerection

(D) Nausea/Vomiting

(E) Profuse sweating

(F) Dizziness

(G) Hyperirritability, Anxiety

(H) Chills

(I) Heat cramps

3. Clinical Findings:

(A) Tachycardia

(B) Hypotension

(D) Hyperventilation

(E) Syncope

(C) Core temperature may be normal or elevated. Spontaneous cooling may occur, usually with profuse sweating. This will not happen with Heat stroke.

(4) Treatment is based on symptoms and focuses on reduction of body temperature and rehydration. (FMST.07.35c)

(A) Lay down in a cool, shaded area.

(B) If body temperature is elevated, lower by any means available.

(C) Sprinkling with water and fanning.

(D) Ice bags over superficial great vessels and fanning.

(E) Periodic toweling of the skin renews the evaporative process.

(F) If patient is alert, orally rehydrate with cool water or commercially available rehydration mixture. Goal is 2L over 1st 2 hours.

(G) If patient is not alert, rehydrate via IV with LR.

(H) Victim must rest and continue to be rehydrated over next 24 to 36 hours.

d. Heatstroke (Hyperthermia). " A true medical emergency". This is a failure of the body's cooling mechanisms that rid the body of excessive heat build-up. Prior to 1950 the mortality rate was 40 to 75%. Long term survival of victims is directly related to the speed of effective treatment measures. Traditionally, the diagnosis of Heatstroke required the following 3 signs: a core temperature > 104°F(41°C), altered mental status, and cessation of sweating. Although, this symptom complex does represent full-blown Heatstroke adherence to this strict criteria may delay critical interventions. Unless an alternative etiology is obvious, the previously healthy person who collapses after physical exertion in hot weather should be considered to have exertional Heatstroke. Treat for Heatstroke whenever you suspect it and always suspect it in the heat.

1. Symptoms:

   (A) Headache
   (B) Shortness of breath
   (C) Nausea/ vomiting
   (D) Low BP
   (E) Skin may be wet or dry
   (F) Constricted pupils
   (G) Seizures
   (H) Dry mouth
   (I) Core temp > 104°F (41°C)
   (J) Weakness/ dizziness
   (K) Hot, red skin
   (L) Tachycardia
   (M) Confusion, Aggression, Combative
   (N) Coma

2. Clinical Findings.

   (A) CNS: altered mental status, agitation, ataxia, delirium, hallucinations, convulsions, seizures, coma, pupillary constriction.

   (B) Cardiovascular: hypotension (a late and ominous finding), shock, sinus tachycardia with ST segment & T wave abnormalities.

   (C) Pulmonary: hyperventilation, pulmonary edema.

   (D) Renal: acute renal failure, hematuria, pyuria, proteinuria, UA casts.

   (E) Gastrointestinal: nausea, vomiting, +/- disseminated intravascular coagulation, hematemesis, melena.

   (F) Hematologic: WBC 20-30 thousand, < platelets, < clotting factors.

   (G) Hepatic: increased liver function tests.

3. Emergency Treatment. (FMST.07.35d)

(A) Rapid reduction of body core temperature by any means available until core temperature is 102°F.

1) Ice water bath (controversial).
2) Cool water bath.
3) Pouring or hosing cool water over victim and fanning.
4) Placing ice bags at neck, armpits, groin, forehead, chest

etc..

(B) Maintain ABC's.

(C) Monitor core temperature.

(D) IV access infuses RL or normal saline 500cc/hr x 4hrs. (watch for pulmonary edema).

(E) Control seizures with valium.

(F) Transport STAT. Cool enroute.

UNITED STATES MARINE CORPS
Mountain Warfare Training Center
Bridgeport, California 93517-5001

FMST.07.15
04/01/02

STUDENT HANDOUT

COMBAT CASUALTY CARE

TERMINAL LEARNING OBJECTIVE: Given a simulated casualty in a simulated combat environment, treat the casualty in accordance with the references. (FMST.07.15)

ENABLING LEARNING OBJECTIVES:

(1) Without the aid of references, choose from a given list the three correct phases of care of tactical casualty management, in accordance with the references. (FMST.07.15a)

(2) Without the aid of references, given a simulated casualty and the necessary equipment, conduct combat casualty care, in accordance with the references. (FMST.07.15b)

OUTLINE

1. Summary of Findings and Recommendations.

   a. Pre-hospital care of combat casualties is divided into three phases: (FMST.07.15a)
      1. Care under Fire.
      2. Tactical Field Care
      3. Combat Casualty Evacuation (CASEVAC) Care.

   b. Suppression of hostile fire may take temporary precedence over the rendering of care in the Care under Fire phase.

   c. Cervical spine immobilization is not required for victims of penetrating head or neck trauma.

   d. CPR should not be attempted on the battlefield for victims of blast or penetrating trauma who have no pulse, respirations, or other signs of life.

   e. The use of tourniquets is encouraged in the Care under Fire phase. The tourniquet is removed and bleeding controlled with direct pressure in subsequent phases when feasible.

f. The nasopharyngeal airway is the airway of first choice for unconscious patients until the CASEVAC phase.

g. For patients with airway obstruction in the Tactical Field Care phase unrelieved by a nasopharyneal airway, the next airway maneuver is cricothyroidotomy.

h. Endotracheal intubation is not the standard of care until the CASEVAC phase.

i. Evaluation of the laryngeal mask airway and the Combitube for use by SOF corpsmen and medics is recommended.

j. Progressive, severe respiratory distress in the setting of unilateral blunt or penetrating chest trauma on the battlefield should result in a presumptive diagnosis of tension pneumothorax and decompression of that hemithorax with needle thoracostomy.

k. Chest tube insertion is not indicated until the CASEVAC phase of care.

l. Oxygen is not expected to be carried onto the battlefield, but CASEVAC assets should have it.

m. No IV fluids are to be given until the CASEVAC phase unless the patient is in hypovolemic shock from a bleeding site which has now been controlled. Fluid resuscitation is not necessary for patients who are not in shock and not appropriate for patients in hypovolemic shock from uncontrolled hemorrhage (penetrating wounds of the chest and abdomen).

n. Saline locks may be used instead of IV's if fluid resuscitation is not required.

o. 18 gauge catheters may be used instead of 14 or 16 gauge.

p. Hespan (1000cc) is the fluid of choice for initial resuscitation of patients with controlled hemorrhagic shock.

q. Morphine is to be used IV (5mg) instead of IM.

r. IV cefoxitin is to be used as soon as possible for patients with penetrating abdominal trauma, grossly contaminated wounds, massive soft tissue trauma, open fractures, or any patient in whom a long delay until definitive treatment is expected.

s. Casualties should not be completely undressed for a secondary survey in the field. Removal of clothing should be limited to that necessary to expose known or suspected wounds.

t. More specific combat casualty care planning should be carried out and based on specific mission-oriented scenarios.

u. The establishment of Combat Casualty Transport Teams and their use on CASEVAC assets is recommended.

v. Electronic monitoring should be routinely used on CASEVAC assets unless conditions prohibit.

w. MAST trousers are not recommended as a standard of care in any phase.

2. Phases of Care:

a. Phase One: Care Under Fire

1. Return fire as directed or required.

2. Keep yourself from getting shot.

3. Try to keep the casualty from sustaining additional wounds.

4. Stop any massive external hemorrhage with a tourniquet.

5. Take the casualty with you when you leave.

b. Phase Two: Tactical Field Care

1. Airway management

- Unconscious casualty without airway obstruction: > Nasopharyngeal airway.

- Unconscious casualty with airway obstruction: > Cricothyroidotomy.

- Cervical spine immobilization is not necessary for casualties with penetrating head or neck trauma.

2. Breathing

- Consider tension pneumothorax and decompress with needle thoracostomy if a casualty has unilateral penetrating chest trauma and progressive respiratory disstress.

3. Bleeding

- Control any remaining bleeding with a tourniquet or direct pressure.

4. IV

- Start an 18 gauge IV or saline lock.

5. Fluid Resuscitation

- Controlled hemorrhage without shock: > No fluids necessary.

- Controlled hemorrhage with shock: > Hespan 1000cc

- Uncontrolled (infra-abdominal or thoracic) hemorrhage: > No IV fluid resuscitation.

6. Inspect and dress wound.

7. Check for additional wounds.

8. Analgesia as necessary: > Morphine (5mg) IV, wait 5minutes, repeat as necessary.

9. Splint fractures and recheck pulses.

10. Antibiotics

Cefoxitin 2gm slow IV push (over 3-5 minutes) for penetrating abdominal trauma, grossly contaminated wounds, massive soft tissue trauma, open fractures, or any patient in whom a long delay until CASEVAC or definitive treatment is expected.

11. Cardiopulmonary Resuscitation

CPR on the battlefield for victims of blast or penetrating trauma who have no pulse, no respirations, and no other signs of life will not be successful if attempted.

c. Phase Three: Combat Casualty Evacuation (CASEVAC) Care

1. Airway management

Unconscious casualty without airway obstruction: > Nasopharyngeal airway, endotracheal intubation, Combitube or laryngeal mask airway.

Unconscious casualty with airway obstruction: > Cricothyroidotomy if endotracheal intubation and/or other airway devices are unsuccessful.

2. Breathing

Consider tension pneumothorax or hemothorax and decompress with needle thoracostomy if a casualty has unilateral penetrating chest trauma and progressive respiratory distress.

Consider chest tube insertion if a suspected tension pneumothorax is not relieved by needle thoracostomy

Oxygen.

3. Bleeding

Consider removing tourniquets and using direct pressure to control bleeding if possible.

4. IV

Start an 18 gauge IV or saline lock if not already done.

5. Fluid resuscitation

No hemorrhage or controlled hemorrhage without shock: > Lactated Ringer's at 250cc/hr.

Controlled hemorrhage with shock: > Hespan 1000cc initially.

Uncontrolled (infra-abdominal or thoracic) hemorrhage: > No IV fluid resuscitation.

Head Wound patient: > Hespan at KVO unless there is concurrent controlled hemorrhagic shock.

UNITED STATES MARINE CORPS
Mountain Warfare Training Center
Bridgeport, California 93517-5001

FMST.07.37
04/01/02

STUDENT HANDOUT

BURN MANAGEMENT

TERMINAL LEARNING OBJECTIVE. Given a casualty in a wilderness environment, and necessary equipment and supplies, treat burn injuries to prevent death or further injury per the reference. (FMST.07.37)

ENABLING LEARNING OBJECTIVES

1. Without the aid of references, given a list of the depths and appearances of burns, match them to the type of burn, in accordance with the references. (FMST.07.37a)

2. Without the aid of references, given a list choose the major burns that require rapid casualty evacuation, in accordance with the references. (FMST.07.37b)

3. Without the aid of references, given a list choose the first step in the treatment of burn injuries, in accordance with the references. (FMST.07.37c)

OUTLINE

1. **GENERAL:** Each year there are about 2 million individuals burned severely enough to seek medical attention. Around 70,000 require hospitalization. Deaths are on the decrease largely due to the increase in the use of smoke detectors. It has been estimated that 90% of all burn injuries are preventable. Children and young adults are frequently affected by burn injuries.

2. **PHYSIOLOGY:** For non-chemical burns, the primary events of injury occur during the time of heat contact. Coagulation necrosis takes place within cells and denaturation of collagen in the dermis. Blood vessels are completely destroyed or endothelium damage is severe enough to cause clotting which leads to ischemic necrosis of remaining viable cells. Burn wounds are not static. Surrounding the "coagulation zone" is a zone of capillary and small vessel stasis. Circulation becomes stagnant from a strip formed by red blood cells, and aggregation of platelets and white blood cells. The fate of the burn wound depends on the progression of this zone of stasis.

7

3. In patients with burns less than 10% of the total body surface area (TBSA), the inflammatory process is generally limited to the burn site itself. As burns approach 20% TBSA, the local

response becomes systemic. Burns of 30% TBSA or greater, often lead to multiple organ system failure, which in a patient with a severe burn almost invariably leads to a fatal outcome, often due to acute renal failure.

Cardiovascular changes begin immediately after a burn. The degree of these changes depends initially on the size of the burn and to a lesser extent the depth of the burn. Burns less than 15% TBSA can be fluid resuscitated with crystalloid fluids. As burns pass 20% TBSA, massive shifts of fluid and electrolytes occurs. This results in massive fluid shift from the intravascular to the extravascular space. Unless intravascular volume is repleted, classic hypovolemic shock occurs.

3. **TYPES OF BURNS**

a. Scald Burns: Scalding is the most common cause of burns, usually resulting from hot water. Water at 140°F creates a deep partial-thickness or full-thickness burn in three seconds. At 156°F the same burn occurs in one second. Freshly brewed coffee is generally about 180°F. Boiling water always causes a deep burn. Soups and sauces are thicker, remaining in contact with the skin longer. Exposed areas tend to burn less deeply than areas covered with a thin layer of clothes. Clothing retains heat and keeps liquid in contact with the skin longer. Immersion scalds are deep and cause severe burns resulting from longer contact time. These burns are more common in children and elderly patients because of their thin skin. Burns from hot oils are generally deep partial-thickness or full-thickness. Temperatures for cooking oils reach around 400°F, and tar can be up to 500°F.

b. Flame Burns: The next most common burn injuries are flame burns. Although household injuries have gone down with the advent of smoke detectors, a high number of burn injuries still occur from careless smoking, improper use of flammable liquids, automobile accidents, and clothing ignited from stoves or space heaters.

c. Flash Burns: Flash burns result from intense heat for a brief time caused by explosions of natural gas, propane, gasoline, and other flammable liquids.

d. Contact Burns: These burns usually result from direct contact of hot metals, plastics, glass, or coals. Contact burns are usually limited in size, but are deep wounds. With the increase of wood stove usage, this type of burn injury is on the rise in children. Contact burns, dealing with an unconscious patient or involving molten materials, are usually fourth degree. The most common contact burn in the wilderness is from hot coals.

e. Electrical Burns: Electrical burns are actually thermal burns from very high-intensity heat. When electricity encounters resistance from the body it turns into heat. Severity of the burn depends on the portion of the body affected relative to the amperage. The smaller the body part, the more intense the heat is, and the part is less able to dissipate the heat. Consequently the fingers and toes are almost always destroyed with severe damage to the foreams. Larger areas can dissipate the current enough to prevent extensive damage. Arc burns take the most direct route, not the path of least resistance. These are deep and destuctive burns which mostly occur at the joints.

Electrical burns may have other associated injuries that should not be overlooked on the initial assessment. Fall related injuries are common. The intense muscle contraction resulting from electrocution can cause fractures of the lumbar vertebrae, humerus, or femur, and also may cause dislocation of the hips and shoulders. The casualty may have cardiac symptoms of a myocardial contusion or infarction. There can be a conduction system failure or an actual rupture of the heart wall. If there are no signs present of cardiac problems intially after a shock of 110 or 220 volts there is a minute chance they will appear later. Nervous system damage is possible whenever a current passes from one side of the body to the other more than likely affecting the spinal cord.

f. Chemical Burns: Usually caused by strong acids or alkalis and are most often the result of industrial accidents, home use of drain cleaners, assault, and improper use of harsh solvents. Chemical burns continue to burn until they are inactivated by reaction with tissue or dilution by flushing with water. Acids tend to tan the skin, creating an impermeable barrier that limits futher penetration. Alkalis combine with cutaneous lipids and saponify[1] the skin until they are neutralized. The appearance of the wound can be misleading. Unless the observer can be 100% sure, all chemical burns should be considered full thickness.

g. NBC Burns: Nuclear weapons cause burns in two ways: by direct absorption of thermal radiation through exposed surfaces (flash burns); or by the indirect action of fires caused by the weapon (flash burns). The first step in treatment is to prevent contamination to the rescuers and other patients.Vesicants such as mustard gas cause skin burns with edema and blister formation. Burns resulting from vesicants should be flushed with copious amounts of water. Absorbent powder (flour, talcum powder, Fuller's earth) can be used if water is scarce. The powders should be wiped away with a moist towel. The military uses M258A1 kits for skin decontamination. White Phosphorus ignites spontaneously when exposed to air. Remove the air with an air-tight seal; mud is optional. A brief rinse with 1% copper sulfate[2] will impead further oxidation.

4. **CLINICAL PRESENTATION:** The severity of the burn injury is related to the size of the burn, the depth of the burn, and the part of the body that is burned.

a. Burn Size: Burns are the only quantifiable form of trauma. The burn size is the single most important factor in predicting mortality, need for specialized care, and the complications expected from the burn. We can measure burn size with the "rule of nines". The "rule of nines" is where areas of the body represent approximately 9% of a person's total body surface area (TBSA), each leg 18% (the front of the leg 9%, and the back of the leg 9%), the front of the torso represents 18%, the back of the trunk is 18%, the head represents 9%, and the perineum 1%. For infants and small children under four years of age, the head represents a larger percentage 18%, and the legs a smaller percentage 13%, each arm 9%, front troso 13%, back troso 13%, and the perineum 1%. The percentage of the burn area and that of the unburned area need to add up to equal 100%. For smaller burn areas, use the Rule of Palmar Surface: the patient's palmar surface equals about 1% TBSA.

[1] Saponify: to convert fat into soap by treating it with an alkali.
[2] Navy uses sodium bicarbonate instead of copper sulfate

b. Burn Depth: The depth of the burn is either described in degree: first, second, third or fourth, or by the depth of the injury. The following describes burns in terms of partial-thickness or full-thickness. While these descriptions appear to separate burns into defined categories, many burns have a mixture of characteristics making a precise diagnosis difficult.

(1) First-Degree: These burns involve the epidermis only and do not blister. They are most commonly caused by ultraviolet light. An example would be a sunburn. The burned skin is painful and red. It should heal in 7 days without scarring.

(2) Superficial Second-Degree, or Superficial Partial-Thickness: These burns include the epidermis and upper layers of the dermis. They characteristically form blisters, under which the skin is red and moist, these burns are painful to the touch. Blisters can take up to two days to appear. These burns are usually caused by hot liquids. Wounds should heal in 14 to 21 days; they may scar depending on the extent of the burn.

(3) Deep Second-Degree, or Deep Partial-Thickness: Burns that involve deeper layers of the dermis. There is damage to hair follicles and sweat glands. They are characterized with a molted pink and white color with blisters forming immediately. These wounds may be less sensitive to touch than the surrounding normal skin, or maybe tender to touch. The patient often complains of discomfort rather than pain. Capillary refill may be slow or absent, when pressure is applied to the wound. These burns are caused by hot liquids, oil, steam, or flame. They may be difficult to distinguish from Third-Degree/Full-Thickness burns. Healing takes 3-9 weeks. Scarring is probable, the degree is related to the amount and the depth of dermal injury. Surgical grafting may be required.

(4) Third-Degree, or Full-Thickness: Involves the entire thickness of the skin, epidermis through the dermis, down to subcutaneous fat. All structures of the epidermis and dermis are destroyed. The wound is classically described as leathery, firm, and depressed when compared to normal skin. The tissue is charred, pale, and insensitive to light and touch. It is often misdiagnosed as a Deep Second-Degree/Deep Partial-Thickness due to the similar clinical findings. This burn will not spontaneously heal. Surgical repair and/or skin grafts are necessary. There is significant scarring.

(5) Fourth-Degree: These burns involve all layers of the skin, subcutaneous fat, muscle, and bone. These are devastating, life threatening injuries. They almost always have a charred appearance, and often only the cause of the burn gives a clue to the amount of underlying tissue destruction. (FMST.07.37a)

c. Burn Classification: The American Burn Association has devised a classification of burns, dividing them into major, moderate, and minor burns. Patients are placed in groups related to their risk. Low-risk patients are between the ages of 10 and 50 years old. High-risk patients are less than 10 and greater than 50 years old. Poor-risk are patients with underlying medical illnesses such as heart disease, diabetes, or chronic pulmonary problems.

(1) Minor Burns: Imply outpatient treatment. Minor burns involve TBSA of less than 15% in the low-risk group, or less than 10% TBSA in the high-risk group; full-thickness burns of less than 2% TBSA in anyone; and no other injuries.

(2) Moderate Burns: Are partial-thickness burns of 15-25% TBSA in the low-risk group, 10-20% TBSA in the high-risk group; full-thickness burns of equal to or less than 10% TBSA in anyone; burns not involving the hands, face, feet, perineum, or circumferential limbs.

(3) Major Burns: Defined as partial-thickness burns greater than 25% TBSA in the low-risk group, or greater than 20% TBSA in the high risk group; full-thickness burns of greater than 10% TBSA in anyone; burns involving the hands, face, feet, or perineum; burns crossing major joints; circumferential limb burns; burns complicated by inhalation injury; electrical burns; burns complicated by fractures or other forms of trauma; and burns in poor-risk patients. These burns require rapid casualty evacuation. (FMST.07.37b)

5. **TREATMENT:**

a. Initial Care:

(1) The first step in the treatment of burn injuries is to stop the burning process. (FMST.07.37c) Remove patients from heat, remove all their clothes that are affected by hot liquids, shut off electricity (very important), remove chemicals by flushing the affected area for 10 to 15 minutes, and decontamination of NBC. Heat can continue to injure tissue for a surprisingly long time. No first aid will be effective until the burning process has stopped.

(2) Manage the ABC's. Once an airway has been established give oxygen. If qualified, consider intubation earlier rather than later.

(3) Assess for associated injuries such as fractures or lacerations and inhalation injury.

(4) Remove clothing and jewelry from the burn area. This will help prevent constriction from swelling of the burn tissue. Do not try to remove anything that has adhered to the wound.

(5) Evaluate the Burn (depth, extent, pain).

b. General Treatment for the Patient:

(1) Gently wash the burn with tepid water and mild soap to remove any debris and to clean the skin surface around the burn site. Pat dry. Remove the skin from blisters that have popped open (do not open blisters unless necessary for function of hands or feet).

(2) Dress burn with a thin layer of antibiotic ointment.

(3) Cover the burn with Second Skin (if the burn is small enough) or cover with a thin layer of gauze, or clean, dry clothing. Burns on the face, neck and hands may be left open to the air after applying silvadene ointment. Covering the wound reduces pain and evaporative losses, but do not use an occlusive dressing.

(4) When evacuation is imminent, do not redress or reexamine the injury. If evacuation is prolonged, redress once a day. Remove old dressings, reclean (removing the old ointment), and apply fresh ointment and a clean dry covering. (Note: if stuck, soak off old dressings with clean, tepid water.)

(5) Do not pack wounds, which are larger than 20% of the body surface area, in ice. Do not leave wet coverings on burns for more than two hours at a time to reduce the risk of hypothermia.

(6) Stabilize the body temperature. When skin is lost, so is the patient's ability to thermoregulate.

(7) Have the patient drink as much fluid as he or she can tolerate without vomiting. Include some salt in the oral fluids, but do not make these solutions stronger than .9%. This is the equivalent of a pinch of salt per 8 ounce glass.

(8) IV therapy as follows, Baxters (Parkland) Formula: In the first 24 hours give Ringers Lactate, 4ml times the weight in kilograms times the percent TBSA burned. Half of the solution is given in the first eight hours. The second half is given over the next 16 hours. The second 24 hours should be given fluids to maintain blood pressure, colloids (usually in the form of albumin) are the preferred choice.

4 x wt (kg) x TBSA= amount LR given in the first 24 hours

Example a 70kg patient with 50% TBSA this is the formula for IV therapy for the first 24 hours.

4 x 70kg x 50% = 14,000ml in 1ST 24 hours
7,000ml hours 1-8
3,500ml hours 9-16
3,500ml hours 17-24

c. General Management of the Burn: Caring for the wound itself is often the least important aspect of burn care. All burn wounds are sterile for the first 24 to 48 hours. Burn management is aimed at keeping the wound clean and reducing the pain.

(1) Elevate burned extremities to minimize swelling. Swelling retards healing and encourages infection. Have the patient gently and regularly move burned areas as much as possible.

(2) Burns are tetanus-prone wounds. Check on the last known tetanus shot. The patient may need a renewal.

(3) Ibuprofen is probably the best over the counter analgesic for burn pain (including sunburn).

(4) If you have no ointment or dressings, leave the burn alone. The burn's surface will dry into a scab like covering that provides a significant amount of protection.

6. **INHALATION INJURIES:** Half of all fire-related deaths are from smoke inhalation. Smoke inhalation doubles the mortality rate for any burn. Smoke inhalation is a general term for carbon monoxide poisoning, thermal airway injury, and smoke poisoning.

a. Carbon Monoxide Poisoning: Carbon monoxide (CO) is a colorless, odorless, tasteless gas that has an affinity for hemoglobin 200 times greater than that of oxygen. Carboxyhemoglobin (COHb) levels are measured as the percentage of hemoglobin bound to carbon monoxide. The following COHb levels correlate with clinical symptoms:

<10%	No symptoms
20%	Headache, nausea, vomiting, loss of dexterity
30%	Confusion, lethargy, ECG ST depression
40%	Coma
50%	Death

The half-life of COHb is 4 to 5 hours at room air, 90 minutes with 100% oxygen, and 20 to 25 minutes at 3 ata in a hyperbaric chamber. Treatment for less than 40% COHb is a 100% oxygen. At 40% or greater the hyperbaric chamber should be used. If patient is still in a coma after COHb levels are normal, prognosis is poor, and it has been experience that they rarely awaken.

b. Thermal Airway Injury: Injury to the respiratory tract from inhalation of hot gas or steam. The lower respiratory tract is usually not damaged, unless there was an inhalation of flame or steam. Any patients who were in an explosion, with burns of the hands, face, and upper torso are at risk. Signs of carbon such as soot in the pharnyx should alert the caregiver to airway injury. Maintaining the airway is the important issue here.

c. Smoke Poisoning: Is the inhalation of noxious gases that are the products of combustion. Cyanide poisoning is known to occur in victims of smoke inhalation. Other toxic gases released are sulfur dioxide, hydrogen chloride, phosgene, and ammonia. These gases cause damage to the respiratory tract similar to chemical burns. Physical signs include facial burns, intraoral or pharyngeal burns, singed nasal hairs, soot in mouth or nose, hoarseness, carbonaceous sputum, and expiratory wheezing.

UNITED STATES MARINE CORPS
Mountain Warfare Training Center
Bridgeport, California 93517-5001

WMC
04/09/02

LESSON PLAN

NOMENCLATURE AND CARE OF MOUNTAINEERING EQUIPMENT

LESSON PURPOSE. The purpose of this period of instruction is to introduce you to the types of equipment used here and how to care for it so as to prevent its untimely failure. This lesson relates to all climbing and installation work that you perform here.

1. **ROPES.** All ropes used in the military must meet UIAA standards or U. S. Federal Test Standard 191A. Most ropes have a 5-year shelf life and maximum 2-year service life.

   a. Static. Black in color.

      (1) Construction. Kernmantle

      (2) Minimum tensile strength. 7500 lbs.

      (3) Maximum elongation. 1.5%

      (4) Diameter. 11mm

      (5) Sizes.

         (a) 165 ft ± 5 ft

         (b) 300 ft ± 10 ft

      (6) Usage. Rescue operations and bridging where a low amount of elongation is desirable under a working load.

   b. Dynamic. Olive Drab in color.

      (1) Construction. Water-resistant treated Kernmantle to reduce friction.

      (2) Minimum tensile strength. 6500 lbs.

      (3) Maximum elongation. 6%

(4) Diameter. 10.5mm and 11mm

(5) Size. 165 ft ± 5 ft.

(6) Usage. For lead climbing/party climbing.

c. Maxim Dry Rope. Olive Drab or Multi-Colored.

(1) Construction. Water-repellent treated Kernmantle.

(2) Minimum tensile strength. 3472 lbs.

(3) Maximum elongation. 6%

(4) Diameter. 9mm

(5) Size. 150 ft ± 5 ft

(6) Usage. For glacier travel/ice climbing

d. Gold Line II.

(1) Construction. Eight strand braided nylon plymor.

(2) Minimum tensile strength. 4500 lbs.

(3) Maximum elongation. 20%

(4) Diameter. 11 mm

(5) Size. 300 ft or 600 ft spools

(6) Usage. Sling Ropes and litters only.

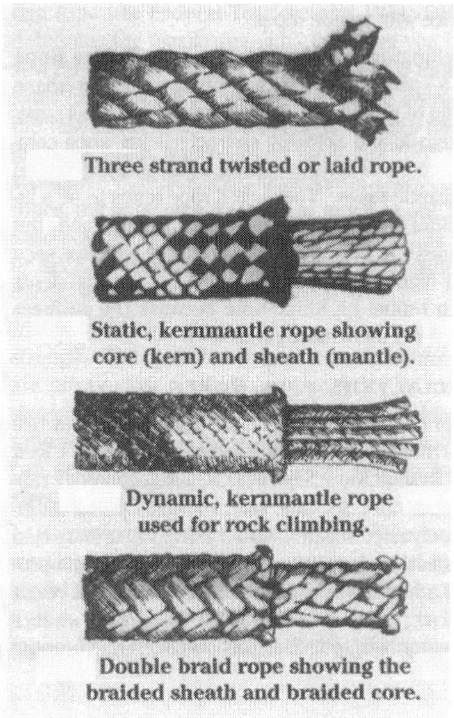

INSIDES AND OUTSIDE SHEATHS OF VARIOUS TYPES OF ROPES

NOTE: Sling ropes are made from 15 foot lengths of plymor or dynamic rope ONLY. Twenty-five foot practice coils should be constructed with static rope, but dynamic rope can be used.

2. (5 Min) **ADVANTAGES/DISADVANTAGES**

a. Advantages of Nylon Rope

(1) High strength to weight ratio.

(2) Good energy absorption in dynamic ropes.

(3) Flexible.

(4) Rot resistant, not affected by frost.

b. Disadvantages of Nylon Rope

(1) Low melting point. Nylon fuses at 400°F and melts at 480°F.

(2) Susceptible to abrasions and cuts.

(3) Affected by chemicals and light.

c. Advantages of Manila Rope

(1) Easily gripped.

(2) Hard wearing.

(3) Does not deteriorate in heat.

d. Disadvantages of Manila Rode

(1) Heavy, kinks, especially when wet. Absorbs water and swells.

(2) Burns at +300°F.

(3) Edible by rodents.

3. (3 Min) **GENERAL INFORMATION**

a. Nylon rope stretches under tension and will rupture at between 30% and 70% elongation, depending on construction.

b. Nylon rope loses as much as 30% strength when wet.

c. Temperatures as low as 250°F will damage a nylon rope.

4. (5 Min) **NYLON WEBBING**

a. The type of nylon webbing available is tubular. Tubular nylon webbing is very strong and flexible. All rules that apply to nylon rope apply to tubular nylon webbing. The size of nylon webbing used is:

(1) 1 inch tubular nylon. Tensile strength approximately 4,000 - 4,500 lbs., depending on the manufacturer.

b. Pre-sewn Spectra Runners. Tensile strength approximately 5,500 lbs.

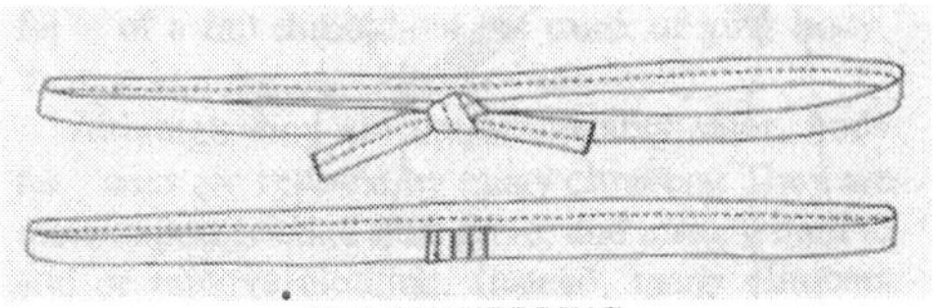

NYLON WEBBING

NOTE: These are minimum strengths. Some manufactures make even stronger webbing.

TRANSITION: We have just discussed general information in nylon webbing, are there any questions? Not only do we use ropes, but we also use carabiners in our installations, we will discuss the types of carabiners used:

5. (5 Min) **CARABINERS**. Also commonly known as snaplinks. Both locking and non-locking are used.

   a. Purpose. Carabiners are used for the following purposes:

      (1) To attach ropes or runners to pieces of protection.

      (2) To attach the rappel rope to the rappel seat for seat-hip rappels or for crossing rope bridges.

      (3) To attach the individuals safety rope to a safety line on a rope installation.

      (4) To form field expedient pulley systems.

   b. Nomenclature of a non-locking carabiner

(1) Gate

(2) Gate pivot pin.

(3) Locking pin.

(4) Body.

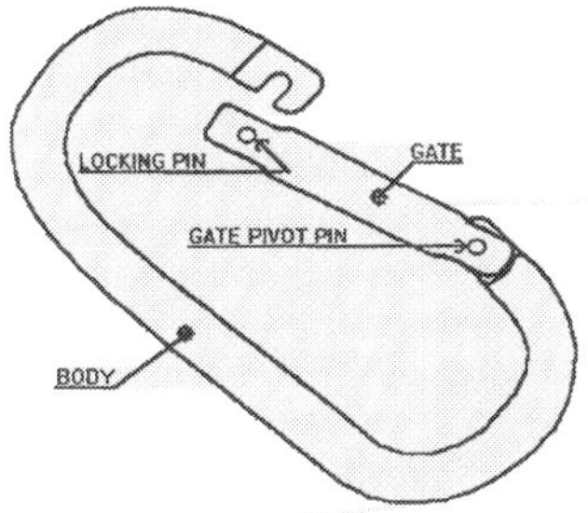

c. Nomenclature of a locking carabiner

NON LOCKING ALUMINUM CARABINER

(1) Gate.

(2) Gate pivot pin.

(3) Locking notch.

(4) Locking nut.

(5) Body.

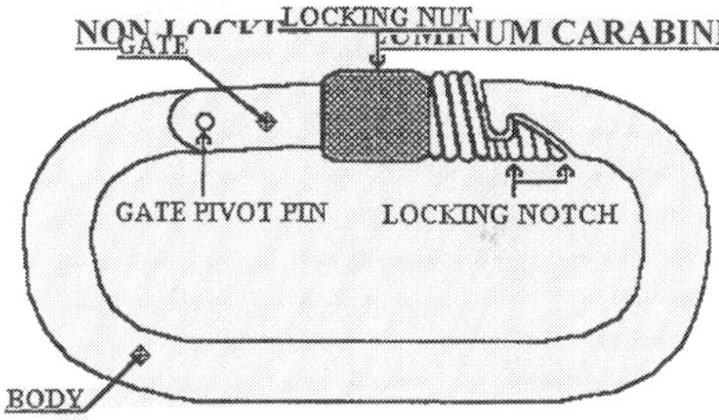

STEEL LOCKING CARABINER

d. There are two types of carabiners used. The two types and their characteristics are:

(1) Steel locking carabiners

(a) Large steel locking "D" (various manufacturers): Minimum tensile strength of 5,500 lbs.

(b) Steel-locking oval Stubai 82 is not in the MAC Kit and obsolete. However, it is being used at MWTC to save money, even though they are beyond the service life. Tensile strength of only 3,300 lbs.

(2) Aluminum non-locking carabiners

(a) Aluminum non-locking oval (various manufacturers): minimum tensile strength of 4,200 lbs.

e. Serviceability Check for a Carabiner. The following steps are used for you to check a carabiner for serviceability:

(1) The gate snaps shut with no friction and with no gap between the locking pin and locking notch.

(2) There is no excessive side to side movement of the gate.

(3) The pivot pin is tight.

(4) The locking pin is tight.

(5) The locking nut travels freely and locks securely.

(6) There are no cracks or flaws in the metal.

NOTE: The weakest part of a carabiner is the gate. If an engraver is used to mark a carabiner, it should be applied to the gate and not the load bearing side.

f. Preventive Maintenance for a Carabiner.

(1) Remove all dirt, moisture and grime.

(2) Lubricate with tri-flow graphite and clean off thoroughly.

NOTE: Whenever you use a locking carabiner ensure that the locking nut is always locked down (tightened).

6. (2 Min) **CARE OF THE CARABINER**. Do not drop the carabiner as this may result in either actual damage to the carabiner or in dirt getting into the workings of the carabiner and damaging it.

UNITED STATES MARINE CORPS
Mountain Warfare Training Center
Bridgeport, California 93517-5001

FMST.07.11
05/01/01

STUDENT HANDOUT

HYPOTHERMIA/REWARMING

TERMINAL LEARNING OBJECTIVE: Given a hypothermia casualty in a cold weather environment and necessary equipment and supplies, manage a hypothermia casualty to prevent death or further injury per the reference. (FMST.07.11)

ENABLING LEARNING OBJECTIVES:

(1) Without the aid of references, select from a given list the definition of hypothermia, in accordance with the references. (FMST.07.11a)

(2) Without the aid of references, select from a given list three of the five mechanisms by which the body loses heat, in accordance with the references. (FMST.07.11b)

(3) Without the aid of references, select from a given list the reason why gentle handling of a hypothermia victim is important, in accordance with the references. (FMST.07.11c)

(4) Without the aid of references, select from a given list the four rewarming techniques used to rewarm a hypothermia victim, in accordance with the references. (FMST.07.11d)

(5) Without the aid of references, select from a given list the proper steps of the sleeping bag rewarming method of rewarming a hypothermia victim, in accordance with the references. (FMST.07.11e)

(6) Without the aid of references, select from a given list the temperature of the water when rewarming a hypothermia casualty in a warm water bath, in accordance with the references. (FMST.07.11f)

8

1. **HYPOTHERMIA**:

a. Definition: A fall in core body temperature to 95°F (35°C) or less. (FMST.07.11a)

b. Hypothermia is potentially lethal. The mortality rate is greater than 50% in severe cases and cases complicated by trauma where the mortality can approach 100%. The most common cause of death in hypothermia is ventricular fibrillation.

c. When speaking of hypothermia, core temperature values must be used. In a controlled environment of an MTF or ER the gold standard is a core temperature taken by rectal or esophageal thermometry. The term "exposure" has often been used to speak of hypothermia but this term is inexact and has numerous other connotations. It should not be used as a synonym for hypothermia.

d. Incidence.

(1) Classically, hypothermia has been thought to occur in extremely cold environments. However, it can occur in temperate climates and at Southern latitudes. In fact, greater than 50% of cases of hypothermia occur at ambient temperatures of 50°F or greater. Contrary to popular belief, the highest incidence of hypothermia is seen in urban areas.

(2) Hypothermia tends tends to occur more frequently in males with a male to female ratio of 2.8:1.0. It's also more common in the elderly, those with drug and alcohol dependency and those with psychiatric diagnoses.

Many standard clinical thermometers don't register a low enough temperature to accurately diagnose hypothermia. Currently, a low-reading rectal thermometer is available through the Federal Stock System (NSN 6515-00-139-4593). All medical personnel attached to cold weather units or deploying to cold weather environments should carry this thermometer. In a high altitude cold weather environment medical providers should be adept at correlating signs and symptoms with specific temperature ranges so as not to unduly expose the patients to the elements. The more accurate, and invasive temperature determination should only be performed when the patient can be adequately protected from the elements.

2. **PATHOPHYSIOLOGY**:

a. Heat Conservation/Loss.

(1) The body is a furnace that generates heat. The ability of the body to thermo regulate depends on the ratio of heat production to heat loss. An imbalance of these two processes can lead to hypothermia. About 75% of food energy consumed goes toward generating body heat while the other 25% goes toward maintaining basic metabolic function.

(2) Thermogenesis can be divided into two main categories:

(A) Shivering thermogenesis creates heat by muscle activity with subsequent breakdown of glycogen. Shivering usually begins at a core temperature of around 97°F (36°C) and finally ceases at core temperatures of 85°- 90°F (30°- 34°C).

(B) Non-shivering thermogenesis creates heat by hormonal interaction with target organs such as the heart. Epinephrine, norepinephrine, growth hormone and thyroid hormone have all been implicated in playing a role in non-shivering thermogenesis.

(3) Changes in hormone secretion patterns are extremely important in acclimatization to cold weather.

(4) Once heat is generated it is retained by three main mechanisms: Peripheral vasoconstriction, abolition of sweating and piloerection. Peripheral vasoconstriction is by far the most important and efficient mechanism of heat conservation. The main internal temperature sensors in the body are located in the hypothalamus in the brain. When these areas sense a drop in core temperature they initiate peripheral vasoconstriction at the surface of the skin to prevent a loss of core body heat.

(5) The body loses heat by five main mechanisms: (FMST.07.11b)

(A) **Radiation** is heat transferred to the surrounding environment via infrared radiation.

(B) **Convection** is the loss of heat via air/water current.

(C) **Conduction** is the direct contact-transfer of heat from a hot to a cold object.

(D) **Evaporation** is the conversion of water from a liquid phase to a gaseous phase (perspiration).

(E) **Respiratory heat loss** takes place when warm humidified air from the lungs is expelled into the atmosphere.

(F) The mechanism accounting for the greatest percentage of heat loss is highly variable and depends on ambient conditions, i.e., a nude man standing in a cold room will lose approximately 65% of his heat by radiation. The same man standing in calm cold water will lose most of his heat by conduction and at a rate 25 times faster.

**NOTE: The liver and brain are the most thermogenic organs on a per weight basis. However, due to their greater mass, skeletal muscle and skin produce the largest percentage of body heat.

(6) Numerous factors predispose to the development of hypothermia. These include but are not limited to:

(A) Malnutrition.
(B) Lack of adaptation.
(C) Inactivity.
(D) Sedatives.
(E) Physical exhaustion.
(F) Burns.
(G) Extremes of age.
(G) Dehydration.
(H) Comorbid illness.

3. **SIGNS AND SYMPTOMS**:

a. Clinical manifestations of hypothermia reflect CNS, and cardiorespiratory involvement.

(1) 98°-95°F (37°-35.6°C): Mild shivering, cold sensation and impaired fine motor coordination.

(2) 95°- 92°F (35°-32.8°C): Violent shivering, difficulty speaking, sluggish thinking, and amnesia. Large muscle in coordination.

(3) 92°-86°F (32.2°-30°C): Shivering is replaced by muscular rigidity. Exposed skin is blue or puffy. Movements are jerky. Dulled sensorium, but victim may be able to maintain posture and the appearance of being in contact with surroundings. Possible atrial arrhythmias.

(4) 85°-81°F (29.4°-27.2°C): Coma, lack of reflexes, possible ventricular arrhythmias. Cessation of shivering below 86oF.

(5) Below 78°F (25.6°C): Failure of cardiac and respiratory centers, pulmonary edema, ventricular fibrillation, and asystole.

b. Some patients with hypothermia will exhibit a phenomenon that is called, "paradoxical undressing."

(1) In these instances the hypothermia victim is found nude with their cloths beside them. If the victims are women they are often misdiagnosed as sexual assault cases. This phenomena results from loss of hypothalamic control over peripheral vasoconstriction which leads to flushing of the periphery with relatively warm blood causing the individual to feel hot. Subsequently the individual undresses to cool down.

c. Again it must be stated that the only way to correctly diagnose hypothermia is with a core temperature reading.

4. **FIELD TREATMENT**:

a. ABC's, C-Spine precautions, remove wet clothing, replace with dry clothes and insulate with a vapor barrier system. (Wrapping the victim in two plastic garbage bags or bubble wrap is ideal. Further insulation may be necessary). Handle the victim carefully to prevent ventricular fibrillation caused by unnecessary jarring and rough handling. (FMST.07.11c) CASEVAC.

b. Victims with altered consciousness should be NPO. Under no circumstances should ETOH be given. This is a peripheral vasodilator and will only enhance heat loss.

c. If CASEVAC is not feasible for a long period of time, or if the victim is only mildly hypothermic, it may be necessary to rewarm in the field by using a proper sleeping bag rewarming technique. Additionally, insulated hot packs placed at the high heat loss areas will help.

d. The hypothermic victim often has no discernible vital signs.

   (1) Before beginning CPR thoroughly assess pulses for 30-45 seconds to detect viable cardiovascular activity.
   (2) Severely hypothermic victims in ventricular fibrillation and asystole have been successfully resuscitated even after periods as long as 4 hours.

5. **REWARMING**: The re-warming of a hypothermia casualty in a BAS setting is less than optimal since the AMAL for a BAS does not include monitoring devices, laboratory facilities, etc., but rewarming of a casualty can be successfully conducted even in these fairly primitive conditions. The following are three basic techniques of re-warming:

   - Passive external re-warming: a sleeping bag or blanket that is not pre-warmed.
   - Active external re-warming: a pre-warmed sleeping bag, bear hugger or warm water bath.
   - Active internal or core re-warming: IV fluids, warmed/humidified O2, lavage, extracorporal re-warming.

a. Four ideal re-warming strategies that may be available in the field include: (FMST.07.11d)

   (1) Sleeping bag rewarming.

   (2) Water bath/counter Current rewarming.

   (3) Warm/humidified airway rewarming.

   (3) Warmed I.V. solutions. NS is preferable as a cold liver does not like to metabolize Lactate.

6. **REWARMING TECHNIQUES (BAS)**:

a. Sleeping Bag Rewarming. (FMST.07.11e) This is the simplest, but least effective method.

(1) Make the diagnosis. Take a rectal temperature to determine if the casualty is clinically hypothermic (95°F or less). Use this temperature as a baseline to determine stabilization.

(2) Warm sleeping bag (pre-warm sleeping bag with two volunteers). Placing a hypothermic casualty into a cold sleeping bag will cause further heat loss by conduction.

(3) Strip casualty. Remove all wet clothing avoiding unnecessary handling. Muscular movement will pump cold blood to the core.

(a) Place the casualty between the two volunteers, if space permits. Their body heat is transferred from the two volunteers to the cold casualty.

(4) Monitor core temperature frequently throughout transport to ensure that the casualty's core temperature is not continuing to drop.

(5) Adequate insulation is required to prevent further heat loss to the environment.

(6) Augment heat by placing insulated heating pads in the high heat loss areas: head, axilla, groin, popliteal region and antecubital fossa. (Be careful not to cause burns)

**NOTE: It should be noted that an evacuation bag can be used; this will provide more room for the casualty and volunteers.

b. Water Bath Re-warming. (FMST.07.11f) This method has a long history of success in rewarming even severely hypothermic casualties. The casualty is placed in a portable/field expedient bathtub or a life raft. The extremities and head are kept out of the water to avoid significant afterdrop. The water bath temperature should be 104°-108°F. An anesthesia temperature probe should be used to constantly monitor the core temperature. Since the casualty will cool the bath water around him, it will be necessary to continue to add warm water to maintain the proper water bath temperature. Keep in mind however that this technique has definite drawbacks. In the event of ventricular fibrillation caused by re-warming shock, unsynchronized cardioversion can be very dangerous in a wet environment. Furthermore, if the casualty in the water is not on a firm surface CPR, if needed, can be very difficult to perform ineffective.

**NOTE: Individual in heat distress can be rapidly cooled using this method with "cool" water instead.

c. Warm Airway Re-warming. A warmed and humidified air/oxygen mixture is used to provide a warming media within the lungs. These devices consist of a method of generating warm humidified air and are usually portable. The warm humidified air modestly increases the amount of heat that can be delivered to core. There should be a thermometer in the airway tubing to monitor the temperature. To prevent injury to the bronchi, the temperature should not exceed 115°F (108° to 115°F is ideal). Warm airway rewarming should be used in conjunction with other rewarming methods. This method, by itself, will probably not deliver sufficient heat to rewarm a severely hypothermic casualty by itself and may in fact blunt the shivering response robbing the body of an important endogenous mechanism for producing heat.

Caveat: When possible, active external re-warming should always be combined with active core re-warming techniques so as to minimize the effects of afterdrop. This occurs as a result of peripheral vasodilatation causing warm core blood to move to the surface while relatively cold surface blood moves to the core and may lower the fibrillation threshold.

**NOTE: Hypothermic submersion incident casualties cannot tolerate humidified air for any length of time.

d. Warmed I.V. Solutions. Hypothermia is a common response to I.V. therapy. Solutions that have been prewarmed have been shown to prevent this complication. There are various methods for warming I.V. solutions:

(1) Crystalloid solutions can be warmed by warm water bath and microwaved with no adverse changes to its integrity.

(2) Fresh Frozen Plasma (FFP; -20°C): Studies have shown that microwaving for 30 seconds followed by gentle manipulation of the bag for 10 seconds, repeated five times will result in only 2.6% cell destruction.

(3) Packed Red Blood Cells (PRBC 4°C): Using conventional rewarming methods, studies have shown that the outer five millimeters of the bag can have isolated areas of hemolysis. Until a device is made that can uniformly shake up PRBC during warming, we do not recommend microwaving of blood. Rewarming of blood by dilution with warm, calcium-free crystalloid solution is useful.

WIND CHILL CHART

WIND	AMBIENT TEMPERATURE (FAHRENHEIT/CENTIGRADE)													
CALM F	35	30	25	20	15	10	5	0	-5	-10	-15	-20	-25	-30
CALM C	27	-1	-4	-7	-9	-12	-15	-18	-21	-23	-26	-29	-32	-34
	EQUIVALENT WIND CHILL FACTOR (FAHRENHEIT/CENTIGRADE)													
5 MPH F	33	27	21	16	12	7	1	-6	-11	-15	-20	-26	-31	-35
8 KPH C	1	-3	-6	-9	-11	-14	-17	-21	-24	-26	-29	-32	-35	-37
10 MPH F	21	16	9	2	-2	-9	-15	-22	-27	-31	-38	-45	-52	-58
16 KPH C	-6	-9	-13	-17	-19	-23	-26	-30	-33	-35	-39	-43	-47	-50
15 MPH F	16	11	1	-6	-11	-18	-25	-33	-40	-45	-51	-60	-65	-70
23 KPH C	-9	-12	-17	-21	-24	-28	-32	-36	-40	-43	-46	-51	-54	-57
20 MPH F	12	3	-4	-9	-17	-24	-32	-40	-46	-52	-60	-68	-76	-81

32 KPH C	-11	-16	-20	-23	-27	-31	-36	-40	-43	-47	-51	-56	-60	-63
25 MPH F	7	0	-7	-15	-22	-29	-37	-45	-52	-58	-67	-75	-83	-89
40 KPH C	-14	-18	-22	-26	-30	-34	-38	-43	-47	-50	-55	-59	-64	-67
30 MPH F	5	-2	-11	-18	-26	-33	-41	-49	-56	-63	-70	-78	-87	-94
48 KPH C	-15	-19	-24	-28	-32	-36	-41	-45	-51	-53	-57	-61	-66	-70
35 MPH F	3	-4	-13	-20	-27	-35	-43	-52	-60	-67	-72	-83	-90	-98
56 KPH C	-16	-20	-25	-31	-33	-37	-42	-47	-51	-55	-58	-64	-68	-72
40 MPH F	1	-4	-15	-22	-29	-36	-45	-54	-62	-69	-76	-87	-94	-101
64 KPH C	-17	-20	-26	-30	-34	-38	-43	-48	-52	-56	-60	-66	-70	-74

7 REWARMING TECHNIQUES (HOSPITAL):

a. Optimal re-warming techniques for higher echelon care facilities still tend to employ rather simple therapeutic modalities. The three safest and most effective are listed below:

- Warmed IVF.
- Warmed/humidified oxygen.
- Bear Hugger.

b. Many texts advocate the use of extracorporal blood re-warming. However, this technique has some disadvantages. It requires a high level of technical and medical support, which are often not available. Since patients undergoing bypass must be anti-coagulated, patients with multiple traumatic injuries may not be good candidates. Lastly, research indicates that anticoagulation with heparin may actually worsen the severity of frostbite that may occur in patients with hypothermia.

c. Lavage: peritoneal, thoracic and mediastinal lavage have all been suggested for use in severe hypothermia. However, the use of these techniques in inexperienced hands tends to carry a much higher complication rate. If performed correctly, peritoneal lavage can be very beneficial with a lower complication rate than thoracic lavage. Gastric, colonic and bladder lavage are indeed safer but are of limited efficacy, as they do not offer access to areas with significant surface area.

d. ACLS: At core temperatures below 86°F attempt three progressive unsynchronized shocks at 200J, 300J, and 360Js. If no response continue CPR and REWARM THE VICTIM. Re-attempt counter-shocks when core temperature rises above 86°F. Above 86°F, perform counter shocks in 2° increments above 86°F, i.e.88°F, 90°F and 92°F. Employ ACLS drug protocols above 93°F along with conventional normothermic ACLS protocol.

UNITED STATES MARINE CORPS
Mountain Warfare Training Center
Bridgeport, California 93517-5001

FMST.07.12
03/21/02

STUDENT HANDOUT

SUBMERSION INCIDENTS

TERMINAL LEARNING OBJECTIVE: Given a casualty in a cold weather environment and the necessary equipment and supplies, manage a submersion incident casualty to prevent death or further injury per the reference. (FMST.07.12)

ENABLING LEARNING OBJECTIVES:

(1) Without the aid of references, given a list of the events, which take place to a submersion victim, select the correct sequence of events, in accordance with the references. (FMST.07.12a)

(2) Without the aid of references, select from a given list, three of the five methods that can be used to increase survival time during cold water immersion, in accordance with the references. (FMST.07.12b)

(3) Without the aid of references, select from a given list, three of the six favorable prognostic signs post submersion incident, in accordance with the references. (FMST.07.12c)

(4) Without the aid of references, select from a given list, three of the four unfavorable prognostic signs post submersion incident, in accordance with the references. (FMST.07.12d)

(5) Without the aid of references, select from a given list the correct field management priorities of a submersion victim, in accordance with the references. (FMST.07.12e)

10

OUTLINE.

1. **TERMINOLOGY**.

a. Definition: A submersion incident is that medical problem that occurs after a casualty has been submersed under water.

(1) Drowning refers to death by suffocation following submersion in water.

(2) Near drowning refers to submersion in water with at least temporary survival (greater than 24 hours).

NOTE: An effort is being made to standardize the terminology used regarding submersion incidents, in order to clarify communication between care-givers. Terms like secondary drowning and other definitions are being discouraged.

2. **PATHOPHYSIOLOGY**.

a. Sequence of events: (FMST.07.12a)

(1) Panic initially, followed by a violent struggle.

(2) Gulping and swallowing air and water to avoid aspiration.

(3) Breath holding until hypoxia leads to unconsciousness.

(4) Once consciousness is lost, the gag reflex relaxes and passive influx of water into the lungs occurs. A small percentage of victims (10-15%) have significant laryngospasm, which prevents any appreciable volume of water to enter the lungs. This is referred to as "dry drowning".

(5) Aspiration is more likely in submersion victims who enter very cold water (10°C). An involuntary gasp reflex occurs, with a marked increase in respiratory rate and depth. This individual is at increased risk of aspiration, and subsequent drowning.

b. Submersion in very cold water will quickly results in hypothermia.

(1) Rapid induction of total body hypothermia (core temperatures of 89.6°F or less) may be protective in submersion incidents. Significant decreases in oxygen requirement result, especially in the case of the brain itself. Marked decreases in cerebral blood flow and oxygen demand have been demonstrated (see Hypothermia lesson). This protective effect of hypothermia has been used to explain the many cases of complete recovery after extended periods of submersion, which have only occurred in cold water.

(2) The "Mammalian Diving Reflex" slows the heart rate, shunts blood to the brain and closes the airway when more primitive marine mammals are exposed to extremely cold water. General consensus holds that this reflex is not a significant response in human beings. However, it has been found to be more prominent in new-born human infants. It has also been speculated that this "Reflex" may be more significant in those individuals who have survived extended submersion times.

c. Fatalities in submersion victims associated with hypothermia are primarily due to three known mechanisms:

(1) Immersion syndrome. Sudden death due to cardiac arrest secondary to massive vagal stimulation. Usually associated with sudden total body exposure to frigid water.

(2) Excessive fatigue and confusion leading to fatal errors of judgment.

(3) Direct hypothermic effects. Thermal conductivity in water is approximately 25 times that of air. Despite the evidence showing hypothermia's protective effects, the complications associated with hypothermia (cardiac arrhythmias, CNS depression, loss of motor control) can lead to submersion and drowning.

d. Some authorities emphasize distinguishing between "wet" and "dry" drowning. In wet drowning, aspiration of water into the lung occurs. (85-90% of casualties). In dry drowning, laryngospasm prevents aspiration of water (10-15% of cases). Ultimately, the end result is the same: obstruction to respiration, oxygen depletion, and asphyxiation. The morbidity and mortality associated with the two types of drowning are not significantly different.

e. Fresh-water and salt-water drowning are somewhat different.

(1) In fresh-water aspiration, the osmotic difference between hypotonic fresh-water and the relatively hypertonic blood leads to net water movement into the vasculature.

This flow across the alveolar membrane also leads to "washout" of surfactant, which leads to collapse of the alveolar spaces. Hypoxemia follows.

(2) In sea-water aspiration, the osmotic shift is reversed, leading to fluid outpouring from the blood stream flooding the alveoli and the interstitium. This leads to decreased lung compliance as well as direct damage to pulmonary capillaries. Again, hypoxemia results.

(3) Hypoxemia is the end result in both types of drowning. No significant difference in morbidity in survivors has been shown between fresh-water and salt-water drowning. The only bodies of water shown to have a worsening effect on prognosis are the hyper-saline bodies, like the Dead Sea and the Great Salt Lake. Otherwise, prognosis is similar.

f. Hypoxemia is the primary insult after submersion. Efforts to correct this problem should be vigorous and sustained. All clinically dead submersion casualties should receive CPR, or ACLS if available, immediately upon removal from the water, regardless of water temperature or duration of submersion. Since the overwhelming majority of naturally occurring water bodies in the world have temperatures below 92°F, all submersion victims should be treated aggressively as hypothermic casualties. Survivals after submersion times up to 66 minutes have been documented. Remember: No one is dead until they are warm and dead.

g. In the case of prolonged submersion or significant aspiration, pulmonary derangements occur rapidly. In lesser insults, the onset of symptoms may be delayed for as long as 24 hours. Physiological changes include:

(1) Cardiovascular. Casualties develop respiratory acidosis and hypoxemia, which may trigger dysrhythmias. Electrolyte changes are not usually significant in survivors of submersion incidents and should correct once spontaneous respirations are re-established.

(2) CNS. C-Spine injuries are common in submersion incidents, most commonly a result of shallow-water diving. Injury patterns include wedge, tear-drop and burst fractures of the upper C-Spine, frequently resulting in tetraplegia. Concussive head injuries also occur. Drug and alcohol use should always be considered.

(3) Metabolic. Respiratory acidosis is the initial problem. Acid/base derangement should correct after resuscitation of the casualty without specific intervention.

3. **SURVIVAL TECHNIQUES IN COLD WATER**.

a. Swimming motion increases heat loss more than it increases heat production. The direct hypothermic effects of cold-water immersion make it impossible to swim for long periods of time. The average swimmer is unlikely to swim more than a kilometer in 50°F water. Swimming is advised only when a rescue vehicle is nearby, or the distance to shore is less than ½ mile.

b. Treading Water and Drown Proofing.

(1) Treading water results in reduced survival time due to increased heat loss secondary to increased activity.

(2) "Drown Proofing" is a technique whereby the swimmer relaxes in the water, floating facedown, arms spread, allowing his face to submerge between breaths. "Drown proofing" is the least desirable survival technique in cold water since the head and neck (high heat loss areas) are maximally exposed.

c. Personal Floatation Device (PFD). Without the aid of the PFD, the swimmer caught in cold water is forced to use less desirable methods to stay afloat.

(1) If wearing a PFD, the individual can utilize the "Heat Escape Lessening Posture" (HELP). This position minimizes the exposure of the swimmer's high heat loss areas because the arms are folded across the chest and pressed to the sides, the knees are drawn up to the chest and the legs are crossed at the ankles. Unfortunately, this position is fatiguing and becomes difficult to hold.

(2) If there is more than one person in the water, the "Huddle" position should be used. Similar to a football huddle, the chest, abdomen and groin area should be pressed together. Children or injured individuals should be placed in the middle of the huddle, since they become hypothermic more quickly.

d. Increasing Survival Time: (FMST.07.12b)

(1) Try to enter the water in a lifeboat or raft.

(2) Locate and don a PFD as quickly as possible.

(3) Wear several layers of clothes, especially to protect high heat loss areas.

(4) If you must enter the water, enter gradually rather than jumping or diving.

(5) If afloat in the water, avoid unnecessary movement. Assume the HELP or Huddle position.

(6) Don't try to swim.

(7) Remember the situation is not hopeless. Have the will to SURVIVE!

4. **RESCUE TECHNIQUES**.

a. Reaching the Casualty. The basic order for water rescue is: Reach and Pull, Throw, Tow and, as a last resort, Go.

(1) Reach and pull. When the casualty is responsive and close to the shore, begin by holding out an object for him to grab then pull him from the water. Floating line is considered the best choice, but if it is not available, a tree branch, fishing pole, oar or other object will do as well.

(2) Throw. Should the casualty be alert, but too far away for you to reach and pull, throw an object that will float to them. A PFD or ring buoy is best, if available. The primary course of action is to throw anything that will float, and to do it as quickly and as accurately as possible.

(3) Tow. Once the conscious casualty has an object to hold onto, try to find a way to tow him to shore. From a secure position, tow the casualty ashore.

(4) Go. Swim to the casualty as a last resort. You should be a strong swimmer, trained in water rescue, in order to avoid the dangers associated with this form of rescue. If the casualty is unconscious or unresponsive, you must go to the casualty.

5. **SIGNS AND SYMPTOMS**.

a. History of Incident. Several questions need to be answered, as they will help determine the approach to the casualty.

(1) When a person is found floating face down on the water surface or is brought to the surface by a rescuer, there usually is little question about the cause of unconsciousness.

(2) When near-drowning is unwitnessed or occurs after a dive or other water activity (surfing), a head injury and a C-spine injury should be presumed, and C-Spine precautions should be instituted.

(3) Determine how long the casualty was submerged. This helps determine prognosis.

(4) How cold was the water? Very cold water (less than 70°F/21°C) may be protective.

(5) What was the casualty's health prior to submersion? Drowning could have been secondary to alcohol ingestion, cardiac arrest or other illness/injury.

b. Casualties may be symptomatic or asymptomatic at the scene:

(1) Respiratory signs and symptoms may include dyspnea, tachypnea (SOB), sore throat, substemal burning, pleuritic chest pain, persistent cough, frothy sputum, cyanosis, rales, rhonchi and wheezes.

(2) Cardiovascular signs may include tachycardia and hypotension.

(3) Neurological findings may include anxiety, restlessness, lethargy, confusion, convulsions, incontinence, hyporeflexia and coma.

(4) Abdominal distention is common as is fever and hypothermia.

c. Favorable prognostic signs: (FMST.07.12c)

(1) Casualty is alert.

(2) Basic or advanced life support is received at the scene.

(3) Good response to initial resuscitation.

(4) Hypothermic.

(5) Older child or adult.

(6) Brief submersion time.

d. Unfavorable prognostic signs: (FMST.07.12d)

(1) Age under three years.

(2) Fixed dilated pupils.

(3) Submersed five minutes or longer.

(4) No resuscitation attempts for the first ten minutes.

(5) PH less than 7.10.

(6) Cardio-pulmonary arrest as initial presentation.

(7) Resuscitation time greater than 30 minutes.

6. **FIELD MANAGEMENT**.

a. The scene of a submersion incident is frequently chaotic and first consideration should be the safety of the rescuers, then retrieval of the casualty from the water.

b. Priorities: (FMST.07.12e)

(1) Establish an airway and check breathing and circulation (ABC's). Rescue breathing, can be performed by a trained rescuer, while the casualty is still in the water.

(2) Observe C-Spine precautions, since cervical spine and head injuries are common, especially in diving accidents.

(3) Begin CPR as soon as possible, if necessary, when the casualty reaches the shore, or suitable rescue platform.

(4) The symptomatic patient should be given high-flow 02 and IV access should be established, if available.

(5) Medevac ASAP.

(6) It is recommended that all victims of submersion be observed in a medical facility for a minimum of 6-8 hours. It is not uncommon for asymptomatic near-drowning victims to refuse further care. All that can be done is attempt to get a contact number in order to follow-up with the victim via telephone 8 hours later.

c. Vomiting is common and should be anticipated by turning the patient's head and having suction ready, if available. Frequently, the cause of vomiting is gastric distention.

(1) The American Heart Association recommends that gastric distention be relieved only when it interferes with CPR.

(2) Gastric decompression can be achieved by "breaking" the patient. This maneuver is performed by placing the patient in the prone position and lifting upwards at the abdomen.

(3) Another method is to turn the patient to a lateral recumbent position and compress the abdomen with the palm of one hand.

(4) The preferred method of gastric drainage is to pass an NG tube and relieve distention with gentle suction.

(5) Gastric drainage should not be attempted without suction equipment available, due to the risk of aspirating vomitus.

7. **<u>PROGNOSTIC PREDICTORS</u>**.

a. 0 - 2 "Unfavorable Prognostic Indicators" yield a greater than 90% survival rate (see above).

b. 3 or more "Unfavorable Prognostic Indicators" yields a less than 5% survival rate.

c. Corm's Criteria: Based on the status of the patient 1 hour after initial resuscitation.

(1) A - Awake	All survive (99+%)
(2) B - Blunted	Most survive (95+%)
(3) C – Comatose	
C1 - Flexor Response	Good survival with ICU care (80+%)
C2 - Extensor Response	Good survival with ICU care (80+%)
03 - Flaccid	Poor survival
C4 - Arrest	Dismal survival

UNITED STATES MARINE CORPS
Mountain Warfare Training Center
Bridgeport, California 93517-5001

FMST.07.14

04/02/02

STUDENT HANDOUT

WILDERNESS ORTHOPAEDIC/ TRAUMA INJURIES

TERMINAL LEARNING OBJECTIVE: Given a casualty in a mountainous enviornment and, necessary equipment and supplies, manage common orthopedic injuries in a mountainous enviornment to prevent death or further injury per the reference. (FMST.07.14)

EABLING LEARNING OBJECTIVES:

(1) Without the aid of references, select from a given list the phases of clearing a cervical spine in the field in accordance with the references. (FMST.07.14a)

(2) Without the aid of references, select from a given list, the correct treatment for an open book pelvic fracture in accordance with the references. (FMST.07.14b)

(3) Without the aid of references, select from a given list, six considerations for orthopedic patient evacuation in accordance with the references. (FMST.07.14c)

OUTLINE

1. INTRODUCTION.

Orthopedic injuries occuring in the wilderness are not significantly different from those suffered in garrison. Bones break, ligaments are torn and strained, and joints are dislocated in both settings. The key to preparing for and managing wilderness orthopedic injuries is understanding that the provider will have significant limitations with regard to diagnostic tools and initial management resources at his/her disposal. The personal risks each expedition member is exposed to can rise dramatically if the team is needed to transport a victim out of the back-country. Determination of a casualty's mobility becomes a major consideration for the wilderness medical provider. Assessment of the terrain, creative use of available natural resources, and effective leadership are all necessary components of caring for orthopedic injuries in the wilderness.

10

←

2. GENERAL: Making the diagnosis.

The wilderness "doc", whatever his/her training background, needs to have good physical exam skills. NO X-RAY and NO LABS will be available in the back-country. These basic tools are not available in the field. The provider will need to rely on his/her observation, palpation, and auscultation skills to make a diagnosis. This person will also need to use the patient, if possible, to uncover the likely injury by taking a thorough history which includes careful review of the mechanism of injury.

A. HISTORY: Critical information can be gained from talking to your patient.

- Mechanism of Injury - Determine:
  1. High vs. low-velocity accident
  2. Direction of force
  3. Nature of activity
  4. Loss of consciousness

- Ask the casualty:
  1. Did they hear a pop, snap, crack, or breaking sound?
  2. How long ago did the injury occur?
  3. Can they move the injured body part?
  4. Was there swelling within 15 minutes of injury or did it take longer?
  5. Was there immediate disability or was there a gradual loss of function?
  6. Was there immediate discoloration of a joint?
  7. Can they bear weight on the injured extremity?

- Other important information:
  1. Evaluate the environment. (i.e. Hot/Cold, Wet/Dry)
  2. Allergies.
  3. Medications currently being taken.
  4. Previous injuries/surgeries to the site in question.

B. PHYSICAL EXAMINATION

Your powers of observation become crucial.

- Look for:
  1. Swelling.
  2. Discoloration and bruising.
  3. Obvious deformity/angulation.
  4. Open wounds. (with or without protruding bone fragments)
  5. Differences right vs. left.

6. Muscle spasm surrounding injured site.
7. Point tenderness.
8. Crepitus.

- Beware:
  Consider the environment while assessing your patient
  - Don't undress the casualty in the cold!
  - Protect your patient from the elements. (Sun, Rain, Cold)
  - Talor your physical exam to meet the constraints of the environment.
    * palpate under clothing.
    * visualize one region at a time, then re-dress.
    * set up temporary shelter from the elements.

3. Head Injuries:
   a. A blow to the head may lead to increased intracranial pressure (ICP) or intracranial bleeding neither of which are managable in the wilderness. The job of the clinician is to diferentiate a serious life threatening injury from a minor one.

      1. Minor injury: No loss of consiousness (LOC), or LOC of less than 15 seconds with immediate return to full alertness. The casualty can not be on medications which increase risk of bleeding or have a history of bleeding disorders. Patient may be monitored every two hours for mental status changes, lethargy, irritibility, persistent nausea and vomiting, changes in speech or visual changes.

      2. Serious Injury: LOC greater than 15 seconds, and/or persisitent confusion or memory loss; signs or symptoms of increased ICP: Debilitating headache, mental status changes, persistent nausea and vomiting, appearance of clear fluid in external auditory canal, Battle sign, raccoon eyes, or seizures.

      3. Field treatment of serious head injury: Suspect injury of C-spine, manage airway-be able to clear vomit, elevate head 30 degrees. Evacuate to treatment facility ASAP.

4. Spinal Injuries.

   A. Cervical Spine

      1. High Risk Activities:

C-spine injuries in the wilderness usually occur after either a fall from a significant height or high-velocity accident from skiing. Common winter activities predisposing participants to C-spine injuries are skiing, snow-boarding, and snow-mobiling.

      2. Anatomy:

The cervical spine consists of seven cervical vertebrae interposed between the base of the skull and the thoracic spine. The cervical spine has a great deal of mobility to allow maximal range of motion for the skull. This increased mobility comes at a high price; this portion of the spine is less stable than the rest of the vertebral column, and hence more subject to injury. The C-Spine is stabilized by three longitudinal ligaments: The anterior longitudinal ligament (ALL) runs longitudinally along the anterior surface of the vertebral bodies. It is broad and very strong and helps to prevent hyper-extension of the head and neck. The posterior longitudinal ligament (PLL) also runs longitudinally, this time along the posterior surface of the vertebral bodies, within the vertebral canal itself. This ligament is relatively narrow and somewhat weaker than the ALL, and helps to prevent hyper-flexion at the neck. Third is the supraspinous ligament, connecting the spinous processes. This, too, is a strong ligament, acting to inhibit hyper-flexion.

The peripheral nerves innervating the muscles and the sensory nerves of the upper extremities originate from the cervical portion of the spinal cord and these are some of the nerves likely to be affected in the event of a C-Spine injury. Hence, it is important to assess the neurologic status of the upper extremities if a cervical injury is suspected.

3. C-Spine Injury Statistics:

-Most common injury: Flexion injury at C5/6
(so look for a deficit in the C6 distribution)

-28% of C-Spine injuries have another spinal fracture associated with it.
(so examine entire spine)

-10% of Head injuries/Facial fractures also have a C-Spine injury, especially if there was LOC.

4. Remember:

a. Head injury -> Assume C-Spine injury.
b. Perform complete Neurovascular exam
   -Motor, Sensory, Reflexes, Pulses, Babinski reflex.

B. Clear a C-Spine. (FMST.07.14a)

When faced with a possible C-Spine injury in the Wilderness, clearing it without X-ray should be considered only in extreme situations. However, there are situations which demand that every effort be made to make a casualty ambulatory, because the movement of the individual would demand herculean effort from the rest of the team, potentially placing more individuals at risk of injury or prolonging exposure to extreme environmental conditions.

Clearing a C-Spine in the field should follow a three-phased assessment. If the patient fails the assessment at any one of the phases, then the team is obligated to maintain spinal precautions and transport the patient. The phases are outlined below, and expanded upon in the lecture.

PHASE ONE:

-ensure no alteration of mental status.
-maintain in-line traction.
-perform complete neurovascular exam (focusing on the upper extremities)
-palpate spinous processes one-by-one while asking casualty whether palpation causes pain (supraspinous ligament)
-distinguish between pain over spinous process and muscle soreness associated with paraspinal musculature.
-only if palpation was completely pain-free can you move on to ...

PHASE TWO:

-continue to maintain in-line traction
-palpate anteriorly to the left and right of trachea. along the anterior vertebral bodies (anterior longitudinal ligament).
-Assess for bony tenderness under the angle of the jaw to the clavicle.
-Distinguish between tenderness of the sternocleidomastoid muscle and actual vertebral pain.
-only if palpation was pain-free can you progress to...

PHASE THREE:

-while loosely maintaining in-line traction, have patient move head through active range of motion

*first flexion/extension
*followed by rotational movement

-stop test at the first indication of pain with movement

Note: If you are able to move the patient's head through a full range of motion without pain, you have effectively cleared his C-Spine. Have the patient resume activity slowly and cautiously; evaluate any other complaints noted.

B. Thoracic and Lumbar Spine:

-Look for a T-L Spine injury with calcaneal fractures. Approximtely 10% of patients with a calcaneal fracture will also have an associated lumbar fracture.

C. Pelvis:

-Think HEMORRHAGE/SHOCK if faced with a Pelvic Fracture
Up to 6 liters of blood loss possible – (internal iliac areteries)

Place gentle constricting wrap around pelvis if open book fracture is suspected. (FMST.07.14b)

4. FRACTURES.

Some common wilderness fractures:

a. Metacarpal/Phalangeal	-the hands are your means of interfacing with the environment, sometimes it's a forceful meeting.
b. Distal Radius	-Very common wilderness fracture.
c. Scaphoid	-check for snuffbox tenderness, pain with axial load to thumb.
d. Lunate	-also common, pain dorsally at base of 3rd MC.
f. Clavicle	-need good NV exam, check for pneumothorax.
g. Long bone	-Radius/Ulna: if proximal, check elbow -Humerus: radial nerve runs in spiral groove -Tib/Fib: think compartment syndrome- Fib fx may ambulate with cane -Femur: think Hemorrhage, think traction.
f. Ankle	-commonly a fracture/dislocation, check NV
g. Hip	- Leg will typically be externally rotated

REMEMBER:

- Always think HEMORRHAGE with Long Bone fractures
- Always perform NEUROVASCULAR Exam before and after treatment
- Always SPLINT AND PAD for stability, comfort, function and transport.
- Indications for REDUCTION are: NV deficit , severe angulation, severe pain, if angulation prediposes to open fx or makes transport difficult.
- OPEN FRACTURES: Have 6-8 hours to get to surgery. Field Tx- Gently wash off with betadine/ iodine solution to get off obvious dirt- don't scrub. Wrap in sterile gauze. Don't place exposed bone back under skin. Give antibiotics: Augmentin, 2nd or 3rd generation cephalosporin, a quinolone or tetracycline.

THE REASONS WE PLACE FEMUR FRACTURES UNDER TRACTION.

1. Re-establishing length tightens fascia and tamponades bleeding.
2. Dramatic pain relief.
3. Helps prevent open fractures.

4. Helps reduce secondary soft tissue damage.
5. Increases ease and safety of transport.

SOME BASIC RULES OF EXTREMITY SPLINTING:

- Splint all fractures before moving the casualty, unless in imminent danger.
- Splint all fractures as they are found, unless severe angulation complicates transport or is causing neurovascular deficits.
- Splint to include the joint above and below.
- Construct splints using uninjured extremity, then splint in the position of function, comfort or stability.

5. DISLOCATIONS.

a. Rapid Diagnosis and Reduction is imperative, if evacuation time is > 1-2 hours.

- Easier to reduce immediately after injury. (muscle spasm)
- Makes transport much easier. (increased patient comfort)
- Dramatic pain relief.
- Early reduction reduces risk of Neurovascular injury.
- Reduction could make difference between an ambulatory vs. litter patient
  the safety of the entire expedition could be placed at risk

b. Signs and Symptoms: Remember, NO X-RAYS!

- Decreased Range of Motion.
- Obvious Deformity. (compare right vs. left)
- Usually NO Crepitus.
- Typical Posture.

c. Reduction Techniques:

- Phalanges	Linear traction, buddy tape. Can not reduce volar displacement of 1st phalanyx.
- Shoulder	Bedsheet method- counter traction. Prone with weights method. External rotation. Snowbird technique.
- Patella	Extension with medially directed pressure.
- Ankle	Linear traction with knee flexion.

- Hip — Anchor pelvis, hip flexed at 30 degrees, upward traction.

- Sternoclavicular — Roll between scapulae, and apply forceful downward pressure.

-Knee — Should only be reduced as an absolute last resort. Apply anterior traction gently.

REMEMBER: Analgesia and Sedation should be utilized when available before attempting joint or fracture reduction, unless multiple traumatic injuries are present.

6. EVACUATION CRITERIA:

Conditions demanding expeditious evacuation: (FMST.07.14c)

1. Suspected cervical, thoracic, and/or lumbar spine injuries.
2. Pelvic injury with instability and/or significant blood loss.
3. Any open fractures. (6-8 hour window, antibiotics early)
4. Compartment syndromes. (pain with passive motion, pallor, pulseless, paresthesia)
5. Hip or Knee dislocations.
6. Any other injury the medical provider feels he/she is not prepared to manage.

7. OTHER TRAUMA.

Heavy bleeding: Usually can be treated by direct pressure, if a tourniquet is required,release pressure every five minutes to reassess.

Contusion: Apply ice 1st 48 hours which will provide pain relief and limit expansion of hematoma.

Lacerations and avulsions: High pressure irrigation with clean water, may remove debris with flame treated forceps. Use steri-strips for minor clean lacs. Keep dirty or puncture wounds open and give antibiotics.

Animal Bites: Copious irrigation, and antibiotic, do not close. Consider rabies if bite was unprovoked and animal was acting strange. (bats, raccoons, skunks, canines, and felines) Has never been seen in livestock, rabbits, squirrels, rats or mice. Most common zoonotic source in North America is the skunk.

Severed part: Can be reattached up to 24 hours after injury if the cut is clean. Gently clean part, wrap in gauze and keep cool without freezing.

UNITED STATES MARINE CORPS
Mountain Warfare Training Center
Bridgeport, California 93517-5001

FMST.07.38
04/02/02

STUDENT HANDOUT

REPTILE AND ARTHROPOD ENVENOMATION BITES AND STINGS

TERMINAL LEARNING OBJECTIVE: Given a simulated casualty, in a wilderness environment, treat reptile and arthropod envenomation, in accordance with the references. (FMST.07.38)

ENABLING LEARNING OBJECTIVES:

1) Without the aid of references, select from a given list the first aid measures used for pit viper envenomation, in accordance with the references. (FMST.07.38a)

2) Without the aid of references, select from a given list the first aid measures used for coral snake envenomation, in accordance with the references. (FMST.07.38b)

3) Without the aid of references, select from a given list the first aid measures used for widow spiders (lactordectus species) bites, in accordance with the references. (FMST.07.38c)

4) Without the aid of references, select from a given list the first aid measures used for brown recluse spider (loxosceles species) bites, in accordance with the references. (FMST.07.38d)

5) Without the aid of references, select from a given list choose the first aid measures used for bee, wasp, and hornet stings, in accordance with the references. (FMST.07.38e)

12

OUTLINE.

1. **PIT VIPERS**. Crotalidae have a triangular head, cat-like vertical pupils, hinged fangs and a heat-sensitive "pit" on each side of the head between the tip of the nose and the eye. Rattlesnakes have a variable number of rattles depending upon age and number of molts. They sometimes strike without rattling. About 60% of this country's venomous bites are attributed to rattlesnakes. Cottonmouths (water moccasins) and copperheads are the other two commonly encountered North American pit vipers. Copperhead and cottonmouth venom are quite similar but weaker than most rattlesnake venom.

a. Anatomy of Pit Vipers.

(1) Venom apparatus: Venom gland; Compressor glandular muscle; Primary duct; Accessory gland; Secondary duct; Fang sheath; and fangs.

(2) Facial pit organs: Highly sensitive paired receptors of infrared radiation that can detect temperature changes of less than 0.2°C; used to detect warm-blooded prey/predators and aim strike; also pit organs may have a role in determining volume of venom injected.

(3) Rattles or buttons: Interlocking keratin rings at tip of tail in rattlesnakes; a new rattle is added with each molting cycle (every 50-400 days); some are lost due to trauma.

(4) The strike: Aimed primarily by facial pits; rarely strike farther than 1/2 their body length; speed - approximately 8 feet/second.

(a) Rattlesnakes generally inject 25-75% of their venom when they bite humans.

(b) Can bite without injecting any venom ("dry bite").

1. 20% - no envenomation.
2. 20-30% - mild envenomation.
3. 30-50% - moderate to severe envenomation.

b. Signs and Symptoms of Snake Venom Poisoning:

(1) Pit Vipers:

(a) Local:

1. Puncture wounds/scratches:

A. Pattern can be misleading (venomous vs. nonvenomous).

B. Must differentiate from other animal/insect bites or plant puncture wounds- usually venomous snakebite wounds are larger, with more bleeding.

C. May be one or many fang marks present; range from a few mm to 4cm apart.

2. Pain, edema, erythema:

A. Usually presents within 30 minutes if envenomation.

B. More severe envenomation generally yields more rapid progression and more severe pain.

C. Pain is usually characterized as burning and immediate in onset.

D. Pain usually confined to bite site.

E. Edema limited to subcutaneous tissues (with no increase in intra-compartmental pressures).

F. Usually starts within 5 minutes (if none in 10 minutes, probably no significant envenomation unless dealing with Mojave rattlesnake).

G. Gradually spreads up extremity for 36 hours.

H. Danger of airway obstruction in bites to head and/or neck.

3. Ecchymosis:

A. Ecchymosis starts within several hours.

B. May involve entire extremity.

4. Lymphangitis:

A. Venom absorbed rapidly through lymphatic system results in lymphatic inflammation.

5. Petechiae, vesicles, hemorrhagic bullae:

A. Will occur in untreated rattlesnake envenomation.

B. Rarely seen when treated early with adequate antivenin.

C. Usually occurs 6-36 hr after bite.

6. Necrosis, tissue destruction:

A. Due to direct venom effects.

B, Can be prevented if adequate antivenin is given within 2 hours of bite.

(b) Systemic:

1. Nausea, vomiting:

A. Common (early onset may indicate severe envenomation).

2. Weakness, diaphoresis, chills, dizziness/vertigo.

A. Syncope is common with all pit viper bites.

B. Frequency is proportional to severity of bite.

3. Change in taste:

A. May complain within minutes of rubbery or metallic taste.

4. Increased salivation.

5. Fever.

6. Tingling, numbness in scalp/face/fingers/toes:

A. Can occur within 10 minutes and indicates moderate to severe envenomation.

7. Fasciculations in face/neck/back/ or other involved extremity:

A. Can occur early and indicates severe envenomation.

8. Visual disturbances:

A. Blurred vision, yellowing of vision and blindness.

9. Tachycardia, bradycardia.

10. Hemorrhage, hemolysis, thrombosis and DIC:

A. Bleeding from wounds, gingival bleeding, epistaxis, Hematuria, hematemesis, Melena, lower GI bleeding, hemoptysis, peritoneal hemorrhage and cerebral hemorrhage.

B. Can occur as early as 6 hours; more common 12-72 hours after bite.

C. Systemic bleeding only occurs in moderate to severe envenomation.

D. Most coagulopathic effects secondary to proteases acting at various sites in the coagulation cascade.

11. Hypotension, shock:

A. Can occur rapidly in severe envenomation.

B. Early, due to pooling of blood.

C. Late may be due to loss of volume.

D. Generally little effect from decreased cardiac contractility.

12. Pulmonary edema:

A. Common in severe envenomation.

B. Secondary to toxin and inflammatory injury of pulmonary capillary membranes and pooling of blood in major vessels and capillary beds of the lung.

C. Compounded by direct cardiodepressant factor in some venoms.

13. Oliguria, anuria:

A. Hypotension with resultant decreased GFR. (#1).

B. Hemoglobinuria and myoglobinuria with renal tubular obstruction.

C. Direct toxic effect of venom on kidneys.

D. Acute tubular necrosis from rhabdomyolysis.

14. Paresis, Paralysis:

A. Seen with Eastern Diamondback and Mojave rattlesnakes.

B. Uncommon with other rattlesnakes.

15. Respiratory failure:

A. Uncommon with pit viper bites (except Mojave).

B. Complicated by cardiovascular failure.

16. Seizures:

A. Probably due to hypertension and hypoxia.

17. Coma:

A. Usually sensorium is clear.

B. When coma occurs, it is secondary to cerebral anoxia.

18. Death:

A. Generally occurs in 6 to 48 hours in untreated severe envenomation.

B. Can occur sooner with IV envenomation.

c. First Aid Measures for Pit Viper Envenomation.
(FMST.07.38a)

(1) Do not try to kill the snake and bring it in for identification.

(a) Only vital to identify the snake when Mojave rattlesnake or coral snake bites are a possibility, as management is altered.

(b) Risk of a second bite to the victim or rescuer.

(c) Should never delay transportation of the victim.

(2) Remove any jewelry, which could become a tourniquet with progression of edema.

(3) Calm and reassure patient to decrease heart rate and circulation of venom.

(4) Constriction band.

(a) Most widely accepted first aid measure in the literature.

(b) Should be more than 1/2 inch wide.

(c) Apply to proximal end of extremity affected and wrap to approximately 5 cm above swelling (site of envenomation). Then wrap back towards proximal end of extremity from where the wrap started.

(d) Apply only tight enough to occlude lymphatic and superficial venous return.

(e) Probably no benefit when applied 30 min or more after bite.

(f) Maintain until antivenin is started.

(5) Incision and suction.

(a) Very controversial; should only be used by experienced medical personnel.

(b) Do not incise wound.

(c) Suction best applied by mechanical means (Sawyer extractor).

(d) Do not use mouth (unknown oral lesions may expose rescuer to risk).

(e) Should be started within 30 minutes of envenomation.

(f) Continue suction for 30 to 60 min.

(6) Rest and immobilization.

(a) Put victim at rest and splint extremity in position of function, at or just below heart level.

(b) Allow room for swelling in splint.

(7) Ice.

(a) Avoid any method of cooling.

(8) Watch for adverse reactions.

(a) Rare documented cases of anaphylaxis caused by snake venom in patients previously bitten or otherwise exposed to snake venom.

(b) Some evidence that venom can activate the alternate-complement system directly and cause a similar reaction.

(9) Rapid transport to a medical facility.

(a) The most important measure in first aid.

(b) No measures should delay this; antivenin is the only proven definitive therapy.

2. **CORAL SNAKES**. The elapidae species of coral snakes are brightly colored, with black noses and alternating red-yellow-red-black bands around their bodies (remember "red on yellow, kill a fellow"). They have relatively small mouths with fixed fangs. From southern Mexico through tropical South America, the rules for distinguishing coral snakes are highly unreliable. Unless you are a knowledgeable herpetologist, it is best not to pick up colorful snakes in the tropics.

a. Signs and Symptoms of Coral Snake Envenomation:

(1) Fang marks may be hard to see.

(2) Frequent delay in onset of symptoms, followed by extremely rapid progression.

(3) Little or no pain at bite site.

(4) No local necrosis.

(5) Earliest evidence may be drowsiness or euphoria.

(6) N/V, increased salivation, paraesthesias at bite site.

(7) Bulbar paralysis progressing to peripheral paralysis.

(8) Paraesthesias, and fasciculation.

(9) Occasional seizures.

(10) Mild Hypotension.

(11) Death is due to respiratory and cardiovascular failure.

b. First Aid Measures For Coral Snake Envenomation. (FMST.07.38b)

(1) None of proven benefit except rest, reassurance and rapid transportation to medical facility for antivenom administration.

3. **GILA MONSTERS**. No direct injection mechanism for their venom, but have powerful jaws that allow them to chew and tear at their victims. Drooling venom produces a substantial amount of pain and possible complications resembling pit viper envenomation. Follow the recommendations for pit viper envenomation. Fatal Gila monster encounters are extremely rare.

4. **VENOMOUS ARTHROPODS**. Arthropods represent the largest phylum in the animal kingdom. It contains spiders, scorpions, insects, ticks, kissing bugs, water

bugs, caterpillars, moths, butterflies, grasshoppers, centipedes, and millipedes, among others. Some arthropods sting (bees, ants, scorpions, etc.), others bite (spiders, centipedes, kissing bugs, etc.), while still others discharge a secretion that is toxic (millipedes, caterpillars, etc.). Although the number of arthropods that are sufficiently venomous to endanger humans is not known, these animals are implicated in far more bites and stings than all other phyla combined. Almost all of the 20,000 species of spiders are venomous, but luckily for, man, only relatively small numbers have fangs long and strong enough to penetrate the human skin. There are some 500 species of scorpions and all are venomous, although only a small number pose a significant danger to humans.

a. The number of deaths from arthropod stings and bites is not known, nor do most countries keep records of the incidence of such injuries. In Mexico, parts of Central and South America, and in North Africa, deaths from scorpion stings may exceed several thousand per year. Spider bites probably account for no more than 200 deaths per year, worldwide. The number of deaths from arthropod bites or stings in the temperate countries is far greater than the number of deaths from snakebite. Almost all of these deaths, however, are from anaphylactic reactions.

5. **SPIDERS**. -There are at least 200 species of spiders that have been implicated in significant bites to humans (Russell, 1988a). Spiders got the misplaced blame for the bites seen in one large series of case studies. Of some 600 suspected spider bites, 80% were found to be caused by other arthropods. The arthropods most frequently involved in these mis-diagnoses were ticks (including their embedded mouth parts), kissing bugs, mites, bedbugs, fleas, flies, beetles, water bugs and various other stinging arthropods.

a. Lactordectus species (Widow spiders): It is most commonly referred to as the Black Widow, Brown Widow or Frog-legged spider depending on the species. There are many other commons names such as: hourglass, poison lady, deadly spider, red-bottom spider, T-spider, Gray Lady spider, or shoe-button spider. Although both male and female widow spiders are venomous, only the latter have fangs large and strong enough to penetrate the human skin. Mature females range in body length from 10 to 18mm, whereas males are from 3 to 5mm.

(1) Signs and Symptoms of spider envenomation.

(a) Sharp, pinprick-like bite, but in some cases the incident is so minor that it goes unnoticed.

(b) Initial bite-pain is often followed by dull-numbing pain in the affected extremity, and cramps in one of the large muscle masses.

(c) Muscle fasciculations can be seen within 30 minutes of the bite.

(d) Sweating is common.

(e) Complaints of weakness and pain in the regional lymph nodes.

(f) Pain in the lower back, thighs, or abdomen is common and rigidity of the abdominal muscles is seen in severe envenomations.

(g) Severe paroxysmal muscle cramps may occur accompanied by arthralgias.

(h) Tissue reaction around the bite may vary, depending on the species involved. In most cases in the United States, there is no reaction at the bite site and the puncture wounds may be impossible to find. There may be an immediate 2 to 4 mm blanched area around the puncture wounds, surrounded by a slightly erythematous area, which in time becomes pallid. This may persist for 20 to 30 minutes.

(i) Hypertension is a characteristic finding in moderate to severe poisonings. It usually appears 2 to 3 hours post envenomation.

(j) Elevation of body temperature.

(k) Increased salivation.

(l) Anorexia.

(2) Treatment for Widow Spider envenomation.
(FMST.07.38c)

(a) First aid measures are supportive in nature. If the pain is intense, ice can be placed over the wound until the patient arrives at a medical facility.

(b) Antivenin is given in severe cases at medical facilities.

b. Loxosceles species (Violin spiders): Are known in the United States as fiddle-back, or brown recluse spiders. There are over 100 species of Loxosceles. The abdomen of these spiders varies in color from gray to orange or reddish-brown to dark brown. The violin on the cephalothorax is brown to blackish and distinct from the pale yellow to reddish-brown background of the cephalothorax. This spider has three pairs of eyes, forming a curved row. Females average 8 to 15 mm in body length whereas males average 6 to 10 mm. Both are venomous.

(1) Signs and Symptoms of Violin spider envenomation.

(a) Pain; about the same degree of a bee sting but patients may be completely unaware of the bite.

(b) Localized burning sensation around the site of injury.

(c) Pruritus is often present and the area begins to appear red, with a small-blanched area around the immediate bite site.

(d) The reddened area enlarges during the subsequent 1- 8 hours. It often becomes irregular in shape and, as time passes, hemorrhages may develop throughout the area.

(e) A small bleb or vesicle forms at the bite site and increases in size.

(f) Vesicle subsequently ruptures and a pustule forms.

(g) The whole area becomes swollen and painful.

(h) Lymphadenopathy may develop.

(i) Necrosis can develop from superficial to deep layers.

(j) Rare systemic signs and symptoms include:

1. Malaise.
2. Stomach cramps.
3. Nausea and vomiting.
4. Jaundice.
5. Spleen enlargement.
6. Hemolysis.
7. Hematuria.
8. Thrombocytopenia.

(2) Treatment for Loxosceles spider envenomation: (FMST.07.38d)

(a) There are no first-aid measures of value. In fact, all first-aid measures should be avoided, because the natural appearance of the lesion is most important in determining the diagnosis.

(b) An ice pack may be placed over the wound if the pain is severe.

(c) Excision of the bite site is no longer recommended as this has been demonstrated to cause worsening of necrosis and tissue damage. Definitive therapy includes use of dapsone, antibiotics, tetanus prophylaxis and possible steroid use.

c. There are many other species of spiders that include Steatoda species (cobweb spiders), Phidippus species (jumping spiders), Chiracanthium species (running spiders), Lycosa species (wolf spiders), and the newest member to the United States is the Hobo spider. There have been several reported deaths in the Northwestern United States. Originally a European species of brown spider, the Hobo spider was introduced into the U.S. in 1936, and little research has been done on it to date.

6. **SCORPIONS**. Approximately 75 of the 650 species of scorpions can be considered of sufficient importance to warrant medical attention. In the United States, members of the genera Hadrurus and Vejovis are capable of inflicting a painful and erythematous, but non-lethal lesion. However, species of the genus Centruroides are sufficiently dangerous to warrant definitive medical care. Most stings are inflicted by the Vejovis species.

a. Centruroides species. There are approximately 30 species of this genus confined to the New World. Of these, eight are of considerable medical importance, and most of these species are found in Mexico. In the United States, they are commonly referred to as "bark scorpions", because of the preference for hiding under the loose bark of trees, in dead trees, or in logs. Their general color is straw to yellowish-brown or reddish-brown, and they are often easily distinguishable from other scorpions in the same habitat by their long thin tail with its "telson" on the end, the pedipalps, or pincer-like claws. They can reach lengths up to 55 mm.

(1) Signs and symptoms of envenomation.

(a) Initial pain, rarely severe.

(b) Area becomes sensitive to touch (tap test).

(c) Tachycardia may ensue after one hour.

(d) Respiratory rate increases.

(e) Fasciculations may be seen over the face or large muscle masses.

(f) Complaints of generalized muscle weakness.

(g) Respiratory distress may progress to paralysis.

(h) Slurring of speech may be present.

(i) Convulsions have been reported.

(2) Treatment for scorpion envenomation.

(a) There is no first-aid measure of value for scorpion stings.

(b) Ice may be placed over the wound to reduce pain.

(c) In severe cases, antivenin may be recommended.

b. Vejovis species is closely related to the Urectonus. Both genera are ground scorpions. Whereas Urectonus usually inhabits mountain habitats from southern California to Oregon, Bejovis has a wide distribution from the southern portions of Canada, south through Wyoming and Colorado to Texas and west to California. Some species measure up to 85mm, but most are from 30 to 35mm. Vejovis spinigerus, a common western U.S. species, is often called the striped-tail scorpion.

c. Hadrurus species is native to North America. These are the longest and most stout-bodied of our scorpions, generally referred to as the giant hairy scorpions because of their size and the conspicuous bristles on their legs, pedipalps and caudal segments. Adults may can measure up to 135 mm. These are burrowing species and may be found as deep as 2 feet in sandy soil. They are native to Arizona, California and parts of Utah, Nevada and Idaho, as well as Mexico.

7. **<u>BEES, WASPS, HORNETS AND ANTS</u>**. While it may take over 100 bees to inflict a lethal dose of venom in most adults, one sting can cause a fatal anaphylactic reaction in a hypersensitive person. There are 3 to 4 times more deaths in the United States from bees than from snakebites. In the few fatalities that have resulted from multiple bee stings, death has been attributed to acute cardiovascular collapse.

a. Treatment for bee stings: (FMST.07.38e)

(1) The stingers of many bees may remain in the skin and should be removed by teasing or scraping rather than pulling, to avoid even more venom being injected into the patient.

(2) Ice may be applied to the sting site.

(3) Persons with known hypersensitivity to stings should carry a Bee-Sting Kit, containing an antihistamine and epinephrine when in endemic areas.

(4) Desensitization can be carried out using whole-body antigens or, preferably, whole-venom antigens.

8. **<u>TICKS AND MITES</u>**. Carry many diseases. In North America, some species of Dermacentor and Amblyomma cause tick paralysis.

a. Signs and symptoms of Tick Paralysis.

(1) Restlessness and irritability.

(2) Lethargy and anorexia.

(3) Generalized muscle weakness.

(4) Incoordination and ataxia.

(5) Nystagmus.

(6) Ascending flaccid paralysis with loss of deep tendon reflexes.

(7) Bulbar or respiratory paralysis may develop.

b. Treatment for Tick Paralysis:

(1) Remove the tick.

(2) Treat symptomatically (ventilator support may be necessary).

(3) Lesions should be cleaned.

(4) Antivenin and/or corticosteroids are used in severe cases.

9. **LEPIDOPTERA (CATERPILLARS)**. Caterpillars commonly cause envenomation in humans. There are at least ten families of caterpillars that are venomous.

a. MEGALOPYGE OPERCULARIS (Puss Caterpillar or Woolly Slug) The most commonly seen venomous caterpillar in the United States.

(1) Description: Hairy, flat, ovoid shaped: 30-35 millimeters in length.

(2) Location: Texas north to Maryland and Missouri.

(3) Stinging apparatus: Consists of spines intermingled with the hairs of the body:

b. SIGNS AND SYMPTOMS.

(1) Needling pain (Can be intense).

(2) Redness.

(3) Swelling.

(4) Nausea.

(5) Headache.

(6) Fever.

(7) Vomiting.

(8) Lymphadenopathy.

(9) Hypotension.

(10) Shock.

(11) Bleeding of the mucus membranes.

a. Treatment

(1) Apply contact tape to affected area, and remove tape. This will effectively remove spines from affected area.

(2) Group I corticosteroid cream and ointment.

(3) Codeine or Demerol may be required for pain.

(4) Antiemitics may be needed to control vomiting.

Note: Anaphylactic reactions have been reported, but are rare.

UNITED STATES MARINE CORPS
Mountain Warfare Training Center
Bridgeport, California 93517-5001

FMST.07.10

04/02/02

STUDENT HANDOUT

FREEZING AND NEAR FREEZING TISSUE INJURIES

TERMINAL LEARNING OBJECTIVE: Given a casualty in a cold weather environment, and necessary equipment and supplies manage common cold weather injuries in a cold weather environment to prevent death or further injury per the reference. (FMST.07.10)

ENABLING LEARNING OBJECTIVES.

1) Without the aid of references, from a given list select the correct definition of frostbite, in accordance with the references. (FMST.07.10a)

2) Without the aid of references, from a given list select the correct two mechanisms of injury for frostbite, in accordance with the references. (FMST.07.10b)

3) Without the aid of references, from a given list select the correct signs and symptoms of frostbite, in accordance with the references. (FMST.07.10c)

4) Without the aid of references, from a given list select the correct favorable post thaw signs of a frostbite injury, in accordance with the references. (FMST.07.10d)

5) Without the aid of references, from a given list select the correct unfavorable post thaw signs of a frostbite injury, in accordance with the references. (FMST.07.10e)

6) Without the aid of references, from a given list select the correct field treatment for frostbite, in accordance with the references. (FMST.07.10f)

12

7) Without the aid of references, from a given list select the correct primary field treatment consideration when treating frostbite in the field, in accordance with the references. (FMST.07.10g)

8) Without the aid of references, from a given list select the correct four methods to reduce the chance of becoming a frostbite casualty, in accordance with the references. (FMST.07.10h)

9) Without the aid of references, from a given list select the correct definition of immersion foot, in accordance with the references. (FMST.07.10i)

10) Without the aid of references, from a given list select the correct mechanism of injury for immersion foot, in accordance with the references. (FMST.07.10j)

11) Without the aid of references, given a list of categories and characteristics match the category of immersion foot to its characteristics, in accordance with the references. (FMST.07.10k)

12) Without the aid of references, from a given list select the correct field treatment of immersion foot, in accordance with the references. (FMST.07.10l)

OUTLINE.

1. **FROSTBITE**

a. Definition: Frostbite is the actual freezing of tissue. (FMST.07.10a)

(1) Frostbite usually occurs as a result of exposure to subfreezing temperatures for a long period of time. However, brief exposure to high wind-chill or very cold temperatures can also cause frostbite. Frostbite usually occurs after more than twelve hours of exposure to subfreezing temperatures. Skin temperature of the extremities is dependent on heat production and is more a function of underlying muscle mass than of circulation. Thus, the areas of the body most prone to frostbite (hands, feet, ears, nose, chin and molar regions) have small amounts of underlying muscle tissue.

(2) Incidence. Frostbite, unlike hypothermia tends to spare the extremes of age. Thus, it is most prevalent in the middle years of life and like hypothermia, the greatest incidence occurs in urban areas. Other factors predisposing a person to frostbite include: poor nutritional status, dehydration, prior cold injury, individuals from the southeastern part of the United States, diabetics, individuals with atherosclerotic disease, and use of tobacco products.

(3) Although the actual numbers are not known, persons born in the south are 3.7 times more susceptible to frostbite. Persons with a history of prior cold injury are 1.7 times more prone to have a second injury. This includes parts of the body other than the originally frostbitten region.

b. Mechanisms of Tissue Injury. Tissue injury resulting from frostbite is produced in two ways: (FMST.07.10b)

(1) *Actual freezing of the tissues.* Ice crystal formation in the extra-cellular space leads to an osmotic gradient across the cell membrane drawing water out of the cell. This process, known as crenation, causes a marked change in intra-cellular electrolyte and enzyme concentrations, which can lead to cell death or recovery depending on the duration of exposure to freezing conditions and post thaw care of the tissue. Intra-cellular ice crystal formation is much more destructive, however, this requires much faster freezing rates.

(2) *Obstruction of blood supply to the tissues.* Cell injury and death caused by ice crystal formation leads to the release of inflammatory mediators, which damage vascular endothelial cells. The net result is increased vascular permeability. Fluid leaks out of the vascular space and into the intercellular space causing tissue edema and increased blood viscosity. Elevated viscosity combined with platelet aggregation, to the walls of damaged blood vessels, lead to vascular obstruction with resultant tissue damage.

c. Signs/Symptoms. (FMST.07.10c) Initially there is discomfort and pain in the affected tissues, which can progress to loss of sensation. The skin appears white and bloodless. There is decreased or no capillary refill. The skin may feel waxy and will not glide freely over a bony prominence. With minor injury, blisters form, usually in the first 24 hours, and edema may also be present. Deep injury is not associated with substantial formation of blisters. There is also little edema seen in deep injury.

(1) Favorable prognostic signs of frostbitten tissue. These signs indicate probably little or no tissue loss, post thaw. (FMST.07.10d)

(A) Warmth of tissue.

(B) Normal tissue color.

(C) Preservation of sensation.

(D) Blister formation. *A* blister extending the full distal length of the injured digit is a good prognostic sign. However, a blister limited to the line of demarcation between normal and frostbitten tissue is a less favorable sign.

(E) Edema persisting more than 24 hours.

(F) Rapid capillary refill.

(2) Unfavorable prognostic signs. These indicate a poor prognosis with probable tissue loss, post thaw: (FMST.07.10e)

(A) Complete absence of edema.

(B) Cyanotic tissue.

(C) Continued loss of sensation.

(D) Affected areas continue to stay cold.

(3) Since it is extremely difficult to predict ultimate tissue loss from initial examination, any classification must be done retrospectively. Classically, frostbite is divided into four degrees of severity:

(A) First degree - Erythema, edema, transient tingling and/or burning.

(B) Second degree - Blisters, edema, anesthesia and/or paresthesias.

(C) Third degree - Involves the entire thickness of skin and extends into subcutaneous tissue.

(D) Fourth degree - Involves the entire thickness of the skin along with muscle and bone.

(F) Frostnip is a completely reversible injury that is technically not considered frostbite. Frostnip classically presents as a white patch and numbness and is readily warmed in the field by placing the affected extremity in one's axilla or groin. No resultant skin changes occur.

d. Field Management (FMST.07.10f)

(1) The correct treatment for frostbite is as follows:

(A) Once an extremity has suffered frostbite, the primary consideration of treatment is not to rewarm the region if there is a chance of refreezing. (FMST.07.10g) This occurs because the amount of tissue damage is greatly increased if the part is thawed then allowed to re-freeze. Generally, the sooner a frostbitten region is thawed the lower the probability and extent of damage. Rapid rewarming is the preferred method of thawing a frostbitten region. This consists of immersion of an affected part in circulating warm water of 104°-108°F until the tissues are soft and pliable, which usually takes about 40-60 minutes.

(B) If slow rewarming methods (such as allowing a part to sit at room temperature) is used, some refreezing of the melted fluid occurs and the new ice crystals are actually larger than those formed during the original freezing causing further tissue destruction.

(C) Out dated methods such as rubbing with snow or rewarming next to an open fire, etc., only enhance tissue damage.

(D) Once thawing has taken place no benefit of continued rapid rewarming can be demonstrated.

(E) Cold injuries that are old and have thawed spontaneously should not be rapidly re-warmed again.

(F) Nicotine use should be prohibited since it is a vasoconstrictor and may enhance tissue loss.

(G) There is no need for ointments, salves, etc.

(H) Blisters should be allowed to remain intact.

(I) Medevac all cases of frostbite, as soon as the tactical situation permits.

(J) One final note; in military tactical situations such as ours, the victim often must walk down below the snow line to reach a BAS or base camp where rapid rewarming can be carried out. In this situation let the victim walk if he has frostbitten feet. Once his feet have thawed, he will be unable to walk and will have to be carried out.

(K) Hospital and long-term care: After re-warming the frostbitten tissue, attention must turn to prevention of infection and functional restoration. The patient's tetanus status should be determined and meticulous care should be taken to keep the injury clean. Daily whirlpool therapy for 30-45 minutes has been shown to be beneficial. The use of post-thaw medications is controversial but there is evidence that topical aloe vera and NSAIDS to control inflammation are probably beneficial. Some studies have demonstrated benefit from the use of allopurinol as a free radical scavenger. Also, animal studies have shown a definite benefit from the use of 90% DMSO gel, however, this use is not FDA approved on humans and it may be difficult or impossible to obtain this medication. The wide scale use of heparin and dextran 40 has been suggested but a definitive recommendation that supports their use still cannot be made. Studies suggesting the use of thrombolytics are very small and not well controlled therefore there is no current recommendation for use of thrombolytics. Lastly, it has been shown that there is a clear benefit for the use of early and aggressive physical therapy to preserve function and prevent contracture.

e. Prevention (FMST.07.10h)

(1) Some general principles for the prevention of frostbite are listed below

(a) Dress to keep comfortably cool. The head radiates 30-50% or more of all body heat and so goes the old adage, "If your feet are cold put on your hat".

(b) Nutrition. Eat the right type and quantity of foods.

(c) Avoid tight fitting and constrictive clothing. This decreases blood flow and increases the chance of frostbite. Rings, jewelry, etc., conduct heat away from the skin and increase the chance of frostbite.

(d) Wear mittens instead of gloves. They are much warmer. For work involving a high level of dexterity use contact gloves.

(e) Do not touch metals with bare hands at extremely low temperatures. Your flesh will stick to the surface and this will result in subsequent frostbite and tissue loss. If you find yourself in such a predicament pour a warm fluid over the metal to thaw your flesh off the metal.

(f) Avoid nicotine.

2. **IMMERSION FOOT**.

a. Definition. (FMST.07.10i) Immersion foot (AKA trenchfoot) is a non-freezing injury of the extremities in which the tissues are damaged.

b. Usually a disease of the feet (but it can occur in the hands), it occurs when the extremity is exposed to a cold-wet environment (external temperatures above freezing and less than

$50^{o}F$) for long periods of time (usually greater than 12 hours). Prolonged standing enhances the pathologic potential for injury due to wet and cold conditions.

c. Mechanism of Injury. (FMST.07.10j) Even though this injury is called immersion foot, actual immersion of the feet in water is not necessary. This injury can also occur in the hands.

(1) The circulation of blood to the affected extremity is reduced because the victim's extremity is cold, wet and the peripheral blood vessels have constricted to reduce overall heat loss.

(2) Because prolonged cooling damages sensitive tissues, nerve injuries are also involved. This nerve damage is responsible for the severe pain and paresthesias. Damaged skin and subcutaneous tissues result in erythema, edema and possible blister formation.

d. Signs/Symptoms Early signs include numbness and paresthesias followed by complete numbness and sometimes leg/arm cramps. The skin at first appears erythematous but becomes progressively pale and mottled then grayish-blue as the condition progresses. Often there is a large amount of edema and blister formation. In severe cases the arterial pulses are absent. Immersion foot is classified by severity: (FMST.07.10k)

CATEGORY	CHARACTERISTICS
Minimal	Reddening of skin; slight sensory changes
Mild	Edema; sensory changes (reversible)
Moderate	Edema, redness, blisters, intra-cutaneous hemorrhage; irreversible nerve damage
Severe	Severe edema, massive intra-cutaneous hemorrhage; necrosis, gangrene

e. Field Management (FMST.07.10l) Treatment consists of:

(1) Pat drying of extremity.

(2) Gentle rewarming.

(3) Elevation of affected extremity.

(4) Bed rest.

(5) Treatment after this is supportive. For more severe injuries, debridement or resection of a portion of the foot may be necessary. However, this is best accomplished at a higher echelon of care. Do not be fooled by this injury. These patients can get

extremely ill from wet-gangrene which may necessitate amputation of tissue. However, the initial treatment should be as conservative as possible. DO NOT TREAT THESE PATIENT'S WITH WATER IMMERSION THERAPY. If infection is present obtain surgical consult early, and plan on small sequential amputations.

3. **CHILBLAINS (PERNIO).**

a. Definition. Chilblains are a non-freezing injury that leads to red inflamed, pruritic skin.

b. Mechanism of Injury. Pernio is caused by intermittent exposure to above freezing temperatures and high humidity. (More extreme cold is usually associated with lower humidity). Classically, chilblains occur in young individuals and most commonly on the anterior surface of the legs and the dorsum of the hands. The cause of pernio is not clearly known, but it is thought to reflect an abnormal vascular response to the environment.

c. Signs/Symptoms. Skin is red, tender, and pruritic, edematous and blanches with pressure. Itching may be severe and hyperhydrosis can occur.

d. Management. The field treatment is massage and gentle heat. This treatment is usually self-limiting and will not recur without re-exposure. Topical steroids are effective for itching and burning.

4. **CORNEAL FROSTBITE**

a. Definition. Actual freezing of the cornea.

b. Mechanism of injury. This occurs when large amounts of cold air pass over and cool the cornea, resulting in corneal clouding and loss of vision. This can occur as the result of even short periods of exposure to activities associated with high wind chill i.e. snowmobiling.

c. Signs/Symptoms. On examination the cornea will be clouded over.

d. Treatment and Prevention. Antibiotic ointment and DO NOT PATCH THE EYES. Prevention is the key.

UNITED STATES MARINE CORPS
Mountain Warfare Training
Center Bridgeport, California
93517-5001

FMST.07.22
04/03/02

STUDENT HANDOUT

LAND NAVIGATION REVIEW

TERMINAL LEARNING OBJECTIVE In a mountainous environment, navigate in mountainous terrain, in accordance with the references. (FMST.07.22)

ENABLING LEARNING OBJECTIVES

1) Given a map and compass, orientate the map, in accordance with the references. (FMST.07.22a)

2) Given a map and a protractor locate a six-digit grid coordinate, in accordance with the references. (FMST.07.22b)

3) Given a map, compass, and protractor perform a resection to locate an unknown point from three known points, in accordance with the references. (FMST.07.22c)

4) Without the aid of references, select from a given list the correct definition for true north, grid north, and magnetic north, in accordance with the references. (FMST.07.22d)

OUTLINE.

1. **THE MAP**. Before you can properly use a map, there is some basic information you will need to understand.

a. Definition. A map is a reduced or scale drawing of the ground and important things on the ground as seen from the air. Essentially it is a picture of the surface of the earth, as it would appear looking at it from an aircraft.

b. Marginal Information. As you take a quick look at the borders, you will notice that there is a large amount of information. When using a map, the instructions are placed around the outer edges of the map and are known as marginal information. All maps are not the same, some having more marginal information than others, and the information being located in different places around the border. It is important that you know how to use this information if you plan to get the most out of your map. The different information on the margins is as follows:

(1) Sheet name. The sheet name is found in two places: the center of the upper margin and lower right margin of the map. The map is usually named after it's outstanding cultural or geographical features. Whenever possible, the name of the largest city is used.

(2) The sheet number. The sheet number is found in the upper right margin of your map and is used as a reference number for the map sheet.

(3) Series name and scale.

(a) The series name is found in the upper left margin of your map. A map series usually comprises a group of similar maps on the same scale and on the same sheet lines or format designed to cover a particular geographical area. It may also be a group of maps designed to serve a common purpose. The name given a series is of the most prominent area.

(b) The scale notation represents the ratio of the maps distance to the corresponding (actual) distance on the earth's surface. For example, the scale notation of 1:50,000 indicates that one unit of measure on the map equals 50,000 of the same units of measure on the actual surface of the ground (1 map inch = 50,000 earth inches).

(4) Series number. The series number appears in the upper right margin and lower left margin. It is a comprehensive reference expressed either as a four digit number or as a letter followed by three or four numbers.

(5) Index to adjoining sheets. The index to adjoining sheets is found in the lower right margin of your map.

(6) Bar scales. The bar scale is located in the center lower margin of your map.

(7) Contour interval. The contour interval is located in the center lower margin of your map, just below the bar scales. It states the vertical distance between adjacent contour lines on the map. For example, every contour line represents an increase in elevation as indicated by the distance of the contour interval.

(8) Northern References. There are 3 different northern references located on the declination diagram. It is located in the right portion of the lower margin of your map.

(a) True North. A line from any position on the earth's surface to the North Pole, which is the exact top of the earth.

(b) Grid North. The north direction established by the vertical grid lines on the map.

(c) Magnetic North. The direction to the North Magnetic Pole.

(9) Legend. Is located in the lower left margin. It illustrates and identifies some of the symbols on the map. There is not enough room on a map sheet to show the true outlines of objects and land features so the mapmaker uses a set of standard symbols to represent them. Sometimes the symbols are not the same, or standard types of symbols typically used. This is because of the type of map, the scale, or the origin. Because of this possible difference, to prevent errors in symbol identification, the legend should be referred to every time a map is used. (The example on the legend may state, "unimproved road, hard surface road, jeep trail," etc. The right side of the symbol will have three symbols; the first being "unimproved road", the second "hard surface road", and the third being "jeep trail".)

(10) Map colors. The colors you see on your map are not there to make the map pretty or enjoyable to the user. They have important meaning. The colors vary with different types of maps, but on a standard, large scale map, there are five basic colors:

(a) Black. Represents most of the man-made features.

(b) Red. Represents major roads.

(c) Blue. Represents water.

(d) Green. Represents vegetation.

(e) Brown. Represents all terrain features, such as hills, draws, fingers, depressions, and saddles by the use of contour lines. These contour lines are shown in brown.

(f) Other Colors. Occasionally other colors are used to show special information. These, as a rule, will be indicated in the marginal information. Currently, there are changes being made to the color system, which may result in changes of the marginal information in regards to map colors.

2. **HORIZONTAL SCALES**.

a. Bar Scale. This special ruler, located at the bottom of the map sheet, enables you to measure a distance between 2 points on your map and change that map distance into the actual ground distance.

(1) The primary scale. The primary scale is the part of the scale that starts at the zero (0) and extends to the right.

(2) The extension scale. The second part of the bar scale is the extension scale. It starts at the zero (0) and extends left.

b. Measuring Straight-Line Distance. If you are given a mission that requires you to move from your present position to a different position some distance away and you plan to travel in a straight line between these two points, the following method is used to determine the distance you will have to travel:

(1) First locate your present position and your destination on your map. Then you will lay a straight edged piece of paper on the map so that the edge of the paper touches the location of your present position and the location of your destination. Make a tick mark on the edge of the paper at each point. Now move the paper down to the bar scale. Place the left tick mark on the zero (0) of the scale so that the other tick mark will fall somewhere on the primary scale.

c. Measuring a Curved Line Distance. If you are given a mission that requires you to move from your present position to a different position and this time you choose to move along a road to reach your destination, the following method is used to work out the ground distance you will have to travel:

(1) Find your present position, your destination, and the road you will follow, on your map. Position a straight edged piece of paper at your location and make a tick mark,. Now, rotate the paper so that it runs along the road until the road turns, and make a tick mark on both the paper and the map. Keeping these 2 tick marks together, rotate the paper again until the road turns, and again mark the paper and the map. Continue this procedure until you have reached your destination. Now using your bar scale as previously stated, find the actual ground distance your squad will be traveling.

3. **ELEVATION AND RELIEF**. In order to begin learning the subject of elevation and relief, let's first look at some definitions.

a. Contour lines. Of the several ways to indicate elevation and relief on maps, the most common is through the use of contour lines. A contour line represents an imaginary line on the ground along which all points on this line are the same elevation above sea level.

b. Characteristics of Contour Lines. There are a few characteristics of a contour line that you should know that will help you understand what they represent on your map. These characteristics are as follows:

(1) Contour lines never cross.

(2) Contour lines connect on themselves.

c. Types of Contour Lines. In addition to knowing the definition, purpose, and characteristics of a contour line, you need to realize that there are different types of contour lines.

(1) Index contour lines. Starting at zero elevation, every 5th contour line is drawn with a heavier line.

(2) Intermediate contour lines. These are the contour lines falling between the Index contour lines.

(3) Supplementary contour lines. These lines are shown as dashed brown lines.

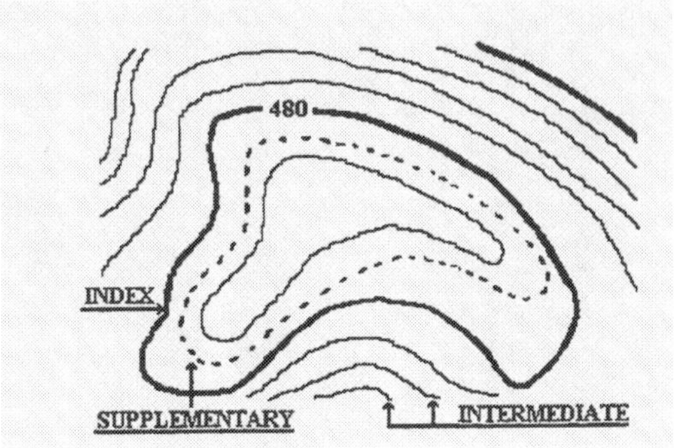

d. Determining Elevation of a Point. Now, you are ready to learn how to determine the elevation of any point of the ground represented on your map.

(1) Given elevation nearest the point, examine the area around your point until you find the nearest point that has a marked elevation. There are a few different ways that contour lines are marked, and they are:

(a) Labeled index contour line. Index contour lines are broken with the elevation of the line printed in the break.

(b) Spot elevations. At various places on your map, usually on prominent landforms such as road junctions or hilltops, a spot elevation will be given. It will be printed in either brown or black.

(c) Bench marks. Similar to Spot elevations, but more accurate. They are marked on your map by a black "x", with the letters "BM", and the elevation.

(d) A point on a contour line. If a point is located on a contour line, its elevation will be the same as that contour line.

(e) A point between contour lines. If a desired point is less than 1/4 the distance between 2 lines, its elevation is considered to be the same as the closest contour line. If the point is from 1/4 to 3/4 the distance, its elevation is considered to be in the middle of the 2 lines.

(f) A point on top of a hill. To estimate the elevation on the top of an unmarked hill, add 1/2 of the contour interval to the elevation of the highest contour line around the hill.)

4. **TERRAIN FEATURES**. There are many types of terrain features that you must be able to identify. By interpreting the arrangement of contour lines, you should be able to get a mental picture of how these terrain features would appear if you were actually out there on the ground looking at them. Some of the features you need to be able to recognize are as follows:

a. Hills. Defined as a point or small area of high ground. A hill is represented on your map by a number of contour lines each of which circles around and connects onto itself. These rough circles get smaller as they progress toward the top of the hill. You need to carefully compare the arrangement of these contour lines to the actual appearance of the hill.

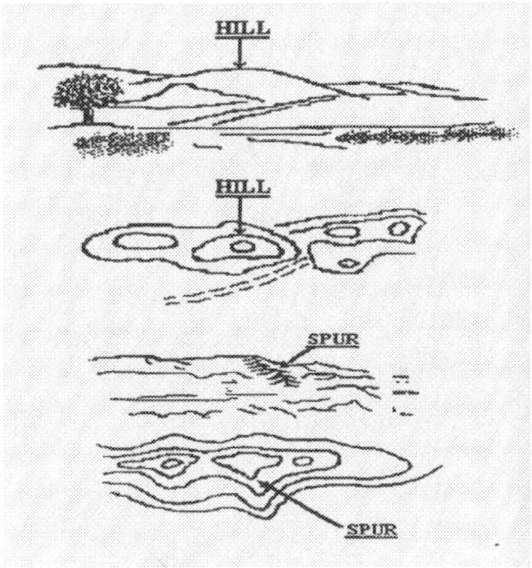

b. Spur. Defined as a line of high ground. The points along the top or crest of the spur are higher than the ground on both sides. A spur is represented on the map by a number of contour lines, each of which form a U-Shaped or V-Shaped design. The end or bottom of the U or V-Shaped lines point down to a lower elevation.

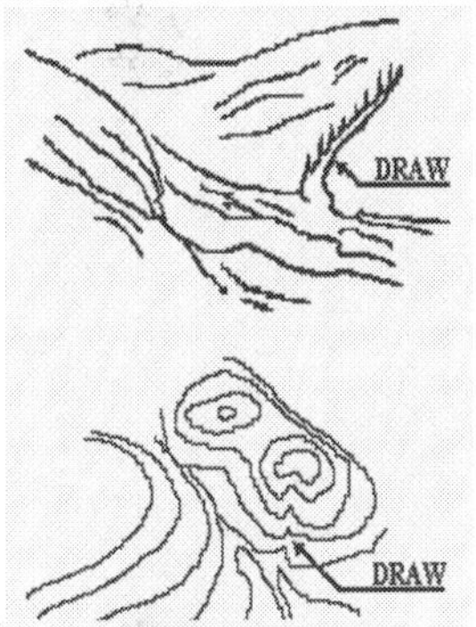

c. Draw. Defined as a line of low ground. The points located in the bottom of a draw are at a lower elevation than the ground to either side. A draw usually runs between 2 spurs. It is sometimes difficult to tell the difference between the contour lines that represent a draw from those that represent a spur because they are both U or V-Shaped lines. Remember that the curved portion representing a spur points down to ground of lower elevation, and the curved portion representing a draw points to ground of higher elevation. Now by looking closely at the area on your map, you will be able to tell which of the 2 you are dealing with. In many cases, draws will contain a stream, while spurs usually never have a stream running down their centers.

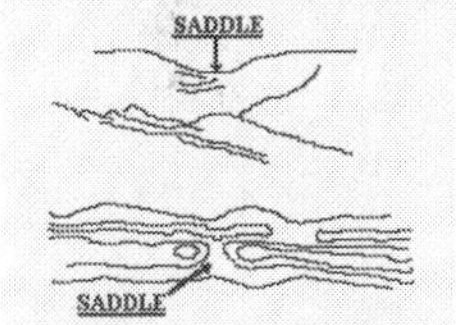

d. Saddle. Defined as the low ground between 2 hilltops, or a dip along the crest of a spur. On your map a saddle is represented by contour lines that form 2 hilltops that are located closely to each other. The area between these 2 hilltops being the location of the saddle.

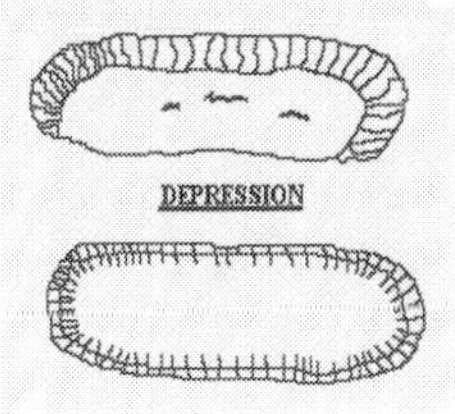

e. Depression. Defined as a point or area of low ground surrounded on all sides by higher ground. A depression is represented by contour lines forming circles within a small area on the map. These rough circles look identical to hilltops except for the presence of hachures. These are short stubby lines that are connected to the contour line at one end and point away from the contour lines at right angles in a downhill direction.

f. Cut. A cut is a man-made feature by which the bed of a road or railroad is leveled by cutting through high ground.

A cut is represented by a sudden ending of contour lines, the ends being connected by a straight line. Hachures are connected to this straight line and point downhill toward the roadbed. This sudden ending of contour lines occurs on both sides of the road bed.

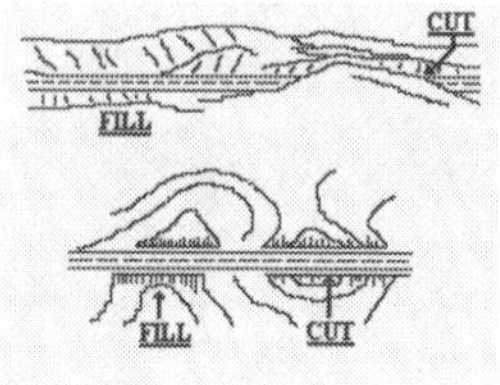

g. Fill. A fill is also a man-made feature by which the bed of a road is raised by filling in the low area. A sudden ending of contour lines can also represent a fill. A straight line connects these ends with Hachures pointing downhill away from the road bed. Again, this occurs on both sides of the road bed.

5. **SLOPES.** The rate of rise or fall of a ground form is known as its slope. In combat, this factor of slope must be considered when making plans for an operation. You must have the ability to recognize the ground forms in your area of operation by studying your map and the spacing between the contour lines on your map.

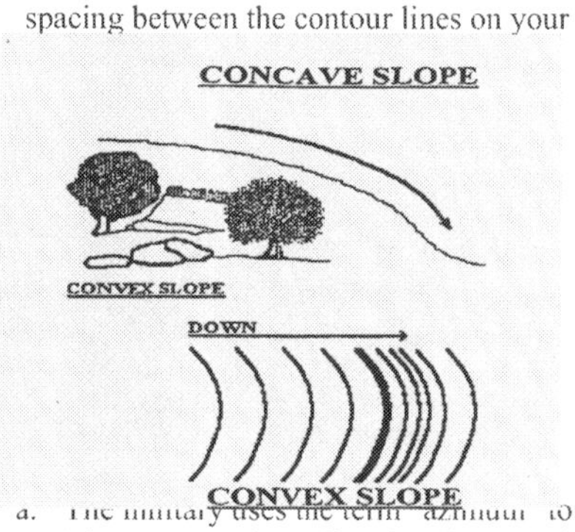

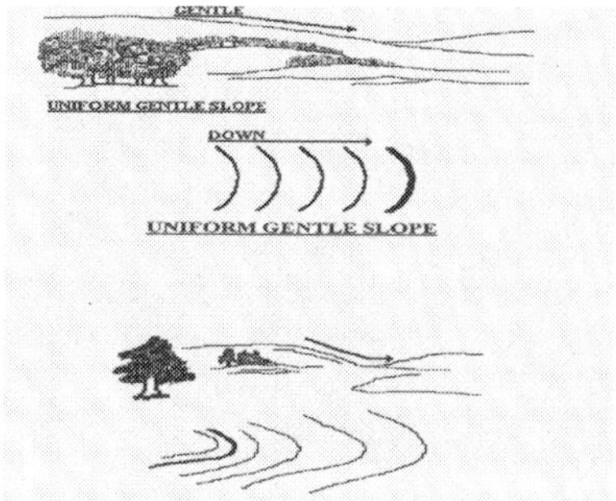

a. The military uses the term "azimuth" to describe the direction from one point to another. There are two base direction lines used to measure the value of an azimuth. They are as follows:

(1) Grid north. It is used when measuring direction from your map.

(2) Magnetic north. Used when measuring direction from your compass.

b. Units of Measurement. Now that you have learned about the two base direction lines that are used to measure direction, we need to learn the units of measurement that are used to measure the angle from one of these direction lines to the direction line.

(1) Degrees. The degree is the unit of measurement that is used most often. Anytime you need an azimuth to move from one point to another, you will use degrees as your unit of measurement. The base direction line, grid north or magnetic north, will be zero (0) degrees and measuring clockwise from this base direction line, the angle increases until you reach 360 degrees which brings you back around in a full circle to the base direction line.

(2) Mils. The mil is also used as a unit of measurement for direction. It is used when working with indirect fire weapons such as mortars and artillery. The azimuth you send will have to

be in mils, not in degrees. Remember, you are still measuring the angle between the base direction line and the line of sight from you to the distant object only you are using a different unit to measure the angle. Moving clockwise from the base direction line, which would be zero (0) mils, the angle back to the direction line. Now you know there are 360 degrees and also 6400 mils in the circle. This means that each degree equals approximately 17.5 mils.

c. Converting Map and Compass Directions. (WMC.5.2d) Earlier you learned that an azimuth was always measured from a base direction line. You also learned that we can use two different base direction lines. Grid north is the base direction line used when you are working with the map and getting a map direction or a grid azimuth. Magnetic north is the base direction line used when you are using the compass to get a compass direction or a magnetic azimuth. Because these two different base direction lines are used and because inmost cases there is a difference of one or more degrees between them, it is necessary for you to learn how to convert from a grid azimuth to a magnetic azimuth or from a magnetic azimuth to a grid azimuth. In order to learn this procedure we must take another look at the declination diagram.

(1) Declination diagram. A diagram that shows the relation between three different norths true north, grid north and magnetic north. It is the difference between grid north and magnetic north that we need to look at. The difference between these two north's is known as the grid-magnetic angle or G-M angle.

(a) Grid-Magnetic angle. The grid-magnetic angle is the angular difference between grid north and magnetic north. As you can see on this diagram a line that connects grid north and magnetic north represents the G-M angle. The size of the angle is stated off to the left side of the diagram in both degrees and mils. The G-M angle on your maps is 18 degrees or 320 mils. It is important to know that the G-M angle will change depending on where you are located on the earth's surface.

(2) Converting. When you receive a mission as a squad leader to move your squad from your present position to some distant point, you take out our map, a protractor and figure out the map direction or the grid azimuth to that distant point. However, before you can use your compass to direct you to your destination, you will have to convert the grid azimuth you have from your map, to a magnetic azimuth that you can set on your compass.

(a) The first step in making this conversion is to examine the declination diagram on your map. The G-M angle on your map is 18 degrees. You know that you will have to add the G-M angle to, or subtract it from the azimuth you measured on your map, or shot on your compass.

(b) The instructions as to whether you will add or subtract this angle are written on the map depending on whether you are going from grid to magnetic or vise-versa.

d. Orientation of a Map. (WMC.6.14a) In order to gain the best use of your map and to avoid confusion while navigating across terrain, you will need to know how to orient your map to the ground. A map is oriented when it is lying flat and level with it's north and south pointing in the same directions as north and south on the ground. There are two methods you can use to orient a map.

(1) Inspection. The first method used to orient the map is by inspection. By carefully examining your map and the ground around the area in which you are located to find linear features such as roads, railroads and powerless, you can align the feature on the map with the same feature on the ground. If there is only one linear terrain feature in the area in which you are located, you will have to be careful not to reverse your directions. Even though you have a road running north-south on your map aligned with the same road on the ground, the north end of your map may be pointed in a southerly direction down the road. To avoid this from occurring, find another terrain feature in the area that you can also identify on the map. Once you have aligned your map along the linear feature, check to insure that the second feature is positioned on the same side of the linear feature on your map as it is on the ground.

(2) Compass. This is the second method of orienting a map. You begin by placing your map on the ground so it lies flat and level. Open your compass so that the cover and eyepiece are completely open. Align the graduated straight edge of the compass along one of the north-south grid lines on your map with the cover end of the compass pointing toward the top of the map. This will place the stationary index line parallel to the grid north. Since the north seeking arrow on the dial of your compass will always point to magnetic north, you set your compass up just like a declination diagram. Grid north being the stationary index line and magnetic north being your north seeking arrow. Now rotate your map, insuring that the graduated straight edge of your compass remains aligned with the north-south grid line, until the face of the compass appears exactly as the declination diagram in the lower margin of the map. Your map is now oriented. For the Sonora Pass map sheet you would rotate the map until the north seeking arrow on the compass lies 16.5 degrees to the right of the stationary index line.

7. **GRID SYSTEM.** (WMC.6.14b) In most combat situations it is very important to be able to pinpoint your location, the location of another unit, or the location of an objective. In order to do this, it is necessary to have a location system that will allow you to accurately determine a specific location and state that location in such a way that will allow others to find that location quickly. This system is known as a grid system.

a. Grid Box. This box gives basic instructions on reading grids in determination of specific points on the map.

b. Grid Lines. The straight black lines that run across, and up and down the surface of your maps form the 1000 meter grid squares. These lines are called grid lines. If you look around the border of your map you will see that both the horizontal and vertical grid lines are identified by a two digit number.

c. Grid Square Identification. Each 1000 meter grid square on your map has it's own unique number. This number is called a grid coordinate. A four digit number is the result of combining the two digit number identification of the vertical grid line and the two digit number identification of the horizontal grid line that cross to form the lower left corner of the grid square. To find the grid coordinate of any grid square on your map you will always read your map right and up. You find the two digit number of that vertical grid line and you will have the first 2 digits of your coordinate. Next, find the two digit number of the horizontal grid line and place it after your vertical number and you now have the 4 digit grid coordinate that identifies your grid square.

d. Locating Terrain Features Within a Grid Square. If you are given a 4 digit grid coordinate to find a certain point on the earth's surface, that grid coordinate would put you within 1000 meters of that point. If you are told to find a certain feature within that grid square and you are good at reading your map, a 4-digit coordinate may get you close enough to your point to enable you to find it. However, there are many situations in which a 4 digit number will not get you close enough to a given position to allow you to accomplish your task. It is necessary to be able to locate points within a grid square and using a six-digit grid coordinate does this.

(1) Six Digit Grid Coordinate. The 1000-meter grid square will have to be broken down further into 100-meter grid squares. Because the surface of your map would become too cluttered by all these additional lines, a special instrument has been made to help you divide up a 1000-meter grid square. This instrument is called a coordinate scale or protractor.

8. **RESECTION**. (WMC.6.14c)

a. You will face a situation where you need to pinpoint the location of your position. This may not be possible by associating the terrain in your area with the map because you may have a poor map or the terrain in your immediate area is flat and has no distinguishing features. When you find yourself in this situation you will need to know how to perform a resection.

(1) Examine the terrain around your position and find two terrain features or objects that you can also identify on your map.

(2) Use your compass to find the magnetic azimuth to one of the distant terrain features you have identified.

(3) Convert the magnetic azimuth you found into a grid azimuth.

(4) Change the grid azimuth you found into a back azimuth.

(5) Drawing a line on your map from a distant point you identified in step one back toward your present position. Place a small mark on your map at this point. Using a straight edge, draw a line from where you plotted the distant object to the mark you made by the edge of your protractor. This line should be extended to insure it continues through the area where you are located.

(6) Repeat, finding the magnetic azimuth to the second distant object. That is at least 60 degrees from the first.

(7) When you have drawn your second line on the map, the point where the second line crosses the first line is the location of your position. The process that has just been covered is called a two-point resection.

(8) If your position is located on a linear line feature such as a road, power line, railroad or river, then it is only necessary to perform a one-point resection in order to pinpoint your location. Instead of identifying two distant points, you will only need to select one. Where the single line that you drew on your map crosses the linear feature on which you are located will be the location of your position.

9. **DETOURING AROUND AN OBSTACLE**.

a. There will be occasions when it will become necessary to move off your plotted route in order to avoid an obstacle such as a small lake or swamp. You must be able to move around these obstacles and return to the route you were following before you took your detour. This can be accomplished by using one of three detour methods: Far side landmark method, nearside landmark method, and the 90-degree offset method.

(1) Far side landmark. Used during hours of daylight when you are able to identify some feature on the far side of the obstacle that lies on your line of march and is obvious enough so that you can find it once you have moved around to the far side.

(2) Nearside landmark. Used during hours of daylight when you cannot find a landmark on the far side. However, you can identify a feature on the nearside that is on your line of march and you will still be able to see once you have moved around to the far side. Again, let us use the example of being on an azimuth of 198 degrees, and it brings you to a small lake. From the nearside you cannot locate any features on the far side. There is, however, a small pier on the nearside that is located on your line of march. You can now move around the lake using the easiest and safest route. Move to an area on the far side that is as close to being opposite of the nearside landmark as possible. You then figure out the back azimuth of your route azimuth, which in this case would be 18 degrees. Shoot an azimuth back across the lake to the pier. Move to the right or left until you are in a position where the pier lies at an azimuth of 18 degrees. This puts you back on your line of march and you can continue on following your original azimuth of 198 degrees.

(3) 90 degree offset. Used during hours of reduced visibility or during daylight hours when there are no visible landmarks along the line of march on either side of the obstacle. This time you are moving along a 360 degree azimuth and you encounter a large, impassable area of thick brush. The brush is too high to see any landmarks from either side of the thicket. The following procedure should be taken to detour around the thicket.

(a) Stop the pace count you have been making for movement along your assigned route.

(b) Decide which way you will detour around the thicket. Let's say in this case the right flank appears to be the easiest and safest. Add 90 degrees to your original azimuth of 360 degrees in order to move your patrol to the right.

(c) Move on this 90 degrees azimuth until you have reached a point where you can clear the right flank of the thicket. You will keep a new pace count starting from the point where you made your 90 degrees turn to the right to the point where you can clear the right flank of the thicket.

(d) Make another change of direction. This time you want to return to a 360 degrees azimuth so you will parallel your original course and move by the right flank of the thicket. In order to return to the 360 degrees azimuth you will now subtract 90 degrees.

(e) Move on the 360-degree azimuth until you have by passed the thicket. At the point of your second 90 degree turn, you will pick up the pace count you were making before you began making your detour around the obstacle. Even though you are not on your original line of march, you are moving parallel to your original line of march and therefore must include the distance you cover during this leg of the offset in your pace count for your route.

(f) Once you have bypassed the thicket, you will make another change of direction. This time, you want to move back toward your original line of march. Since that means you will be turning toward the left, you will subtract 90 degrees from the 360 degrees azimuth you have just completed. Your new direction will be 270 degrees.

(g) Pace back toward the original line of march following the 270-degree azimuth, the same number of paces as you made off the original line of march. In doing this you will have returned to your original line of march.

(h) Once you have moved back to your original line of march you will make your last change of direction. By adding 90 degrees to the original azimuth of the last leg of the offset, which in this case was 270 degrees, you will be back on the original azimuth of 360 degrees.

(i) Return to your original line of march pace count.

UNITED STATES MARINE CORPS
Mountain Warfare Training Center
Bridgeport, California 93517-5001

FMST.07.39
04/03/02

STUDENT HANDOUT

SEARCH AND RESCUE MANAGEMENT

TERMINAL LEARNING OBJECTIVE: In a mountainous environment, execute a search and rescue, in accordance with the references. (FMST.07.39)

ENABLING LEARNING OBJECTIVES:

(1) Without the aid of references and from a given list, choose the definition of a search, in accordance with the references. (FMST.07.39a)

(2) Without the aid of references and from a given list, choose the definition of a rescue, in accordance with the references. (FMST.07.39b)

(3) Without the aid of references and from a given list, choose one the four methods used to determine a search area, in accordance with the references. (FMST.07.39c)

(4) Without the aid of references and from a given list, choose one of the five methods for confining a search area, in accordance with the references. (FMST.07.39d)

(5) Given a lost individual scenario in a mountainous environment, plan a search and rescue in accordance with the references. (FMST.07.39e)

OUTLINE.

1. **GENERAL**. When discussing a search and rescue situation we must first be able to differentiate between the two facets of the operation. A "search" means that you are looking for an overdue missing person (FMST.07.39a). A "rescue" 'is the extraction of a known subject from a known location (FMST.07.39b). In either of these situations time is critical for several reasons:

14

a. Subject may need emergency medical care.

b. Size of the search area is limited.

c. Time and weather destroys clues.

d. Subject may need protection from the environment or from himself.

2. **SEARCH METHODS**. When conducting a search you will be searching for clues, instead of searching for the victim. There are two reasons for this:

a. There are more clues than victims. Anyone moving through an area will leave signs.

b. Clue detection greatly reduces search difficulty by reducing the search area.

c. This determination will assist the search manager in the use of assets, the justification for requesting outside agency assistance, and other management decisions. There are eight points that must be considered in determining the search urgency:

1. Subject profile.
2. Weather profile.
3. Equipment available to subject.
4. Subject's training, experience, and mobility.
5. Terrain hazards.
6. History of incidents in the area.
7. Time.
8. Political sensitivity.

3. **SUBJECT PROFILE**. The subject profile is based on the subject's capabilities and mobility in the terrain and conditions of the search area. The three points are age, medical condition, and number of subjects. Discussing this will assist us is rating the relative urgency:

a. Age: Very young/old, other.

b. Medical condition: Injured, ill, mental problems, healthy, or dead.

c. Number of subjects: Single or group.

d. Subject experience profile: No experience/no knowledge of area, no experience/knows area, experienced/no knowledge of area, or experienced/knows area.

e. Weather profile: Past or current bad weather, bad weather in next 8 hours, bad weather in next 24 hours, or no bad weather predicted. Temperature and visibility.

f. Equipment profile: Inadequate, questionable, or adequate for environment or weather.

g. Terrain/Hazard profile: Known hazardous terrain or, hazards or few or no hazards.

h. History of incidents in the area: Outcome of other searches for individuals with similar profiles in similar situations.

i. Time: Time since the subject went missing and the effects of time on clues.

j. Political sensitivity: External sources; VIPs, politicians, relatives, media, or higher authority in your chain of command.

4. **DETERMINING THE SEARCH AREA**. There are four methods used to determine the search area by rescue managers. The best application is to use a combination of these methods by personnel with search experience. (FMST.07.39c)

a. Theoretical: This is the distance that the victim could have traveled in the elapsed time since he/she was reported missing. (PI x R Squared)

b. Statistical: Historical data on searches for similar individuals in similar conditions.

c. Subjective: Evaluation of limiting factors for this specific incident and location.

d. Deductive reasoning: Sherlock Holmes version, based on analysis of circumstances involving the specific incident.

5. **BEHAVIOR ASPECTS OF A LOST PERSON**. General aspects of lost person behavior will give you an awareness of the actions of the individuals you are searching for by separating them into different categories. Some factors that could effect the search strategy are the general health of the subject, his/her past experiences, physiological effects of the environment, and the victims' mental state or biological cycle. Also, looking at the subject's ability to assist in his/her own survival with simple skills such as fire, shelter, signaling, and land navigation will determine strategy. The psychological impact on the victim should also be taken into account. Will the subject be overcome by Stress? Fear? Panic? Embarrassment? Or will the subject be calm and develop a positive "I'll be OK attitude? A further point to consider is the circumstances under which the subject became lost. Was the subject at a known location such as a campground? Enroute to a specific location or on a trail? Traveling in a wilderness area with no specific location or destination in mind? Or does the subject fall under one of the following categories:

a. Children age 1 to 3: Unaware of concept of being lost, no navigation skills or sense of direction, wander aimlessly, might seek out a convenient place to take a nap. i.e. hollow log, under bush/rock/picnic table.

b. Children age 3 to 6: More mobile, more inquisitive and easily distracted, will attempt to "go home" once they decide that they are lost, taught to avoid strangers and may hide from searchers.

c. Children 6 to 12: Good navigation skills and sense of direction in familiar terrain but become disoriented in unknown surroundings, may have intentionally run away and do not want to be found, may not answer when called, same fears as adults but no experiences to fall back on, once night sets in more likely to respond to searchers and except assistance.

d. Mentally retarded (all ages): Act similar to children 6 to 12, generally will not respond to name, will hide from searchers or seek shelter, once shelter found might remain in place, will do little or nothing to assist in their own survival or rescue.

e. Elderly (above 65): Possibly suffering from senility or Alzheimer's, easily confused or attracted to something that strikes their fancy, may not be oriented to the present, likely to over-extend themselves physically, and often have hearing impairments.

f. Despondent: Often are seeking solitude and will not respond to searchers, usually found near prominent terrain/scenic views, within sight and sound of civilization, rarely found in the underbrush.

g. Hikers: Rely on trails and have a set destination in mind, have poor land navigation once they loose the trail, often get confused at trail intersections or in springtime when trails are covered with snow, hiking parties of different ability levels get separated, will miss trails and head down wrong drainages. Hunters: Concentrate on game more than navigation, often pursue game into drainages/deadfalls/underbrush/deep snow, tend to over-extend themselves, typically not prepared for foul weather.

h. Cone pickers/Berry pickers/Photographers/ rockhounds/etc: Intention to stay or go to a specific location, carry little provisions or survival gear, go in good weather so they don't carry foul weather gear, often are not aware of surrounding as they are intent on pictures or the ground if pickers, these people are an extremely high risk group.

i. Fishermen: Well oriented once they arrive at lake or river, most often late due to accident, these are often recovery operations.

j. Climbers: Generally well equipped and self sufficient, tend to remain on marked routes and leave good plans with friends, primary factors are weather and hazardous conditions such as falling debris or avalanches, once located these searches often become technical rescues. 1. Marines: Subject profile will depend on individual's MOS, mission, training, and abilities learned as a civilian.

6. **BEGINNING THE SEARCH**. Once we have determined that a subject is missing we must determine whether or not he/she is in our search area. The sheriffs department will initiate a "Bastard Search", this is to ensure that the subject is not simply at the local tavern, shacked up with an old friend, or was not in an automobile accident enroute to the area. Next we will begin our search at one of the four following area:

a. Last Known Point (LKP): This is the last place that you have physical evidence to support that the subject was there. i.e. Subject's vehicle at the trailhead, name in a logbook, campsite with gear, etc.

b. Place Last Seen (PLS): The last place he was physically seen. i.e. On the trail by other hikers, at the hunting cabin, at a restaurant or gas station, or back home in San Diego.

c. Departure Point: The place where the subject was supposed to enter the search area. Usually a trailhead or specific point relating to the individual's activity.

d. Clues: As clues are discovered, and you confirm that they belong to the subject, you will update or adjust the search area.

7. **MATTSON**. Once we have determined that we do have an overdue subject and can assume that he/she is in our search area we need to break the search area into manageable sized segments. Depending upon the area to be searched, special training required to search high hazard areas, and assets available, we will divide the area by hazard or by terrain features. The lines of division should follow terrain features so that the searchers on the ground can easily navigate and stay within their zones. We will then determine the POA using the Mattson approach. This is done with each of the five or six members of the management team with the most search experience assigning a percentage to each of the segments using his individual judgment on the likelihood of the subject being in that area. Always include an extra column to represent the "rest of the world". Once all personnel have made their determinations, divide the totals and this will give you your areas to search in order of precedence:

NAME	A	B	C	D	E*	TOTAL
DUNSTON	40%	20%	10%	20%	10%	100%
KEOGH	50%	30%	5%	10%	5%	100%
KOFFEL	40%	25%	15%	10%	10%	100%
AYRES	60%	20%	10%	5%	5%	100%
TOTAL	190%	95%	40%	45%	30%	400%
AVERAGE	47.50%	23.75%	10.00%	11.25%	7.5%	
PRIORITY	1 ST	2ND	4TH	3RD	5TH	

8. **COMMAND SYSTEM**. The Office of Emergency Service (OES), has developed a command system that they use for all emergency situations. This system is called the

Incident Command System (ICS). The breakdown is very similar to the military chain of command and the duties correspond to those of the various "S" shops in the battalion. The personnel in each duty area will complete their assigned duties as you would in a military unit with the Incident Commander (IC) being overall responsible for the entire action. Normally, in California, the IC is the County Sheriff.

a. Incident Commander. CO

b. Operations-Tactics, coordination, insert resources. S-3

c. Plans-Strategy, intelligence, briefing and debriefing teams. S-2

d. Logistics-Beans, Band-Aids, beds, heads, motor-T, Special equipment. S-4

e. Communication-COC and field, landlines, frequency coordinator. S-6

f. Finance-Documentation of all man-hours, damages, legal issues.
S-1

g. Liaisons-inter agency, media, safety, family. PAO/Chaplain

9. **CONFINING A SEARCH AREA**. There are several reasons to confine the search area. The most important is that the longer the subject has free movement, the farther he/she can travel. (Remember the Theoretical Method) Additionally, this will ensure that the subject does not walk out without you being aware of it. It's real dumb to be conducting a search for a subject that is on his/her way home. The use of confinement will not interfere with the other searchers, decreases the chances of a more massive search and if promptly initiated has been shown to be highly effective. There are several methods that we can use. (FMST.07.39d)

a. Sign-cutters - Searchers that hike or ski the perimeter of the entire search area, then the perimeter of the segments. This will let you know if he/she moves from one area to another and allows you to concentrate limited assets in a specific area.

b. Trail-watchers/Road Blocks/Camp-ins - Searchers that camp at trailheads. They can alert you if the subject walks out and interview other individuals concerning possible sightings of the subject.

c. Look-outs - These personnel are already in place in fire towers. Simply notify them of the situation.

d. Track Traps - Clearing areas of the trail that are checked periodically for fresh tracks from the subject.

e. String-Lines - Spools of line with paper arrows attached that point to a safe area of a trail-watcher camp.

10. **USING CLUES**. The use of clues cannot be underestimated. They can tell you if a subject was or was not in the area, by freshness you can tell how long ago the subject was there, by the type of clue you can determine physical and mental condition, and by following clues you can determine a direction and speed of travel. Once you have located a CONCLUSIVE clue, you will re-mattson the area, using the clue as the LKP. The clues we will be looking for fall under four categories:

a. Physical - Footprints, equipment, candy wrappers, cigarette butts, etc.

b. Recorded - Summit logs or trail registers.

c. People - Witnesses who have seen the subject.

d. Event - Mirror flashes, camp fires, whistle signals, etc.

11. **ADDITIONAL ASSETS**. There are a variety of assets available to you in the search and rescue business. Some are volunteers, some charge money and some (Like the military) require higher authority to grant permission for their use. Remember that in addition to active searchers you can also use passive measures such as lights, noise, balloons, fires, smoke generators, sirens, or P.A. systems. Some examples of active searchers are:

a. Scent dogs.

b. Tracking dogs.

c. Human trackers.

d. Hasty teams.

e. Grid teams.

f. Helicopters.

g. Civil Air Patrol.

h. Snowmobile clubs.

i. Horse teams.

UNITED STATES MARINE CORPS
Mountain Warfare Training Center
Bridgeport, California 93517-5001

FMST.07.16
04/03/02

STUDENT HANDOUT

TRIAGE

TERMINAL LEARNING OBJECTIVE: Given simulated casualties, in a wilderness environment, triage casualties, in accordance with the references. (FMST.07.16)

ENABLING LEARNING OBJECTIVES:

1) Without the aid of references and from a given list, choose when actual triage most often begins, in accordance with the references. (FMST.07.16a)

2) Without the aid of references and from a given list, identify one of the four-triage categories used by NATO, in accordance with the references. (FMST.07.16b)

3) Without the aid of references and from a given list, choose the primary cause of death in a mass casualty situation, in accordance with the references. (FMST.07.16c)

OUTLINE.

1. **DEFINITION**: Triage or sorting (often used interchangeably) is the evaluation and classification of casualties for the purpose of establishing the priority for treatment and evacuation.

NOTE: Casualty sorting is one of the most important tasks for the entire medical service. No other job requires more informed judgment, hard work and courage. The Officer and the Corpsman responsible for triaging wounded Sailors and Marines must exercise sound judgment so that casualties receive proper, expedient care. Triage is a continuing process and any individual assigned should be the most capable and experienced available.

a. Decision Factors. The factors concerning casualty triage are based upon the following:

(1) The need for immediate resuscitation.

15

(2) The need for emergency surgery.

(3) The futility of surgery due to obviously lethal wounds.

(4) Time needed to perform surgery compared with:

(a) Probability of success.

(b) Number of other casualties needing treatment.

b. The actual triage begins most often with the casualty himself. (FMST.07.16a). This usually occurs when the injured individual determines that "Buddy-Aid" or "Self-Aid" will not be adequate.

(1) The corpsman, once presented with the casualty, must make several determinations:

(a) Is the casualty a walking-wounded or a litter case?

(b) If severely wounded, are resuscitative efforts required immediately?

(c) Does the casualty need any other immediate treatment (bleeding, asphyxia or pain control)?

(2) Once the casualty has been transported to a Battalion Aid Station (BAS), triage continues. The dental officer will commonly perform the duties of the Triage Officer at the echelon I or echelon II level, if present.

2. **NATO TRIAGE CATEGORY CODES**. (FMST.07.16b)

a. At each medical treatment facility (MTF) in the area of operations (AO), incoming casualties are classified by level of treatment required. Four triage category groups have been universally adopted for use by both United States and NATO Forces. These categories are listed and defined as they appear in standardization agreement (STANAG) No 2879:

(1) Immediate Treatment (Group T1) (RED Tag). To include those requiring emergency life-saving surgery. These procedures should not be time consuming and should concern only those patients with high chances for survival. Some examples are:

(a) Respiratory obstruction.

(b) Accessible hemorrhage.

(c) Emergency amputation.

(2) Delayed Treatment (Group T2) (YELLOW Tag). To include those badly in need of time-consuming major surgery, but whose general condition permits delay in surgical treatment without unduly endangering life. To mitigate the often, critical effects of delay in surgery, sustaining treatment, such as stabilizing I.V. fluids, splinting, administration of antibiotics, catheterization, gastric decompression, and relief of pain will be required. Examples are:

(a) Large muscle wounds.

(b) Fractures of major bones.

(c) Intra-abdominal and/or thoracic, head or spinal injuries.

(d) Uncomplicated major burns.

(3) Minimal Treatment (Group T3) (GREEN Tag). To include those with relatively minor injuries who can effectively care for themselves or who can be helped by untrained personnel. Some examples are:

(a) Minor lacerations.

(b) Abrasions.

(c) Fractures of small bones.

(d) Minor burns.

(4) Expectant Treatment (Group T4) (BLUE Tag). This group comprises patients who have received serious and often multiple injuries, and whose treatment would be time-consuming and complicated with a low chance of survival. If fully treated, they may make heavy demands on medical manpower and supplies. Until the mass casualty situation is under control, they will receive appropriate supportive treatment. The extent of treatment depends on available supplies and manpower, and may involve the use of large doses of narcotic analgesics. These patients should not be abandoned, and every effort should be devoted to their comfort. The possibility of their survival, despite alarming injuries, must always be kept in mind. Examples are:

(a) Severe multiple injuries.

(b) Severe head or spinal injuries.

(c) Large doses of radiation.

(d) Widespread severe burns.

3. **TREATMENT CONCEPTS**.

a. The triage officer must be familiar with all aspects of combat casualty care (airway management, hemorrhage control, use of anesthesia, surgical procedures, fluid resuscitation and post-operative care, to name a few), but ever-present in his mind should be the following concepts:

(1) Asphyxia and hemorrhage are initially the primary causes of death in any mass casualty situation. (FMST.07.16c)

(2) Once at a definitive care facility, uncontrolled hemorrhage is the leading cause of death.

(3) Life has priority over limb, and preserving function overrides the correction of superficial anatomic defects.

(4) A casualty is in a constant state of change until the wound or injury has been repaired.

(5) Systemic disturbances caused by the wound continue until healing is complete.

(6) Combat conditions will dictate the modifications necessary in triage and care of mass casualties.

b. Priorities of Treatment. With the six concepts of treatment in mind, we can now address three basic priorities for treatment.

(1) First Priority:

(a) Asphyxia:

1. Mechanical obstruction of the airway.

2. Sucking chest wound.

3. Tension pneumothorax.

4. Maxillofacial trauma.

5. Massive hemo-pneumothorax.

(b) Shock:

1. Exsanguinating hemorrhage.

2. Major internal hemorrhage (thoracic, abdominal, and pelvic).

3. Visceral injuries/evisceration.

4. Cardio-pericardial injuries.

5. Massive muscle injury leading to Fat Embolization to the lung.

6. Full-thickness or deep-partial burns of 10 % or greater.

(2) Second Priority:

(a) Visceral injuries:

1. Perforation of G.I. Tract.

2. Pancreatic and Biliary system.

3. Genitourinary tract.

4. Thoracic wounds w/o asphyxia.

(b) Vascular injuries requiring repair:

1. Tourniquet cases.

(c) Closed head injuries:

1. Increasing loss of consciousness.

(d) Major burns:

1. Partial-thickness burns of greater than 25% TBSA in the low risk group, 20% TBSA in the high risk group.

2. Burns involving the poor risk group.

3. Burns involving the face, hands, feet, genitalia/perineum.

4. Circumferential limb burns.

(3) Third Priority:

(a) Brain and spinal injuries - requiring burr hole decompression.

(b) Soft tissue injuries.

(c) Debridement necessary with less than major muscle damage.

(d) Lesser fractures and dislocations.

(e) Eye injuries.

(f) Maxillofacial injuries w/o asphyxia.

(g) Burns under 20% in other locations other than priority 2.

4. **<u>TRIAGE TOOLS</u>**.

a. Trauma Score/Glasgow Coma Scale. In recent years, many methods have been proposed to speed up the classification and sorting of casualties. One method that has gained wide acceptance is the Trauma Score/Glasgow Coma Scale. Several studies have been performed using various parameters, i.e., respiratory rate, chest wall expansion, BP, capillary refill, etc. By assigning a number scale to each specific response, the likelihood of survival as a percentage can be calculated.

1. Glasgow Coma Scale.

a. Eye opening:

1) Spontaneous = 4.
2) To voice = 3.
3) To pain = 2.
4) None = 1.

b. Verbal Response.

1) Oriented = 5.
2) Confused = 4.
3) Inappropriate words = 3.
4) Incomprehensive words = 2.
5) None = 1.

c. Motor Response.

1) Obeys Command = 6.

2) Localized Pain = 5.
3) Withdraw Pain = 4.
4) Flexion (pain) = 3.Extension (pain) = 2.
5) None = 1.

d. Total the points up and they equal:

1) 13-15= 4.
2) 9-12 = 3.
3) 6-8 = 2.
4) 4-5 = 1.
5) <3 = 0

2. Trauma Scale.

a. Respiratory Rate:

1) 12-29 min = 4.
2) 10-12 min = 3.
3) 6-9 min = 2.
4) 1-5 min = 1.
5) None = 0.

b. Systolic BP:

1) >89 mmHg = 4.
2) 76 mmHg = 3.
3) 50-75 mmHg = 2.
4) 1-49 mmHg = 1.
5) None = 0.

c. Total the points.

3. Total trauma Scale + GCS = % survival (1-12)

12 = 99
10 = 88
8 = 67
6 = 63
4 = 33
2 = 29

1 = 25
0 = 4

UNITED STATES MARINE CORPS
Mountain Warfare Training Center
Bridgeport, California 93517-5001

FMST.07.08
04/03/02

STUDENT HANDOUT

PREVENTIVE MEDICINE AND WATER PURIFICATION

TERMINAL LEARNING OBJECTIVE. Given a unit in a mountainous environment and the necessary equipment and supplies, perform preventive medicine, in accordance with the references. (FMST.07.08)

ENABLING LEARNING OBJECTIVES.

1) Without the aid of references, from a given list select the five important areas of personal hygiene, in accordance with the references. (FMST.07.08a)

2) Without the aid of references, from a given list select the correct steps taken to ensure proper immunization prior to deployment, in accordance with the references. (FMST.07.08b)

3) Without the aid of references, from a given list select the three methods of water purification, in accordance with the references. (FMST.07.08c)

4) Without the aid of references, from a given list select the four forms of halogens used for water purification, in accordance with the references. (FMST.07.08d)

5) Without the aid of references, from a given list select the definition of giardia, in accordance with the references. (FMST.07.08e)

6) Without the aid of references, from a given list select the most common drug used to treat giardia infection in the United States, in accordance with the references. (FMST.07.08f)

7) Without the aid of references, from a given list select the three important reasons for proper waste disposal, in accordance with the references. (FMST.07.08g)

8) Without the aid of references, from a given list select the correct methods of waste disposal, in accordance with the references. (FMST.07.08h)

16

1. **PERSONAL HYGIENE** Your body's defense mechanisms break down due to the physical demands of the arduous terrain, environment, stress and metabolic changes. Personal

hygiene becomes the key to prevent the spread of communicable diseases. Five important areas are: (WMC.6.17a)

a. Body.

(1) The body should be washed frequently in order to minimize the chances of small cuts and scratches developing into full blown infections and as a defense against parasitic infections.

(2) A daily bath or shower consisting of soap and hot water is ideal. However, when this is not possible, you should:

(a) Give yourself a "sponge bath" using soap and water, making sure particular attention is given to body creases, i.e., armpits, groin area, face ears, hands and feet.

(b) If water is in extremely short supply, you should take an "air bath" by:

(1) Remove all clothing and hang it up to air.

(2) Expose the body for two hours to the sunlight, which is ideal, but the effects will basically be the same if done indoors or during an overcast day. BE CAREFUL NOT TO SUNBURN!

(3) Shaving. Shaving at night before bedtime is preferable because shaving will remove facial oils, which aid in protecting the face from a harsh cold environment during the day.

b. Hair.

(1) Clean frequently.

(2) Inspect at least once a week for parasites. Have your mountain buddy inspect the areas not within your view. This is particularly important during the summer season. Any time the area you are operating in an area that contains wild life there is the possibility of parasites like fleas and ticks.

c. Hands. Hand washing with soap and water will cut down in the transmission of viral and parasitic infections via the person-person and fecal-oral route.

(1) Keep fingernails trimmed and clean to prevent accidentally scratching yourself and to prevent harborage for bacteria.

d. Feet. Are your prime sources of transportation, so take care of them!

(1) Inspect frequently for:

(a) Blisters.

(b) Infections, bacterial and fungal.

(2) Preventive measures:

(a) Keep your feet dry, change socks frequently.

(b) Use antifungal foot powder.

(c) Antiperspirants. Pretreatment for 2 weeks prior to field with an antiperspirant containing Aluminum Chlorhydrate. This will aid in the prevention of athlete's foot.

e. Oral Hygiene. The mouth and teeth should be cleaned at least daily to prevent tooth decay and gum disease.

(1) Ideally with:

(a) Tooth brush.

(b) Tooth paste.

(c) Dental floss.

(2) If these items are not available:

(a) A "chew stick" made from a clean twig about 8" long and finger width. Chew one end until it becomes frayed and brush-like, then use as a toothbrush.

(b) Field expedient dental floss can be pulled from the center lines of paracord.

(c) The gums should be stimulated at least once daily by rubbing them vigorously with a clean index finger.

2. **IMMUNIZATIONS (NAVMEDCOMINST 6230.3)**. In addition to being required, immunizations are your best front line defense against communicable disease.

a. Eight weeks prior to deployment: (WMC.6.17b)

(1) Ensure Basic Series Completed (BSC) on required immunizations.

(2) Determine if additional immunizations/chemophylaxis will be required in that geographical area of operations.

(3) Identify and flag health risk personnel: i.e. TB converters (NAVMEDCOMINST 6224.1), hepatitis (NAVMEDCOMINST 6230.1A), mononucleosis, parasitic infections and anemic personnel.

(4) Review medical intelligence reports:

(a) EPMU, Navy Environmental Preventive Medicine Unit.

(b) G-2, USMC.

(c) NAVDISVECTOECOLCONCEN, Navy Disease Vector Ecology and Control Centers.

(d) CDC - Center For Disease Control.

(e) WHO - World Health Organization.

3. **WATER.** A satisfactory source is one where quantity and quality is enough to supply the needs of all the troops. Water or ice taken from the environment should be considered contaminated and must be purified with available means. Be aware that cold weather slows the chemical reaction time of purification measures.

a. Sources.

(1) Rivers

(2) Streams

(3) Ponds

(4) Lakes

(5) Ice

(6) Snow

b. Water Source Selection. The choice of a water source is influenced by:

(1) Not contaminated by: (Check all to one mile upstream).

(a) Sewage

(b) Enemy pollution

(c) Chemicals

(d) Biologicals

(e) Radiologicals

NOTE: Check condition of vegetation or for dead animals.

(2) Quantity.

(3) Ease of procurement.

(4) Ease of purification.

(5) Freedom from turbidity.

(6) Excessive organic or non-organic contamination.

(7) Should be easily protectable.

4. **PURIFICATION.** Simply consists of removing or destroying enough impurities to make water safe and pleasant to drink.

a. The first step in purifying is to select a treatment method. (All figures are based on one quart of water.) The three different purifying methods are: (WMC.6.17c)

(1) Boiling. The standard recommendation is to bring the water to a rolling boil for 1 minute at sea level for complete sterilization of the water. Recent research showed that additional boiling time for sterilization was not necessary and that boiling for two minutes at any altitude rendered water safe for consumption.

(2) Halogens. The number of tablets and contact time must be considered. All charts based on one-quart canteen. (WMC.6.17d)

(a) Iodine tablets.

TYPE WATER	TABLETS	CONTACT TIME
CLEAR	1	25 MIN
CLOUDY	2	25 MIN
COLD	2	25 MIN

(b) Chlorine bleach. The strength of the solution, number of drops and contact time must be considered.

PERCENT	CLEAR WATER	CLOUDY WATER	COLD WATER
1%	10GTTS/30 MIN	20GTS/30 MIN	10GTTS/60MIN
4-6%	2GTTS/30 MIN	4GTTS/30 MIN	2GTTS/60 MIN
7-10%	1 GTTS/30 MIN	2GTTS/30 MIN	1 GTTS/60 MIN

(c) Iodine solution. The strength, number of drops and contact time must be considered.

PERCENT	CLEAR WATER	CLOUDY WATER	COLD WATER
2%	5GTTS/30 MIN	10GTTS/30 MIN	8GTTS/60 MIN

(d) Betadine solution.

CLEAR WATER	CLOUDY WATER	COLD WATER
8GTTS/30 MIN	16GTTS/30 MIN	8GTTS/60 MIN

(1) Filtration Most filters are not small enough to completely disinfect water. They remove particulate debris, which allows lower halogen dose and improved appearance and taste of "dirty" water. Filters clog quickly if the water is dirty or has a lot of suspended particles. NEVER USE FILTERS AS YOUR SOLE METHOD OF WATER PURIFICATION.

(a) Maximum Filter Pore Size:

1) Parasitic eggs and larvae: 20 micrometers (um).

2) Giardia, E. histolytica: 5 um.

3) Cryptosporidium: 1 um

4) Enteric bacteria: 0.2 um.

5) Viruses: 0.01 um (too small for field water filtration).

5. **GIARDIA**

a. Definition. (WMC.6.17e) An intestinal flagellate protozoa that can cause clinical illness in humans.

(1) Found worldwide in tropical and temperate areas, the infectious cysts are passed by humans and animals and can be transmitted by direct fecal-oral route or waterborne contamination (cysts maintain viability for months in cold water). Due to these factors, Giardia is both a common cause of diarrhea in travelers to foreign countries, and campers in the United States where it has become endemic and is now considered the most common human parasite.

(2) It is found with increasing frequency in formerly "pristine" mountain streams, rivers and lakes and outbreaks have occurred in remote recreational areas and resorts throughout the western mountain states.

(3) Giardiasis is currently a risk for anyone drinking untreated surface water in the western mountains and for communities with water treatment techniques inadequate for eliminating the cysts.

b. Pathophysiology. The organism has two forms: Trophozoite and cyst.

(1) These may be passed in the stool of an infected person. The trophozoites die within one hour outside the body, however the cysts are very hardy, retaining infectivity more than two months in cold water. If the cysts are ingested, excystation is encouraged by stomach acid, and the motile trophozoite attaches by suction disc to the wall of the duodenum and proximal jejunum, where they may cause symptoms of illness.

(2) Water, ice, or food prepared with contaminated water, are the major sources of infection, but direct person-to-person transmission can occur from sexual contact or hand-to-mouth contamination.

(3) Incubation time: 5 - 25 days (7 - 10 medium). Most infections are thought to be asymptomatic carrier states, however, a wide array of clinical syndromes may occur.

c. Symptoms.

(1) A number of people have abrupt onset of explosive diarrhea accompanied by abdominal cramps, foul gas, vomiting, fever and malaise. This commonly lasts three to four days before developing the more common subacute syndrome. In most individuals, the onset is more insidious, and symptoms are persistent or recurrent.

(2) Stools become mushy, greasy and malodorous. Over a period of weeks, watery diarrhea may alternate with soft stools and even constipation. Upper GI symptoms accompany stool changes, mid and upper abdominal cramping, substernal burning, acid indigestion, sulfurous belching, nausea and foul flatus, distention with an early "full" feeling are all typically increased after a meal.

(3) Symptoms of anorexia, fatigue and weight loss are common, but fever and vomiting are infrequent except during an acute onset of illness.

d. Laboratory Parameters.

(1) The "typical" victim should suggest a diagnosis of giardiasis. This individual may have a history of camping, vacationing at a mountain resort, or foreign travel. Symptoms of upper abdominal discomfort, foul gas and soft stools arose one or two weeks after a short trip or during a lengthy one and have fluctuated for weeks. Sometimes improving, the individual held off visiting the doctor, only to have the symptoms recur. Work or school is not interrupted, however, physical discomfort and embarrassment from the gas are concerns.

(2) Laboratory confirmation of giardiasis can be difficult. Stool samples are the first means of diagnosis. Trophozoites may be found in fresh, watery stools, but disintegrate rapidly. Cysts remain in fresh stools for at least 24 hours. Cysts passage, however, is extremely variable and not related to clinical symptoms. Three samples, taken every other day, should be examined for a definite diagnosis.

(3) Average duration of symptoms has varied from three to ten weeks. Acquired resistance has been shown. Immunity is suspected, but not proven and human reinfection definitely occurs.

e. Treatment. (WMC.6.17f) Complete cure of most Giardia infections can be achieved 90% of the time with *Metronidazole* (flagyl) 250 mg three time s a day for 5 days.

f. Prevention There is no safe or effective chemoprophylactic drug and prevention should be practiced by travelers and campers. The purity of American wild waters can no longer be assumed and in developing countries, even tap water should be suspected. In addition to choosing a technique for individual field water disinfection, individuals must become more responsible for prevention of water contamination. This requires attention to personal latrines and camper trailer waste disposal, a subject that most people do not want to touch.

6. **WASTE**. The importance of proper waste disposal cannot be over emphasized.

a. It serves to: (WMC.6.17g)

(1) Eliminate harborage for rodents and vermin.

(2) Prevents attraction of rodents and vermin.

(3) Prevents a pathogenic contamination source.

b. Two basic types of waste are organic and non-organic. The basic ways to dispose of them are burning, burying, or hauling it away. (WMC.6.17h)

(1) Organic wastes:

(a) Human waste - Burn or haul away to a designated waste pit area.

(b) Urine - Use only assigned marked areas.

(c) Edible garbage - Burn. It should not be left exposed for animals, vermin or the enemy.

(d) Paper -Burn.

(e) Contaminated bandages - Sterilize/burn.

(2) Non-organic waste:

(a) Metals - Haul away or bury.

(b) Liquids - Burn or bury.

UNITED STATES MARINE CORPS
Mountain Warfare Training Center
Bridgeport, California 93517-5001

FMST.07.19
04/03/02

STUDENT OUTLINE

REQUIREMENTS FOR SURVIVAL

TERMINAL LEARNING OBJECTIVE Given a survival situation in any type of environmental condition and minimal equipment and resources, apply the requirements for survival to facilitate individual or group survival and recovery per the references. (FMST.07.19)

ENABLING LEARNING OBJECTIVES

1. Without the aid of references and from a given list, choose one of the survival stressors, in accordance with the references. (WMC.A1.a)

2. Without the aid of references and from a given list, choose one of the priorities of work in a survival situation, in accordance with the references. (WMC.A1.b)

1. REQUIREMENTS FOR SURVIVAL

a. This mental "mind-set" is important in many ways. We usually call it the "will to survive" although you might call it "attitude" just as well. This basically means that, if you do not have the right attitude, you may still not survive.

A1

b. A guideline that can assist you is the acronym " SURVIVAL".

(1) **S**ize up.

(a) Size up the situation.

1. Conceal yourself from the enemy.

2. Use your senses to hear, smell, and see to determine and consider what is developing on the battlefield before you make your survival plan.

(b) Size up your surroundings.

1. Determine the rhythm or pattern of the area.

2. Note animal and bird noises and their movement.

3. Note enemy traffic and civilian movement.

(c) Size up your physical condition.

1. Check your wounds and give yourself first aid.

2. Take care to prevent further bodily harm.

3. Evaluate condition of self and unit prior to developing survival plan.

(d) Size up your equipment.

1. Consider how available equipment may affect survival senses; tailor accordingly.

(2) **U**ndue haste makes waste.

(a) Plan your moves so that you can move out quickly without endangering yourself if the enemy is near.

(3) **R**emember where you are.

(a) If you have a map, spot your location and relate it to the surrounding terrain.

(b) Pay close attention to where you are and where you are going. Constantly orient yourself.

(c) Try to determine, at a minimum, how your location relates to the following:

1. The location of enemy units and controlled areas.

2. The location of friendly units and controlled areas.

3. The location of local water sources.

4. Areas that will provide good cover and concealment.

(4) **V**anquish fear and panic.

(a) Realistic and challenging training builds self-confidence and confidence for a unit's leadership.

(b) The feeling of fear and panic will be present. The survivor must control these feelings.

(5) **I**mprovise and improve.

(a) Use tools designed for one purpose for other applications.

(b) Use objects around you for different needs. (i.e. use a rock for a hammer)

(6) **V**alue living.

(a) Place a high value on living.

(b) Refuse to give into the problem and obstacles that face you.

(c) Draw strength from individuals that rise to the occasion.

(7) **A**ct like the natives.

(a) Observe the people in the area to determine their daily eating, sleeping, and drinking routines.

(b) Observe animal life in the area to help you find sources of food and water.

NOTE: Remember that animal reactions can reveal your presence to the enemy. Animals cannot serve as an absolute guide to what you can eat and drink.

(8) **L**ive by your wits, **but for now**, learn basic skills.

(a) Practice basic survival skills during all training programs and exercises.

2. **STRESS**. Stress has many positive benefits. Stress provides us with challenges: it gives us chances to learn about our values and strengths. Too much stress leads to distress. While many of these signs may not be self-identified, it remains critical that all survivors remain attentive to each other's signs of distress. Listed are a few common signs of distress found when faced with too much stress:

   a. Difficulty in making decisions (do not confuse this sign for a symptom of hypothermia).

   b. Angry outbursts.

   c. Forgetfulness.

   d. Low energy level.

   e. Constant worrying.

   f. Propensity for mistakes.

   g. Thoughts about death or suicide.

   h. Trouble getting along with others.

   i. Withdrawing from others.

   j. Hiding from responsibilities.

   k. Carelessness.

3. **SURVIVAL STRESSORS**. (FMST.07.19a) **Any** event can lead to stress. Often, stressful events occur simultaneously. These events are not stress, but they produce it and are called "stressors". In response to a stressor, the body prepares either to "fight or flee". Stressors add up. Anticipating stressors and developing strategies to cope with them are the two ingredients in the effective management of stress. It is essential that the survivor be aware of the types of stressors they will encounter.

   a. Injury, Illness, or Death: Injury, illness, and death are real possibilities a survivor has to face. Perhaps nothing is more stressful than being alone in an unfamiliar environment where you could die from hostile action, an accident, or from eating something lethal. Illness and injury can also add to stress by limiting your ability to maneuver, get food and drink, find shelter, and defend yourself.

   b. Uncertainty and Lack of Control: Some people have trouble operating in settings where everything is not clear-cut. The only guarantee in a survival situation is that nothing is guaranteed. This uncertainty and lack of control also add to the stress of being ill, injured, or killed.

c. Environment: A survivor will have to contend with the stressors of weather, terrain, and the variety of creatures inhabiting an area. Heat, cold, rain, winds, snow, mountains, insects, and animals are just a few of the challenges awaiting the Marine working to survive. Depending on how a survivor handles the stress of environment, his surroundings can be either a source of food and protection or a cause of extreme discomfort leading to injury, illness, or death.

d. Hunger and Thirst: Without food and water a person will weaken and eventually die. Getting and preserving food and water takes on increasing importance as the length of time in a survival setting increases. With the increased likelihood of diarrhea, replenishing electrolytes becomes critical. For a Marine used to having his provisions issued, foraging can be a significant source of stress.

e. Fatigue: It is essential that survivors employ all available means to preserve mental and physical strength. While food, water, and other energy builders may be in short supply, maximizing sleep to avoid deprivation is a very controllable factor. Training data collected at MCCDC demonstrates that individuals who lack sleep for 24 hours or more suffer a 25% decrease in effective performance. Further, sleep deprivation directly correlates with increased fear. Effective survival is highly unlikely when fatigue builds to a level where staying awake becomes a stressful evolution.

f. Isolation: Although Marines complain about higher headquarters, we become used to the information and guidance it provides, especially during times of confusion. Being in contact with others provides a greater sense of security and a feeling someone is available to help if problems occur. A significant stress in survival situations is that often a person or team has to rely solely on its own resources.

4. **NATURAL REACTIONS**. Man has been able to survive many shifts in his environment throughout the centuries. His ability to adapt physically and mentally to a changing world kept him alive. The average person will have some psychological reactions in a survival situation. These are some of the major internal reactions you might experience with the survival stresses.

   a. Fear: Fear is our emotional response to dangerous circumstances that we believe have the potential to cause death, injury, or illness. Fear can have a positive function if it encourages us to be cautious in situations where recklessness could result in injury. Unfortunately, fear can also immobilize a person. It can cause us to become so frightened that we fail to perform activities essential for survival.

   b. Anxiety: Anxiety can be an uneasy, apprehensive feeling we get when faced with dangerous situations. When used in a healthy way, anxiety urges us to act to end, or at least master, the dangers that threaten our existence. A survivor reduces his anxiety by performing those tasks that will ensure his coming through the ordeal alive. Anxiety can overwhelm a Marine to the point where he becomes easily confused and has difficulty thinking.

c. Anger and Frustration: Frustration arises when a person is continually thwarted in his attempts to reach a goal. The goal of survival is to stay alive until you can reach help or until help can reach you. To achieve this goal, Marines must complete some tasks with minimal resources. One outgrowth of frustration is anger. Getting lost, damaged or forgotten equipment, the weather, inhospitable terrain, enemy patrols, and physical limitations are just a few sources of frustration and anger. Frustration and anger encourage impulsive reactions, irrational behavior, poorly thought-out decisions, and, in some instances, an "I quit" attitude. If the Marine does not properly focus his angry feelings, he can waste much energy in activities that do little to further either his chances of survival or the chances of those around him.

d. Depression: Depression is closely linked with frustration and anger when faced with the privations of survival. A destructive cycle between anger and frustration continues until the person becomes worn down-physically, emotionally, and mentally. When a person reaches this point, he starts to give up, and his focus shifts from "What can I do" to "There is nothing I can do." If you allow yourself to sink into a depressed state, then it can sap all your energy and, more important, your will to survive.

e. Loneliness and Boredom: Man is a social animal and enjoys the company of others. Loneliness and boredom can be another source of depression. Marines must find ways to keep their minds productively occupied.

f. Guilt: The circumstances leading to your survival setting are sometimes dramatic and tragic. It may be the result of an accident or military mission where there was a loss of life. Perhaps you were the only, or one of a few, survivors. While naturally relieved to be alive, you simultaneously may be mourning the deaths of others that were less fortunate. Do not let guilt feelings prevent you from living.

5. **PRIORITIES OF WORK IN A SURVIVAL SITUATION.** (FMST.07.19b) Each survival situation will have unique aspects that alter the order in which tasks need to be accomplished. A general guideline is to think in blocks of time.

a. First 24 hours: The first 24 hours are critical in a survival situation. You must make an initial estimate of the situation. Enemy, weather, terrain, time of day and available resources will determine which tasks need to be accomplished first. They should be the following:

(1) Shelter.

(2) Fire.

(3) Water.

(4) Signaling.

b. Second 24 hours: After the first 24 hours have passed, you will now know you can survive. This time period needs to be spent on expanding your knowledge of the area. By completing the following tasks, you will be able to gain valuable knowledge.

(1) Tools and weapons: By traveling a short distance from your shelter to locate the necessary resources, you will notice edible food sources and game trails.

(2) Traps and snares: Moving further away from your shelter to employ traps and snares, you will be able to locate your shelter area from various vantagepoints. This will enable you to identify likely avenues of approach into your shelter area.

(3) Pathguards: Knowing the likely avenues of approaches, you can effectively place noise and casualty producing pathguards to ensure the security of your shelter area.

c. Remainder of your survival situation. This time is spent on continuously improving your survival situation until your rescue.

6. **GROUP SURVIVAL.** In-group survival, the group's survival depends largely on its ability to organize activity. An emergency situation does not bring people together for a common goal; rather, the more difficult and disordered the situation, the greater are the disorganized group's problems.

a. Groups Morale: High morale must come from internal cohesiveness and not merely through external pressure. The moods and attitudes can become wildly contagious. Conscious, well-planned organization and leadership on the basis of delegated or shared responsibility often can prevent panic. High group morale has many advantages.

(1) Individual feels strengthened and protected since he realizes that his survival depends on others whom he trusts.

(2) The group can meet failure with greater persistency.

(3) The group can formulate goals to help each other face the future.

b. Factors that Influence Group Survival: There are numerous factors that will influence whether a group can successfully survive.

(1) Organization of Manpower - Organized action is important to keep all members of the group informed; this way the members of the group will know what to do and when to do it, both under ordinary circumstances and in emergencies.

(2) Selective Use of Personnel - In well-organized groups, the person often does the job that most closely fits his personal qualifications.

(3) Acceptance of Suggestion and Criticisms - The senior man must accept responsibility for the final decision, but must be able to take suggestion and criticisms from others.

(4) Consideration of Time - On-the-spot decisions that must be acted upon immediately usually determines survival success.

(5) Check Equipment - Failure to check equipment can result in failure to survive.

(6) Acquiring survival knowledge and skills increases survival Knowledge and Skills - Confidence in one's ability to survive.

UNITED STATES MARINE CORPS
Mountain Warfare Training Center
Bridgeport, California 93517-5001

FMST.
07.20
04/03/02

STUDENT OUTLINE

SURVIVAL KIT

TERMINAL LEARNING OBJECTIVE. In a summer mountainous environment, construct a survival kit, in accordance with the references. (FMST.07.20)

ENABLING LEARNING OBJECTIVES

(1) Without the aid of references, choose from a given list in the criteria for a survival kit, in accordance with the references. (FMST.07.20a)

(2) Without the aid of references, choose from a given list one example of each criteria for a survival kit, in accordance with the references. (FMST.07.20b)

1. COMPONENTS FOR A SURVIVAL KIT

a. The environment is the key to the types of items you will need in your survival kit. How much equipment you put in your kit depends on how you will carry the kit. A kit on your body will have to be much smaller than one carried in a vehicle.

b. Always layer your survival kit, keeping the most important items on your body.

c. In preparing your survival kit, select items that can be used for more than one purpose.

d. Your survival kit does not need to be elaborate. You only need functional items that will meet your needs and a case to hold them. The case might be a first aid case, an ammunition pouch, or another suitable case. This case should be:

(1) Water repellent or waterproof.

(2) Easy to carry or attach to your body.

(3) Suitable to accept various sized items.

(4) Durable.

e. When constructing a survival kit, you should have the following components: (FMST.07.20a)

(1) Fire starting items.

(2) Water procurement items.

(3) Food procurement items.

(4) Signaling items.

(5) First aid items.

(6) Shelter items.

2. **ITEMS CONTAINED WITHIN EACH COMPONENT** (FMST.07.20b)

a. Fire Starting Items.

(1) Matches.

(2) Magnifying glass.

(3) Flint and Steel.

(4) Lighter.

(5) Potassium Permanganate, with a container of sugar or anti-freeze.

(6) Prepackaged Tinder.

(a) Commercially Manufactured

(b) Cotton Balls and Petroleum Jelly

b. Water Procurement Items

(1) Water Disinfecting Chemicals.

(a) Iodine Tablets

(b) Betadine Solution

(c) Iodine Solution

(2) Metal Container. (Serves for boiling water)

(a) Canteen Cup

(b) Survival Kit Container

(c) Any suitable can that contained no petroleum products.

(3) Water Carrying Items.

(a) Canteen

(b) Plastic Bag

(c) Plastic/Metal/Glass Container which contained no petroleum products

c. Food Procurement Items

(1) Fish.

(a) Various sized hooks

(b) Various sized sinkers/weights

(c) Metal leaders and swivels

(d) Small weighted jigs

(e) Fishing line

i. Think about the size of fish for that environment when selecting weights and sizes.

(2) Game.

(a) Snares

i. Commercially Manufactured

ii. Aircraft Cable

iii. Tie Wire

(b) Bait

i. MRE Cheese Spread or Peanut Butter Package

(c) 550 Cord for Gill Net and Trap Construction

i. Engineer/Marking Tape

ii. Sling shot rubber and pouch

d. <u>Signaling Items</u>

(1) Day.

(a) Mirror

(b) Whistle

(c) Pyrotechnics (Smoke, Pen Flares)

(d) Air Panels

(2) Night.

(a) Pyrotechnics (Pen Flares, Star Clusters)

(b) Lights (Flashlight, Strobe, Chemlight)

(c) Whistle

e. <u>Shelter Items</u>

(1) Cordage.

(a) 550 Cord

(b) Wire

(c) Communication wire

(d) Tie wire

(2) Finger Saw.

(3) Sewing Kit with Needles for construction/repair of clothing.

(4) Tentage.

(a) Poncho

(b) Tarp

(c) Space blanket

(d) Plastic trash bags

f. First Aid Items

(1) Band-Aids.

(a) Steristrips

(b) Adhesive Tape

(c) Non-stick pads, 4x4's, Gauze, Battle Dressings

(d) Muslin Bandage

(2) Ointments.

(a) Burn

(b) Anti-septic

(3) Miscellaneous.

(a) Salt

(b) Sugar

(c) Eye Wash

(d) Alcohol prep pads

(e) Suture Kit

(f) Scalpel

(g) Vile of Yarrow

g. Miscellaneous items

(1) Fingernail clippers

(2) Compass

(3) Notebook with pen or pencil

(4) Wood eye screws and nails

(5) Surgical tubing

NOTE: It is assumed that the Marine is always carrying a high quality fixed bladed knife, a multi-tool knife, and a sharpening stone.

UNITED STATES MARINE CORPS
Mountain Warfare Training Center
Bridgeport, California 93517-5001

FMST.07.23
04/05/02

STUDENT OUTLINE

SIGNALLING AND RECOVERY

TERMINAL LEARNING OBJECTIVE Given a survival situation in any type of environmental condition and minimal equipment and resources, signal for aid to facilitate individual or group recovery per the references. (FMST 07.23)

ENABLING LEARNING OBJECTIVES

(1) Without the aid of references and from a given list, choose the audio international distress Signal, in accordance with the references. (FMST.07.23a)

(2) Without the aid of references and from a given list, choose the visual international distress signal, in accordance with the references. (FMST.07.23b)

(3) Without the aid of references and given a list of messages, match in writing the international code symbols to each corresponding message, in accordance with the references. (FMST.07.23c)

OUTLINE

1. **SIGNALING DEVICES**. The equipment listed below are items that may be on your body or items inside an aircraft. Generally, these items are used as signaling devices while on the move. They must be accessible for use at any moment's notice. Additionally, in a summer mountainous environment, Marines may experience areas that are snow covered and must be familiar with the effects that snow will have on specific signaling devices.

    a. Pyrotechnics. Pyrotechnics include star clusters and smoke grenades. When using smoke grenades in snow pack, a platform must be built. Without a platform, the smoke grenade will sink into the snow pack and the snow will absorb all smoke. A rocket parachute flare or a hand flare has been sighted as far away as 35 miles, with an average of 10 miles. Pyrotechnic flares are effective at night, but during daylight their detectability ranges are reduced by 90 percent.

A3

b. M-186 Pen Flare: The M-186 Pen Flare is a signaling device carried in the vest of all crew chiefs and pilots. Remember to cock the gun prior to screwing in the flare.

c. Strobe Light: A strobe light is generally carried in the flight vests of all crew chiefs and pilots. It can be used at night for signaling. Care must be taken because a pilot using goggles may not be able to distinguish a flashing strobe from hostile fire.

d. Flashlight: By using flashlights, a Morse code message can be sent. A SOS distress call consists of sending three dots, three dashes, and three dots. Keep repeating this signal.

e. Whistle: The whistle is used in conjunction with the audio international distress signal. It is used to communicate with forces on the ground.

f. AN/PRC-90 & AN/PRC-112: The AN/ PRC 90 survival radio is a part of the aviator's survival vest. The AN/PRC-112 will eventually replace the AN/PRC-90. Both radios can transmit either tone (beacon) or voice. Frequency for both is **282.8 for voice**, and **243.0 for beacon**. Both of these frequencies are on the UHF Band.

g. Day/Night Flare: The day/night flare is a good peacetime survival signal. The flare is for night signaling while the smoke is for day. A red cap identifies the older version flare with three nubbins while the new generation has three rings around the body for identification during darkness. The flare burns for approximately 20 second while the smoke burns for approximately 60 seconds.

NOTE: Once one end is used up, douse in water to cool and save the other end for future use.

h. Signal Mirror: A mirror or any shiny object can be used as a signaling device. It can be used as many times as needed. Mirror signals have been detected as far away as 45 miles and from as high as 16,000', although the average detection distance is 5 miles. It can be concentrated in one area, making it secure from enemy observation. A mirror is the best signaling device for a survivor; however, it is only as effective as its user. Learn how to use one now, before you find yourself in a survival situation.

(1) Military signal mirrors have instructions on the back showing how to use it. It should be kept covered to prevent accidental flashing that may be seen by the enemy.

(2) Any shiny metallic object can be substituted for a signal mirror.

(3) Haze, ground fog, or a mirage may make it hard for a pilot to spot signals from a flashing object. So, if possible, get to the highest point in your area when flashing. If you can't determine the aircraft's location, flash your signal in the direction of the aircraft noise.

AIMING THE SIGNAL MIRROR

2. METHODS OF COMMUNICATION

a. Audio: Signaling by means of sound may be good, but it does have some disadvantages.

(1) It has limited range unless you use a device that will significantly project the sound.

(2) It may be hard to pinpoint one's location due to echoes or wind.

(3) International Distress Signal. (FMST.07.23a) The survivor will make six blasts in one minute, returned by three blasts in one minute by the rescuer.

b. Visual: Visual signals are generally better than audio signals. They will pinpoint your location and can been seen at a greater distances under good weather conditions.

(1) The visual international distress symbol is recognized by a series of three evenly spaced improvised signaling devices. (FMST.07.23b)

3. **IMPROVISED SIGNALING DEVICES**. Improvised signaling devices are generally static in nature. They must be placed in a position to be seen by rescuers. They are made from any resources available, whether natural or man-made.

a. Smoke Generator: The smoke generator is an excellent improvised visual signaling device. It gives the survivor the flexibility to signal in either day or night conditions. This type of signal has been sighted as far away as 12 miles, with an average distance of 8 miles. Smoke signals are most effective in calm wind conditions or open terrain, but effectiveness is reduced with wind speeds above 10 knots. Build them as soon as time and the situation permits, and protect them until needed.

(1) Construct your fire in a natural clearing or along the edge of streams (or make a clearing). Signal fires under dense foliage will not be seen from the air.

(2) Find two logs, 6 - 10 inches in diameter, and approximately five feet long. Place the two logs parallel to each other with 3 - 4 feet spacing.

(3) Gather enough sticks, approximately two inches in diameter and four feet long, to lie across the first two logs. This serves as a platform for the fire.

(4) Gather enough completely dry branches to build a pyramid fire. The pyramid fire should be 4 feet by 4 feet by 2 feet high.

(5) Place your tinder under the platform.

(6) Gather enough pine boughs to lay on top of the pyramid fire. This serves to protect the fire and the tinder.

(7) To light, remove the pine bough and ignite the tinder. If available, construct a torch to speed up the lighting process, especially for multiple fires.

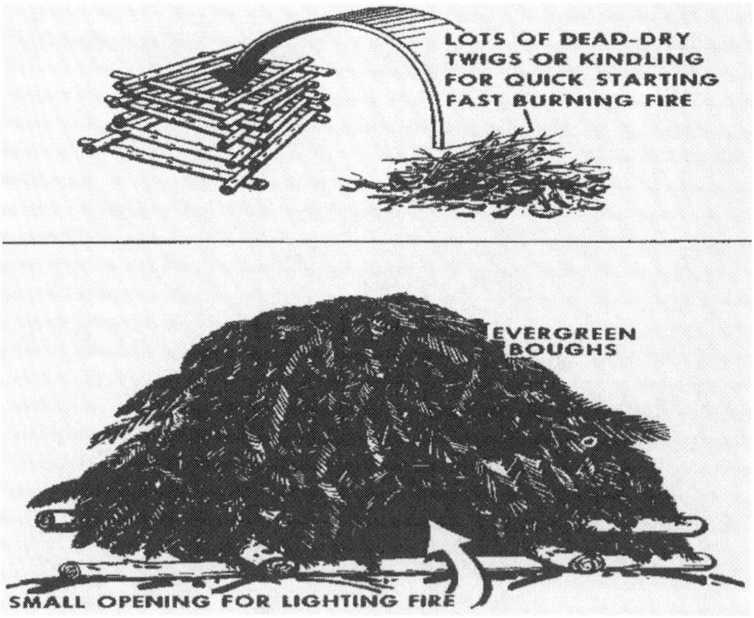

SMOKE GENERATOR

(8) To create a smoke effect during the day light hours, place the pine bough on the ignited fire.

(9) Placing a smoke grenade or colored flare under the platform will change the color of the smoke generated. Remember that you want the fire to draw in the colored smoke, which will create a smoke color that contrasts with the background will increase the chances of success.

b. Arrangement or alteration of natural materials: Such things as twigs or branches, can be tramped into letters or symbols in the snow and filled in with contrasting materials. To attract more attention, ground signals should be arranged in big geometric patterns.

(1) International symbols. The following symbols are internationally known. (FMST.07.23c)

Number	Message	Code symbol
1	REQUIRE ASSISTANCE	V
2	REQUIRE MEDICAL ASSISTANCE	X
3	NO OR NEGATIVE	N
4	YES OR AFFIRMATIVE	Y
5	PROCEED IN THIS DIRECTION	↑

INTERNATIONAL SYMBOLS

(2) Shadows: If no other means are available, you may have to construct mounds that will use the sun to cast shadows. These mounds should be constructed in one of the International Distress Patterns. To be effective, these shadow signals must be oriented to the sun to produce the best shadows. In areas close to the equator, a North—South line gives a shadow anytime except noon. Areas further north or south of the equator require the use of East—West line or some point of the compass in between to give the best result.

(3) Size: The letters should be large as possible for a pilot or crew to spot. Use the diagram below to incorporate the size to ratio for all letter symbols.

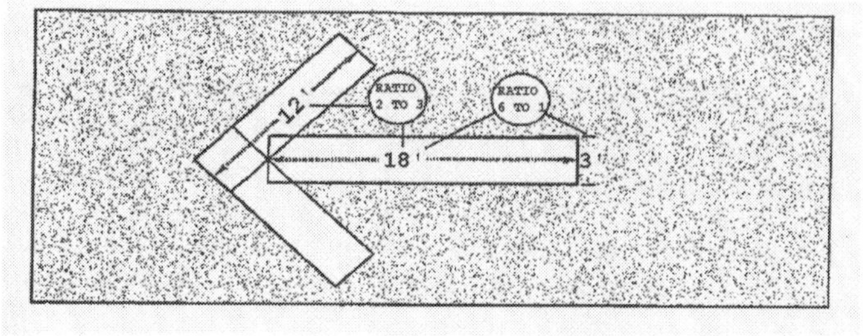

SIZE AND RATIO

(4) Contrast: When constructing letter symbols, contrast the letter from the surrounding vegetation and terrain. Ideally, bring material from another location to build the letter. This could be clothing, air panels, space blanket, etc.

(a) On snow, pile bough or use sea dye from a LPP (Life preserver, personal). Fluorescent sea dye markers have been sighted as far away as 10 miles, although the average detection distance is 3 miles.

4. **AIR TO GROUND COMMUNICATIONS**. Air to ground communications can be accomplished by standard aircraft acknowledgments.

a. Aircraft will indicate that ground signals have been seen and understood by:

(1) Rocking wings from side to side: This can be done during the day or in bright moonlight.

b. Aircraft will indicate that ground signals have been seen but not understood by:

(1) Making a complete, clockwise circle during the day or in bright moonlight.

5. **RECOVERY**. Marines trapped behind enemy lines in future conflicts may not experience quick recovery. Marines may have to move to a place that minimizes risk to the recovery force. No matter what signaling device a Marine uses, he must take responsibility for minimizing the recovery force's safety.

a. Placement Considerations: Improvised signaling devices, in a hostile situation, should not be placed near the following areas due to the possibility of compromise:

(1) Obstacles and barriers.

(2) Roads and trails.

(3) Inhabited areas.

(4) Waterways and bridges.

(5) Natural lines of drift.

(6) Man-made structures.

(7) All civilian and military personnel.

b. Tactical Consideration: The following tactical considerations should be adhered to prior to employing an improvised signaling device.

(1) Use the signals in a manner that will not jeopardize the safety of the recovery force or you.

(2) Locate a position, which affords observation of the signaling device and facilitates concealed avenues of escape (if detected by enemy forces). Position should be located relatively close to extract site in order to minimize "time spent on ground" by the recovery force.

(3) Maintain continuous security through visual scanning and listening while signaling devices is employed. If weapon systems are fire and/or observation should cover available, signaling devices.

(4) If enemy movement is detected in the area, attempt to recover the signaling device, if possible.

(5) Employ improvised signaling devices only during the prescribed times, if briefed in the mission order.

c. Recovery Devices: In mountainous terrain, a helicopter landing may be impossible due to ground slope, snow pack, or vegetation. The survivor must be familiar with recovery devices that may be aboard the aircraft.

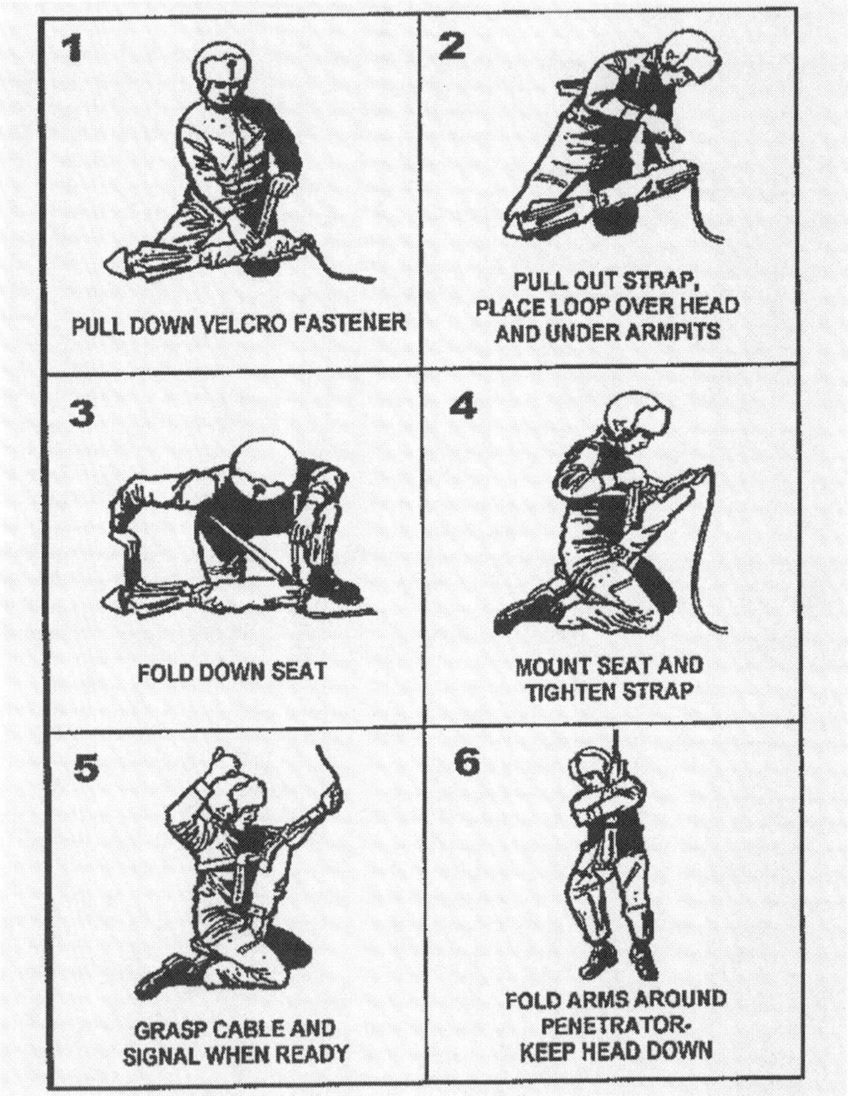

JUNGLE PENETRATOR

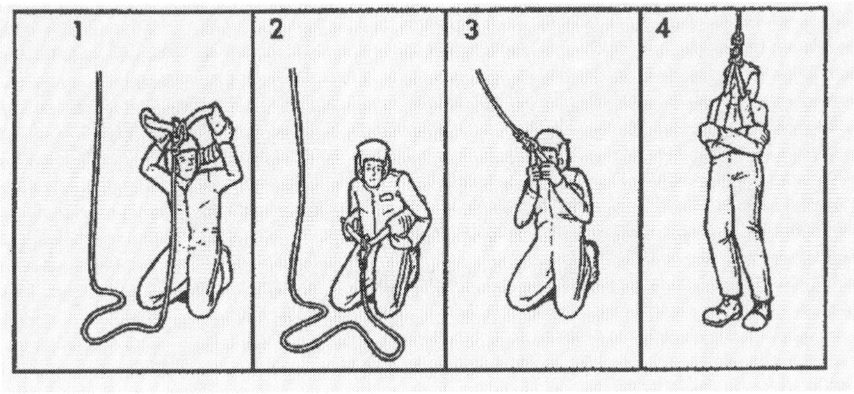

SLING HOIST

d. Recovery by other than aircraft. Recovery by means other than aircraft may occur. Unit Sop's should include signaling and link-up with forces at the following locations:

(1) Border Crossings. The evader who crosses into a neutral country is subject to detention by that country for the duration of the war.

(2) FEBA/FLOT.

(a) Static: Recovery along a static FEBA is always difficult. Under these conditions, enemy and friendly forces can be expected to be densely arrayed and well camouflaged, with good fields of fire. Attempts to penetrate the FEBA should be avoided.

(b) Advancing: Individuals isolated in front of advancing friendly units should immediately take cover and wait for the friendly units to overrun their position.

(c) Retreating: Individuals between opposing forces should immediately take cover and wait for enemy units to pass over their position. After most enemy units have moved on, evaders should try to link up with other isolated friendly elements and return to friendly forces.

(3) Link-up with friendly patrols. Unit authentication numbers and/or locally developed codes may assist the evader to safely make contact in or around the FEBA and when approached by friendly forces.

UNITED STATES MARINE CORPS
Mountain Warfare Training Center
Bridgeport, California 93517-5001

FMST.07.21
FMST.07.04
04/07/02

STUDENT OUTLINE

EXPEDIENT SHELTERS AND FIRES

TERMINAL LEARNING OBJECTIVES

(1) In a summer mountainous environment, construct expedient shelters, in accordance with the references. (FMST.07.21)

(2) In a summer mountainous environment, construct fires, in accordance with the references. (FMST.07.04)

b. **ENABLING LEARNING OBJECTIVES**

(1) Without the aid of references and from a given list, choose the four characteristics of a safe expedient shelter, in accordance with the references. (FMST.07.21a)

(2) Without the aid of references and from a given list, choose the four hazards avoid when using natural shelters, in accordance with the references. (FMST.07.21b)

(3) Without the aid of references and from a given list, choose one of man-made expedient shelters, in accordance with the references. (FMST.07.21c)

(4) Without the aid of references, start a fire using a primitive method, in accordance with the references. (FMST.07.04a)

A4

OUTLINE

1. **BASIC CHARACTERISTICS FOR SHELTER**. (FMST.07.21a) Any type of shelter, whether it is a permanent building, tentage, or an expedient shelter must meet six basic characteristics to be safe and effective. The characteristics are:

    a. Protection from the Elements: The shelter must provide protection from rain, snow, wind, sun, etc.

    b. Heat Retention: It must have some type of insulation to retain heat; thus preventing the waste of fuel.

    c. Ventilation: Ventilation must be constructed, especially if burning fuel for heat. This prevents the accumulation of carbon monoxide. Ventilation is also needed for carbon dioxide given off when breathing.

    d. Drying Facility: A drying facility must be constructed to dry wet clothes.

    e. Free from Natural Hazards: Shelters should not be built in areas of avalanche hazards, under rock fall or "standing dead" trees have the potential to fall on your shelter.

    f. Stable: Shelters must be constructed to withstand the pressures exerted by severe weather.

2. **NATURAL SHELTERS**. Natural shelters are usually the preferred types because they take less time and materials construct. The following may be made into natural shelters with some modification.

    a. Caves or Rock Overhangs: Can be modified by laying walls of rocks, logs or branches across the open sides.

    b. Hollow Logs: Can be cleaned or dug out, then enhanced with ponchos, tarps or parachutes hung across the openings.

    c. Hazards of Natural Shelters: (FMST.04.21b)

        (1) Animals: Natural shelters may already be inhabited (i.e. bears, coyotes, lions, rats, snakes, etc.). Other concerns from animals may be disease from scat or decaying carcasses.

        (2) Lack of Ventilation: Natural shelters may not have adequate ventilation. Fires may be built inside for heating or cooking but may be uncomfortable or even dangerous because of the smoke build up.

        (3) Gas Pockets: Many caves in a mountainous region may have natural gas pockets in them.

(4) Instability: Natural shelters may appear stable, but in reality may be a trap waiting to collapse.

3. **MAN-MADE SHELTERS**. (FMST.04.21c) Many configurations of man-made shelters may be used. Over-looked man-made structures found in urban or rural environments may also provide shelter (i.e. houses, sheds, or barns). Limited by imagination and materials available, the following man-made shelters can be used in any situation.

a. Poncho Shelter.

b. Sapling Shelter.

c. Lean-to.

d. Double lean-to.

e. A-frame Shelter.

f. Fallen Tree Bivouac.

4. **CONSTRUCTION OF MAN-MADE SHELTERS**. To maximize the shelter's effectiveness, Marines should take into consideration the following prior to construction.

a. Considerations.

(1) Group size.

(2) Low silhouette and reduced living area dimensions for improved heat conservation.

(3) Avoid exposed hill tops, valley floors, moist ground, and avalanche paths.

(4) Create a thermal shelter by applying snow, if available, to roof and sides of shelter.

(5) Location of site to fire wood, water, and signaling, if necessary.

(6) How much time and effort needed to build the shelter.

(7) Can the shelter adequately protect you from the elements (sun, wind, rain, and snow). Plan on worst case scenario.

(8) Are the tools available to build it. If not, can you make improvised tools?

(9) Type and amount of materials available to build it.

(10) When in a tactical environment, you must consider the following:

(a) Provide concealment from enemy observation.

(b) Maintain camouflaged escape routes.

(c) Use the acronym BLISS as a guide.

B – Blend in with the surroundings.
L –Low silhouette.
I – Irregular shape.
S – Small.
S – Secluded located.

b. Poncho Shelter: This is one of the easiest shelters to construct. Materials needed for construction are cord and any water-repellent material (i.e. poncho, parachute, and tarp). It should be one of the first types of shelter considered if planning a short stay in any one place.

(1) Find the center of the water-repellent material by folding it in half along its long axis.

(2) Suspend the center points of the two ends using cordage.

(3) Stake the Four Corners down, with sticks or rocks.

c. Sapling Shelter: This type of shelter is constructed in an area where an abundance of saplings are growing. It is an excellent evasion shelter.

(1) Find or clear an area so that you have two parallel rows of saplings at least 4' long and approximately 1 ½' to 2' apart.

(2) Bend the saplings together and tie them to form several hoops, which will form the framework of the shelter.

(3) Cover the hoop with a water-repellent covering.

(4) The shelter then may be insulated with leaves, brush, snow, or boughs.

(5) Close one end permanently. Hang material over the other end to form a door.

d. Lean-to: A lean-to is built in heavily forested areas. It requires a limited amount of cordage to construct. The lean-to is an effective shelter but does not offer a great degree of protection from the elements.

(1) Select a site with two trees (4-12" in diameter) spaced far enough apart that a man can lie down between them. Two sturdy poles can be substituted by inserting them into the ground the proper distance apart.

(2) Cut a pole to support the roof. It should be at least 3-4" in diameter and long enough to extend 4-6" past both trees. Tie the pole horizontally between the two trees, approximately 1 meter off the deck.

(3) Cut several long poles to be used as stringers. They are placed along the horizontal support bar approximately every 1 ½' and lay on the ground. All stringers may be tied to or laid on the horizontal support bar. A short wall or rocks or logs may be constructed on the ground to lift the stringers off the ground, creating additional height and living room dimensions.

(4) Cut several saplings and weave them horizontally between the stringers. Cover the roof with water-repellent and insulating material.

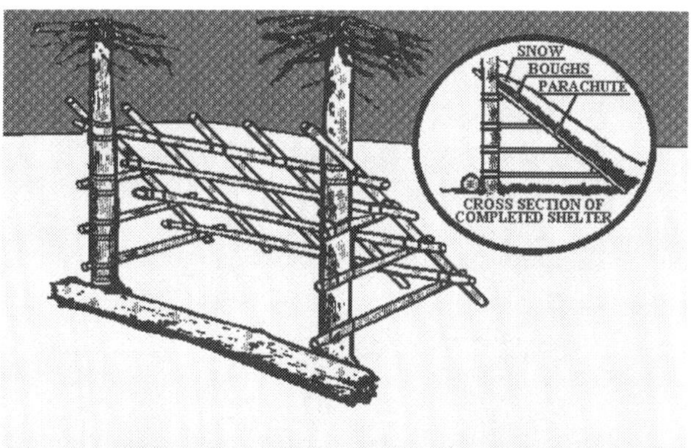

LEAN-TO

e. Double lean-to: The double lean-to shelter is constructed for 2-5 individuals. It is constructed by making two lean-tos and placing them together.

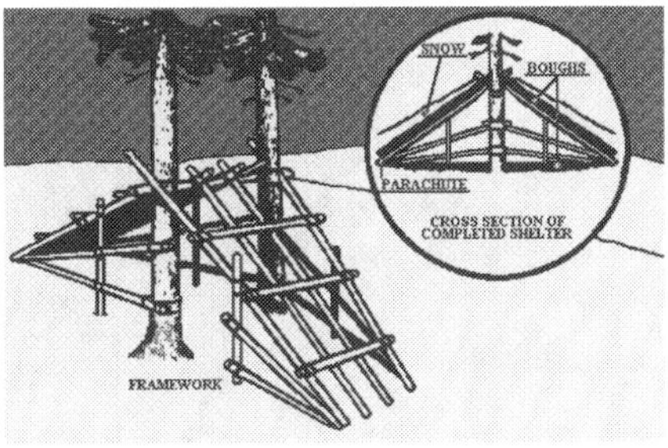

DOUBLE LEAN-TO

f. A-Frame Shelter: An A-Frame shelter is constructed for 1-3 individuals. After the framework is constructed, bough/tentage is interwoven onto the frame and snow, if available, is packed onto the outside for insulation.

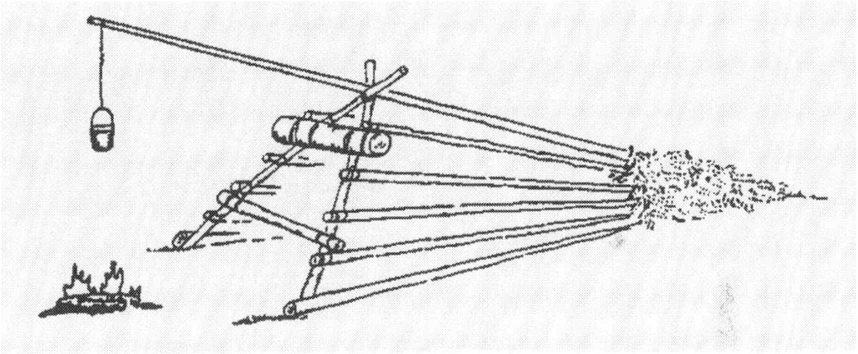

A-FRAME

g. Fallen Tree Bivouac: The fallen tree bivouac is an excellent shelter because most of the work has already been done.

(1) Ensure the tree is stable prior to constructing.

(2) Branches on the underside are cut away making a hollow underneath.

(3) Place additional insulating material to the top and sides of the tree.

(4) A small fire is built outside of the shelter.

FALLEN TREE BIVOUAC

5. **REFLECTOR WALLS.** Heating a shelter requires a slow fire that produces lots of steady heat over a long period of time. A reflector wall should be constructed for all open-ended shelters. A reflector wall is constructed with a flat rock or a stack of green logs propped behind the fire. A surprising amount of heat will bounce back from the fire into the shelter.

6. **FIRES**. Fires fall into two main categories: those built for cooking and those built for warmth and signaling. The basic steps are the same for both: preparing the fire lay, gathering fuel, building the fire, and properly extinguishing the fire.

   a. Preparing the fire lay: There are two types of fire lays: fire pit and Dakota hole. Fire pits are probably the most common.

      (1) Create a windbreak to confine the heat and prevent the wind from scattering sparks. Place rocks or logs used in constructing the fire lay parallel to the wind. The prevailing downwind end should be narrower to create a chimney effect.

      (2) Avoid using wet rocks. Heat acting on the dampness in sandstone, shale, and stones from streams may cause them to explode.

      (3) Dakota Hole: The Dakota Hole is a tactical fire lay. Although no fire is 100% tactical, this fire lay will accomplish certain things:

         (a) Reduces the signature of the fire by placing it below ground.

         (b) Provides more of a concentrated heat source to boil and cook, thus preserving fuel and lessening the amount of burning time.

         (c) By creating a large air draft, the fire will burn with less smoke than the fire pit.

         (d) It is easier to light in high winds.

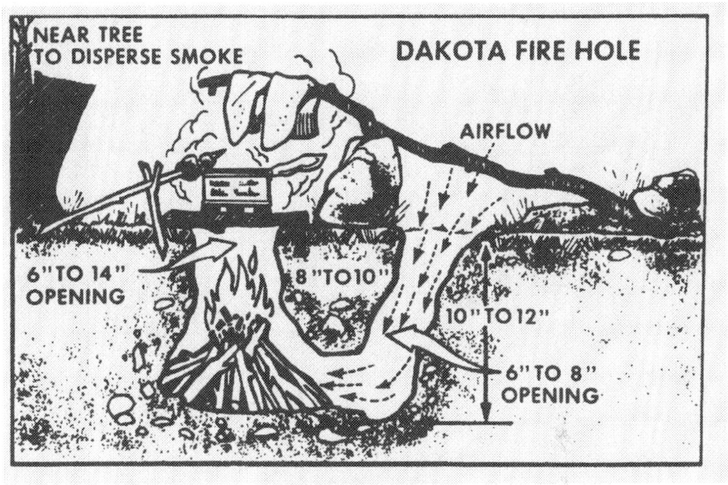

DAKOTA HOLE

b. Gather Fuel: Many Marines take shortcuts when gathering firewood. Taking a few extra minutes can mean the difference between ease and frustration when building a fire.

(1) Tinder: Tinder is the initial fuel. It should be fine and dry. Gather a double handful of tinder for the fire to be built and an extra double handful to be stored in a dry place for the following morning. Dew can moisten tinder enough to make lighting the fire difficult. Some examples are:

(a) Shredded cedar/juniper bark, pine needles.

(b) Dry grass.

(c) Slivers shaved from a dry stick.

(d) Hornet's nest.

(e) Natural fibers from equipment supplemented with pine pitch (i.e., cotton battle dressing).

(f) Cotton balls and petroleum jelly or Char-cloth.

NOTE: Sticks used for tinder should be dry and not larger than the diameter of a toothpick.

(2) Kindling: This is the material that is ignited by the tinder that will burn long enough to ignite the fuel.

(a) Small sticks/twigs pencil-thick up to the thickness of the thumb. Ensure that they are dry.

(b) Due to a typically large resin content, evergreen limbs often make the best kindling. They burn hot and fast, but typically do not last long.

(3) Fuel Wood: Fuel Wood is used to keep the blaze going long enough to fulfill its purpose. Ideally, it should burn slow enough to conserve the woodpile, make plenty of heat, and leave an ample supply of long-lasting coals.

(a) Firewood broken from the dead limbs of standing trees or windfalls held off the ground will have absorbed less moisture and therefore should burn easily.

(b) Refrain from cutting down live, green trees.

(c) Softwoods (evergreens and conifers) will burn hot and fast with lots of smoke and spark, leaving little in the way of coals. Hardwoods (broad leaf trees) will burn slower with less smoke and leave a good bed of coals.

(d) Learn the woods indigenous to the area. Birch, dogwood, and maple are excellent fuels. Osage orange, ironwood, and manzanita, though difficult to break up, make terrific coals. Aspen and cottonwood burn clean but leave little coals.

(e) Stack your wood supply close enough to be handy, but far enough from the flames to be safe. Protect your supply from additional precipitation.

(f) If you happen to go down in an aircraft that has not burned a mixture of gas and oil may be used. Use caution when igniting this mixture.

c. Building the Fire. The type of fire built will be dependent upon its intended use, either cooking or heating and signaling.

(1) Cooking Fires. The following listed fires are best used for cooking:

(a) Teepee Fire: The teepee fire is used to produce a concentrated heat source, primarily for cooking. Once a good supply of coals can be seen, collapse the teepee and push embers into a compact bed.

TEEPEE FIRE

(2) Heating and Signaling Fires.

(a) Pyramid Fire: Pyramid fires are used to produce large amounts of light and heat. They will dry out wet wood or clothing.

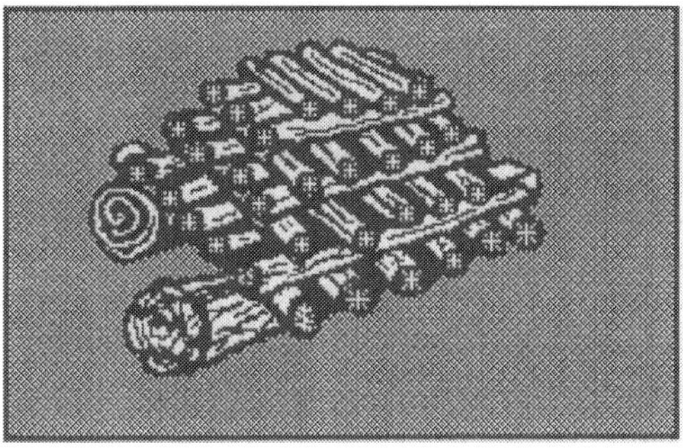

PYRAMID FIRE

d. Starting Fires. Starting a fire is done by a source of ignition and falls into two categories, modern igniters and primitive methods.

(1) Modern Methods: Modern igniters use modern devices we normally think of to start a fire. Reliance upon these methods may result in failure during a survival situation. These items may fail when required to serve their purpose.

(a) Matches and Lighters. Ensure you waterproof these items.

(b) Convex Lens. Binocular, camera, telescopic sights, or magnifying lens are used on bright, sunny days to ignite tender.

(c) Flint and Steel: Sometimes known as metal matches or "Mag Block". Scrape your knife or carbon steel against the flint to produce a spark onto the tinder. Some types of flint & steel designs will have a block of magnesium attached to the device, which can be shaved onto the tinder prior to igniting. Other designs may have magnesium mixed into the flint to produce a higher quality of spark.

(2) Primitive Methods. Primitive fire methods are those developed by early man. There are numerous techniques that fall into this category. The only method that will be taught at MCMWTC is the Bow & Drill.

(3) Bow & Drill. The technique of starting a fire with a bow & drill is a true field expedient fire starting method, which requires a piece of cord and knife from your survival kit to construct. The components of the bow & drill are bow, drill, socket, and fireboard, ember patch, and bird's nest.

(a) Bow: The bow is a resilient, green stick about 3/4 of an inch in diameter and 30-36 inches in length. The bowstring can be any type of cord, however, 550 cord works best. Tie the string from one end of the bow to the other, without any slack.

(b) Drill: The drill should be a straight, seasoned hardwood stick about 1/2 to 3/4 of an inch in diameter and 8 to 12 inches in length. The top end is tapered to a blunt point to reduce friction generated in the socket. The bottom end is slightly rounded to fit snugly into the depression on the fireboard.

(c) Socket: The socket is an easily grasped stone or piece of hardwood or bone with a slight depression on one side. Use it to hold the drill in place and to apply downward pressure.

(d) Fire board: The fire board is a seasoned softwood board which should ideally be 3/4 of an inch thick, 2-4 inches wide, and 8-10 inches long. Cut a depression 3/4 of an inch from the edge on one side of the fireboard. Cut a U-shape notch from the edge of the fireboard into the depression. This notch is designed to collect and form an ember, which will be used to ignite the tinder.

(e) Ember Patch: The ember patch is made from any type of suitable material (i.e., leather, aluminum foil, and bark). It is used to catch and transfer the ember from the fireboard to the bird's nest. Ideally, it should be 4 inches by 4 inches in size.

(f) Bird's Nest: The bird's nest is a double handful of tinder, which will be made into the shape of a nest. Tinder must be dry and finely shredded material (i.e., outer bark from juniper/cedar/sage brush or inner bark from cottonwood/aspen or dry grass/moss). Lay your tinder out in two equal rows about 4 inches wide and 8-12 inches long. Loosely roll the first row into a ball and knead the tinder to further break down the fibers. Place this ball perpendicular onto the second row of tinder and wrap. Knead the tinder until all fibers of the ball are interwoven. Insert the drill half way into the ball to form a partial cylinder. This is where the ember will be placed.

(4) <u>Producing a fire using the bow & drill</u>.

(a) Place the ember patch under the U-shaped notch.

(b) Assume the kneeling position, with the left foot on the fireboard near the depression.

(c) Load the bow with the drill. Ensure the drill is between the wood of the bow and bow string. Place the drill into the depression on the fireboard. Place the socket on the tapered end of the drill.

(d) Use the left hand to hold the socket while applying downward pressure.

(e) Use the right hand to grasp the bow. With a smooth sawing motion, move the bow back and forth to twirl the drill.

(f) Once you have established a smooth motion, smoke will appear. Once smoke appears, apply more downward pressure and saw the bow faster.

(g) When a thick layer of smoke has accumulated around the depression, stop all movement. Remove the bow, drill, and socket from the fire board, without moving the fireboard. Carefully remove your left foot off the fireboard.

(h) Gently tap the fireboard to ensure all of the ember has fallen out of the U-shaped notch and is lying on the ember patch. Remove the fireboard.

(i) Slowly fan the black powder to solidify it into a glowing ember. Grasping the ember patch, carefully drop the ember into the cylinder of the bird's nest.

(j) Grasp the bird's nest with the cylinder facing towards you and parallel to the ground. Gently blow air into the cylinder. As smoke from the nest becomes thicker, continue to blow air into the cylinder until fire appears.

(5) Trouble Shooting the Bow & Drill

(a) Drill will not stay in depression- Apply more downward pressure and/or increase width/depth of depression.

(b) Drill will not twirl- lessens the amount of downward pressure and/or tighten bowstring.

(c) Socket smoking- Lessen the amount of downward pressure. Wood too soft when compared to hardness of drill. Add some lubrication: animal fat, oil, or grease.

(d) No smoke- Drill and fireboard are the same wood. Wood may not be seasoned. Check drill to ensure that it is straight. Keep left hand locked against left shin while sawing.

(e) Smoke but no ember- U-shaped notch not cut into center of the depression.

(f) Bowstring runs up and down drill- use a locked right arm when sawing. Check drill to ensure that it is straight. Ensure bowstring runs over the top of the left boot.

(g) Birds nest will not ignite- Tinder not dry. Nest woven too tight. Tinder not kneaded enough. Blowing too hard (ember will fracture).

e. Extinguishing the Fire. The fire must be properly extinguished. This is accomplished by using the drown, stir, and feel method.

(1) Drown the fire by pouring at water in the fire lay.

(2) Stir the ember bed to ensure that the fire is completely out.

(3) Check the bed of your fire by feeling for any hot spots.

(4) If any hot spots are found, start the process all over again.

UNITED STATES MARINE CORPS
Mountain Warfare Training Center
Bridgeport, California 93517-5001

FMST.07.22
04/07/02

STUDENT OUTLINE

SURVIVAL NAVIGATION

LEARNING OBJECTIVES

TERMINAL LEARNING OBJECTIVE. In a summer mountainous environment, navigate in a survival situation, in accordance with the references. (FMST.07.22)

ENABLING LEARNING OBJECTIVES

(1)Without the aid of references and from a given list, choose one of the considerations for travel, in accordance with the references. (FMST.07.22a)

(2)Without the aid of references and from a given list, choose two methods for locating the North Star, in accordance with the references. (FMST.07.22b)

OUTLINE

1. **CONSIDERATIONS FOR STAYING OR TRAVELLING**. (FMST.07.22a)

   a. Stay with the aircraft or vehicle if possible. More than likely somebody knows where it was going. It is also a ready-made shelter.

   b. Leave only when:

      (1) Certain of present location; have known destination and the ability to get there.

      (2) Water, food, shelter, and/or help can be reached.

      (3) Convinced that rescue is not coming.

A5

c. If the decision is to travel, the following must also be considered:

(1) Which direction to travel and why.

(2) What plan is to be followed.

(3) What equipment should be taken

(4) How to mark the trail.

(5) Predicted weather.

d. If the tactical situation permits leave the following information at the departure point:

(1) Departure time.

(2) Destination.

(3) Route of travel/direction.

(4) Personal condition.

(5) Available supplies.

2. DAYTIME SURVIVAL NAVIGATION

a. Sun Movement. It is generally taken for granted that the sun rises in the east and sets in the west. This rule of thumb, however, is quite misleading. In fact, depending on an observer's latitude and the season, the sun could rise and set up to 50 degrees off of true east and west.

b. The following diagram and terms are essential to understanding how the sun and stars can help to determine direction:

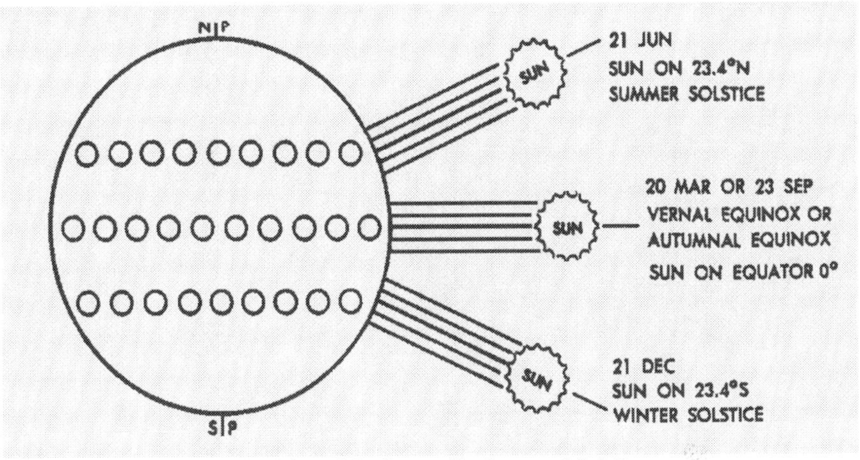

Position of the Sun at Equinox and Solstice

(1) Summer/Winter Solstice: (21 June/21 December) Two times during the year when the sun has no apparent northward or southward motion.

(2) Vernal/Autumnal Equinox: (20 March/23 September) Two times during the year when the sun crosses the celestial equator and the length of day and night are approximately equal.

c. Sun's Movement. As reflected in the diagram above, the sun continuously moves in a cycle from solstice to equinox, throughout each day, however, the sun travels a uniform arc in the sky from sunrise to sunset. Exactly halfway along its daily journey, the sun will be directly south of an observer (or north if the observer is in the Southern Hemisphere). This rule may not apply to observers in the tropics (between 23.5 degrees north and south latitude) or in the Polar Regions (60 degrees latitude). It is at this point that shadows will appear their shortest. The time at which this occurs is referred to as "local apparent noon."

d. Local Apparent Noon. Whenever using any type of shadow casting device to determine direction, "local apparent noon" (or the sun's highest point during the day) must be known. Local apparent noon can be determined by the following methods.

(1) Knowing sunrise and sunset from mission orders, i.e., sunrise 0630 and sunset 1930. Take the total amount of daylight (13 hours), divide by 2 (6 hours 30 minutes), and add to sunrise (0630 plus 6 hours 30 minutes). Based on this example, local apparent noon would be 1300.

(2) Using the string method. The string method is used to find two equidistant marks before and after estimated local apparent noon. The center point between these two marks represents local apparent noon.

e. Sun's Bearing. With an understanding of the sun's daily movement, as well as its seasonal paths, a technique is derived that will determine the true bearing of the sun at sunrise and sunset. With the aid of a circular navigational chart, we can accurately navigate based on the sun's true bearing:

(1) Determine the sun's maximum amplitude at your operating latitude using the top portion of the chart.

(2) Scale the center baseline of the chart where 0 appears as the middle number; write in the maximum amplitude at the extreme north and south ends of the baseline.

(3) Continue to scale the baseline; you should divide the baseline into 6 to 10 tick marks that represent equal divisions of the maximum amplitude.

(4) From today's date along the circumference, draw a straight line down until it intersects the baseline.

(5) The number this line intersects is today's solar amplitude. If the number is left of 0, it is"north" amplitude; if the number is right of 0, it is a "south" amplitude. Use the formula at the bottom of the chart to determine the sun's bearing at sunrise or sunset.

LATITUDE (N OR S)	5	10	15	20	25	30	35	40	45	50	55	60
MAXIMUM AMPLITUDE	24°	24°	24°	25°	26°	27°	29°	31°	34°	38°	44°	53°

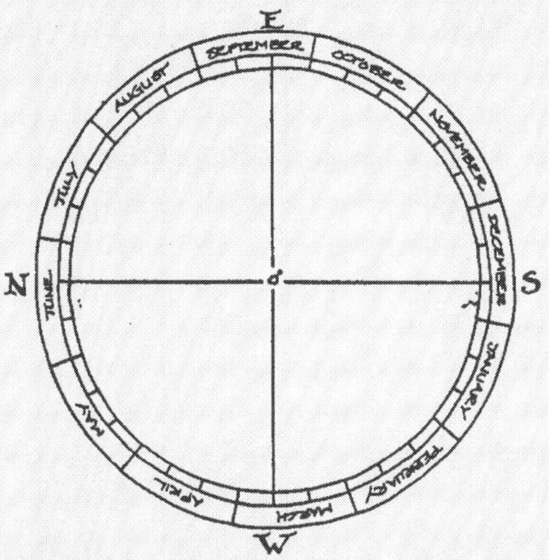

NORTH AMPLITUDES	SOUTH AMPLITUDES
SUNRISE BEARING = 90° − AMPLITUDE	SUNRISE BEARING = 90° + AMPLITUDE
SUNSET BEARING = 270° + AMPLITUDE	SUNSET BEARING = 270° − AMPLITUDE

Circular Navigational Chart

f. Shadow Stick Construction. This technique will achieve a cardinal direction within 10 degrees of accuracy if done within two hours of local apparent noon. Once again, this technique may be impractical near the Polar Regions, as shadows tend to be very long; similarly, in the tropics shadows are generally very small.

(1) Get a straight, 3-6 foot stick free of branches and pointed at the ends and 3-5 small markers: i.e., sticks, rocks, or nails.

(2) Place the stick upright in the ground and mark the shadow tip with a marker.

(3) Wait 10-15 minutes and mark shadow tip again with a marker.

(4) Repeat this until all of the markers are used.

SHADOW STICK METHOD

(5) The markers will form a West—East line.

(6) Put your left foot on the first marker and your right foot on the last marker, you will then be facing north.

3. **POCKET NAVIGATOR** The only material required is a small piece of paper or other flat-surface material upon which to draw the trace of shadow tips and a 1 to 2 inch pin, nail, twig, wooden matchstick, or other such device to serve as a shadow-casting rod.

a. Set this tiny rod upright on your flat piece of material so that the sun will cause it to cast a shadow. Mark the position where the base of the rod sits so it can be returned to the same spot for later readings. Secure the material so that it will not move and mark the position of the material with string, pebbles, or twigs, so that if you have to move the paper you can put it back exactly as it was. Now, mark the tip of the rod shadow.

b. As the sun moves, the shadow-tip moves. Make repeated shadow-tip markings every 15 minutes. As you make the marks of the shadow tip, ensure that you write down the times of the points.

c. At the end of the day, connect the shadow-tip markings. The result will normally be a curved line. The closer to the vernal or autumnal equinoxes (March 21 and September 23) the less pronounced the curvature would be. If it is not convenient or the tactical situation does not permit to take a full day's shadow-tip markings, your observation can be continued on subsequent days by orienting the pocket navigator on the ground so that the shadow-tip is aligned with a previously plotted point.

d. The markings made at the sun's highest point during the day, or solar noon, is the north—south line. The direction of north should be indicated with an arrow on the navigator as soon as it is determined. This north-south line is drawn from the base of the rod to the mark made at solar noon. This line is the shortest line that can be drawn from the base of the pin to the shadow-tip curve.

e. To use your pocket navigator, hold it so that with the shadow-tip is aligned with a plotted point at the specified point. I.e.; if it is now 0900 the shadow-tip must be aligned with that point. This will ensure that your pocket navigator is level. The drawn arrow is now oriented to true north, from which you can orient yourself to any desired direction of travel.

f. The pocket navigator will work all day and will not be out of date for approximately one week.

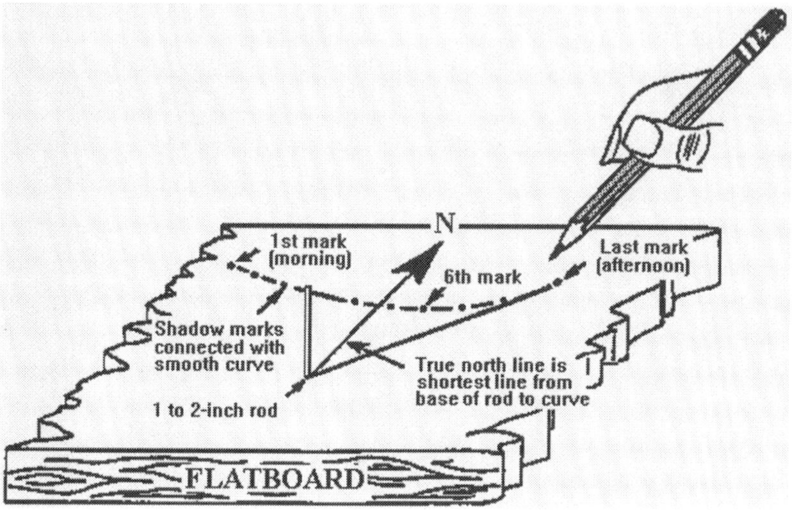

POCKET NAVIGATOR

4. NIGHTTIME SURVIVAL NAVIGATION

a. Mark North. To aid you in navigating at night, it is beneficial to watch where the sun goes down. If you're going to start moving after dark mark the northerly direction.

b. Locating the North Star. There are two methods used in locating the North Star. (FMST.07.22b)

(1) Using the Big Dipper (*Ursa Major*). The best indictors are the two "dippers ". The North Star is the last star in the handle of the Little Dipper, which is not the easiest constellation to find. However, the Big Dipper is one of the most prominent constellations in the Northern Hemisphere. The two lowest stars of the Big Dipper's cup act as pointers to the North Star. If you line up these two stars, they make a straight line that runs directly to the North Star. The distance to the North Star along this line is 5 times that between the two pointer stars.

(2) Using Cassiopeia (*Big M or W*). Draw a line straight out from the center star, approximately half the distance to the Big Dipper. The North Star will be located there.

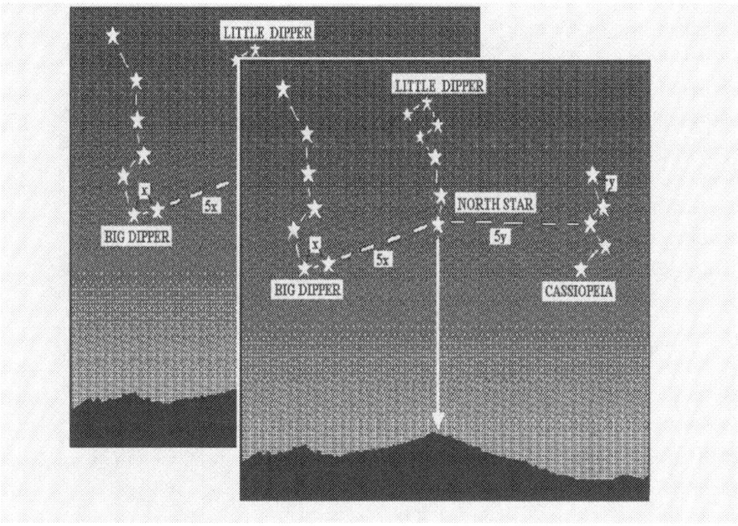

LOCATING THE NORTH STAR

NOTE: Because the Big Dipper and Cassiopeia rotate around the North Star, they will not always appear in the same position in the sky. In the higher latitudes, the North Star is less effective for the purpose of orienting because it appears higher in the sky. At the center of the Arctic Circle, it would be directly overhead and all directions lead south.

c. Southern Cross. In the Southern Hemisphere, Polaris is not visible. There, the Southern Cross is the most distinctive constellation. An imaginary line through the long axis of the Southern Cross, or True Cross, points towards a dark spot devoid of stars approximately three degrees offset from the South Pole. The True Cross should not be confused with the larger cross nearby know as the False Cross, which is less bright, more widely spaced, and has five stars. Two closely spaced, very bright stars that trail behind the crosspiece can confirm the True Cross. These two stars are often easier to pick out than the cross itself. Look for them. Two of the stars in the True Cross are among the brightest stars in the heavens; they are the stars on the southern and eastern arms. The stars on the northern and western arms are not as conspicuous, but are bright.

NOTE: The imaginary point depicted in the picture is the dark spot devoid of stars.

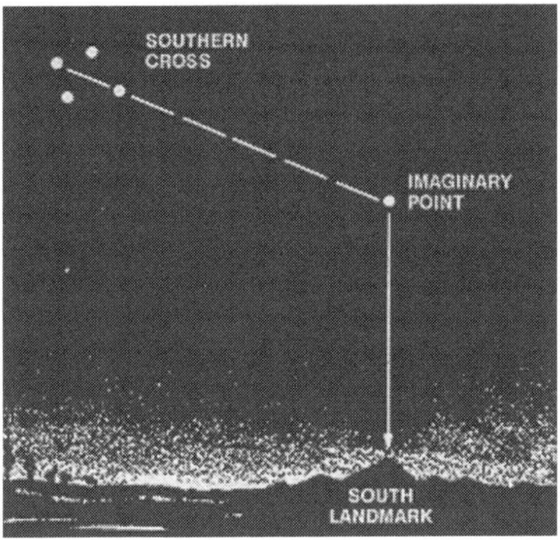

SOUTHERN CROSS

d. Moon Navigator. Like the sun, the moon rises in the east and sets in the west. Use the same method of the shadow stick as you did during the day.

5. **IMPROVISED COMPASSES**. There are three improvised techniques to construct a compass.

a. Synthetic technique. The required items are a piece of synthetic material, (i.e., parachute cloth), and a small piece of iron or steel that is long, thin, and light. Aluminum or yellow metals won't work (only things that rust will do). A pin or needle is perfect, but a straightened paper clip, piece of steel baling wire or barbed wire could also work.

(1) Stroke the needle repeatedly in one direction against the synthetic material. Ensure that you lift the material a few inches up into the air at the end of each stroke, returning to the beginning of the needle before descending for another stroke in the same direction. Do this approximately 30 strokes. This will magnetize the needle.

(2) Float the metal on still water using balled up paper, wood chip, or leaf. Gather some water in a non-magnetic container or a scooped out recess in the

ground, such as a puddle. Do not use a "tin can" which is made of steel. (Aluminum can would be fine.) Place the float on the water, then the metal on it. It will slowly turn to orient itself.

b. Magnet technique. You will achieve the same results by using a magnet. Follow the same steps as you did with the synthetic material. The magnets you are most likely to have available to you are those in a speaker or headphones of a radio.

c. Magnetization through a battery. A power source of 2 volts or more from a battery can be used with a short length of insulated wire to magnetize metal. Coil the wire around a needle. If the wire is non-insulated, wrap the needle with paper or cardboard. Attach the ends to the battery terminals for 5 minutes.

d. Associated problems with improvised compasses. The following are common problems with all improvised compasses.

(1) Soft steel tends to lose its magnetism fairly quickly, so you will have to demagnetize your needle occasionally, though you should not have to do this more than two or three times a day.

(2) Test your compass by disturbing it after it settles. Do this several times. If it returns to the same alignment, you're OK. It will be lined up north and south, though you will have to determine by other means which end is north. Use the sun, stars, or any other natural signs in the area.

(3) Remember that this will give magnetic north. In extreme northern latitudes, the declination angle can be extreme.

6. SURVIVAL NAVIGATION TECHNIQUES

a. Navigator.

(1) Employ a navigation method.

(2) Find the cardinal direction.

(3) Pick a steering mark in the desired direction of travel.

b. Maintain a Log: The possibility may arise when you will not have a map of the area. A log will decrease the chance of walking in circles.

(1) Construction.

(a) Use any material available to you i.e., paper, clothing, MRE box, etc.

(b) Draw a field sketch annotating North, prominent terrain features, and Friendly/enemy position.

(2) Maintenance.

(a) Annotate distance traveled, elevation gained and lost, and cardinal directions.

(b) Maintain and update field sketch as movement progresses.

(c) Ensure readability of your field sketch. (I.e., don't clutter the sketch so much that it can't be read.)

c. <u>During Movement Constantly Refer To</u>.

(1) Log.

(2) Steering marks.

d. <u>Actions If You Become Lost</u>.

(1) Immediate action

(a) Orient your sketch. This will probably make your mistake obvious.

(2) Corrective action

(a) Backtrack using steering marks until you have determined the location of your error.

(b) Re-orient your sketch.

(c) Select direction of travel and continue to march.

UNITED STATES MARINE CORPS
Mountain Warfare Training Center
Bridgeport, California 93517-5001

FMST.07.08
04/07/02

STUDENT OUTLINE

WATER PROCUREMENT

LEARNING OBJECTIVES

TERMINAL LEARNING OBJECTIVE Given a unit in a cold weather environment and the necessary equipment and supplies, perform cold weather preventive medicine, in accordance with the references. (FMST.07.08)

ENABLING LEARNING OBJECTIVES

(1) Without the aid of references from a given list, choose one of the techniques for finding incidental water, in accordance with the references. (FMST.07.08i)

(2) Without the aid of references from a given list, choose one of the four hazardous fluids to avoid substituting for potable water, in accordance with the references. (FMST.07.08j)

(3) Without the aid of references, construct a water-producing device, in accordance with the references. (FMST.07.08k)

OUTLINE

1. WATER INTAKE

a. Thirst is not a strong enough sensation to determine how much water you need.

b. The best plan is to drink, utilizing the OVER DRINK method. Drink plenty of water anytime it is available and particularly when eating.

c. Dehydration is a major threat. A loss of only 5 % of your body fluids causes thirst, irritability, nausea, and weakness; a 10% loss causes dizziness, headache, inability to walk, and a tingling sensation in limbs; a 15% loss causes dim vision, painful urination, swollen tongue, deafness, and a feeling of numbness in the skin; also a loss of more than 15% body fluids could result in death.

d. Your water requirements will be increased if:

(1) You have a fever.

(2) You are experiencing fear or anxiety.

(3) You evaporate more body fluid than necessary. (i.e., not using the proper shelter to your advantage)

(4) You have improper clothing.

(5) You ration water.

(6) You overwork.

2. **<u>INCIDENTAL WATER</u>**. (FMST.07.08.i)

a. During movement, you may have to ration water until you reach a reliable water source. Incidental water may sometimes provide opportunities to acquire water. Although not a reliable or replenished source, it may serve to stretch your water supply or keep you going in an emergency. The following are sources for incidental water:

(1) <u>Dew</u>: In areas with moderate to heavy dew, tying rags or tuffs of fine grass around your ankles can collect dew. While walking through dewy grass before sunrise, the rags or grass will saturate and can be rung out into a container. The rags or grass can be replaced and the process is repeated.

(2) <u>Rainfall</u>: Rainwater collected directly in clean container or in plants that contain no harmful toxins is generally safe to drink without disinfecting. The survivor should always be prepared to collect rainfall at a moment's notice. An inverted poncho works well to collect rainfall.

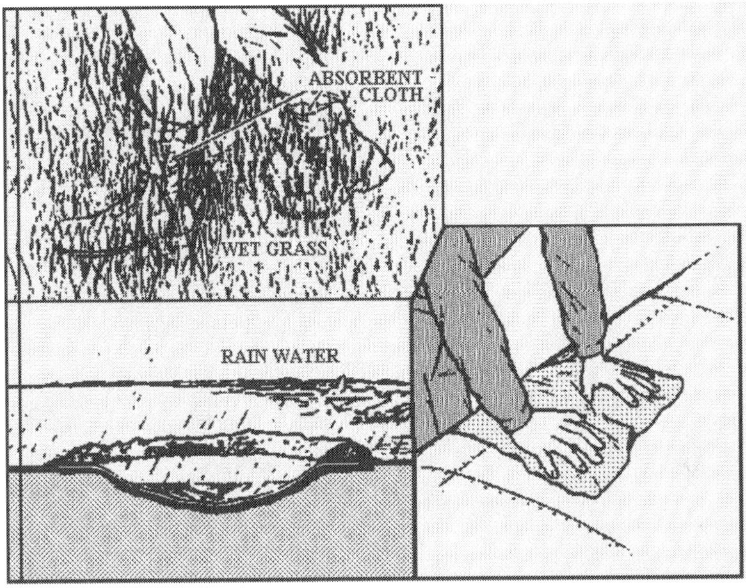

3. **HAZARDOUS FLUIDS** (FMST.07.08j)

a. Survivors have occasionally attempted to augment their water supply with other fluids, such as alcoholic beverages, urine, blood, or seawater. While it is true that each of these fluids has a high water content, the impurities they contain may require the body to expend more fluid to purify them. Some hazardous fluids are:

(1) Seawater: Seawater in more than minimal quantities are actually toxic. The concentration of sodium and magnesium salts is so high that fluid must be drawn from the body to eliminate the salts and eventually the kidneys cease to function.

(2) Alcohol: Alcohol dehydrates the body and clouds judgment. Super-cooled liquid, if ingested, can cause immediate frostbite of the throat, and potential death.

(3) Blood: Blood, besides being salty, is a food. Drinking it will require the body to expend additional fluid to digest it.

(4) Urine: Drinking urine is not only foolish, but also dangerous. Urine is nothing more than the body's waste. Drinking it only places this waste back into the body, which requires more fluid to process it again.

4. **WATER QUALITY**. Water contains minerals, toxins, and pathogens. Some of these, consumed in large enough quantities may be harmful to human health. Pathogens are our primary concern. Pathogens are divided into Virus, Cysts, Bacteria, and Parasites. Certain pathogens are more resistant to chemicals and small enough to move through microscopic holes in equipment (i.e., T-shirt, parachute). Certain pathogens also have the ability to survive in extremely cold water temperatures. Pathogens generally do not live in snow and ice. Water quality is divided into three levels of safety with disinfecting as the most desired level, then purified, followed by potable.

   a. Disinfecting. Water disinfecting removes or destroys harmful microorganisms. Giardia cysts are an ever-present danger in clear appearing mountain water throughout the world. By drinking non-potable water you may contract diseases or swallow organisms that could harm you. Examples of such diseases or organisms are Dysentery, Cholera, Typhoid, Flukes, and Leeches.

   b. Remember impure water, no matter how overpowering the thirst, is one of the worst hazards in a survival situation.

   c. The first step in disinfecting is to select a treatment method. The two methods we will discuss are as follows: (FMST.07.08j)

      (1) Heat. The Manual of Naval Preventive Medicine (P-5010) states that you must bring the water to a rolling boil before it is considered safe for human consumption. This is the most preferred method.

         (a) Bringing water to the boiling point will kill 99.9% of all Giardia cysts. The Giardia cyst dies at 60^{0}C and Cryptosporidium dies at 65C. Water will boil at 14,000' at 86^{0}C and at 10,000' at 90C. With this in mind you should note that altitude does not make a difference unless you are extremely high.

      (2) Chemicals. There are numerous types of chemicals that can disinfect water. Below are a few of the most common. In a survival situation, you will use whatever you have available.

         (a) Iodine Tablets.

         (b) Chlorine Bleach.

         (c) Iodine Solution.

         (d) Betadine Solution.

         (e) Military water purification tablets. These tablets are standard issue for all DOD agencies. These tablets have a shelf life of four years from the date of manufacture, unless opened. Once the seal is broken, they have a shelf life of one year, not to exceed the initial expiration date of four years.

49703 Month / Year / Batch Number

(3) Water Disinfecting Techniques and Halogen Doses.

Iodination techniques Added to 1 liter or quart of water	Amount for 4 PPM	Amount for 8 PPM
Iodine tablets Tetraglcine hydroperiodide EDWGT Potable Aqua Globaline	½ tablet	1 tablet
2% iodine solution (tincture)	0.2 ml 5 gtts	0.4 ml 10 gtts
10% povidone-iodine solution*	0.35 ml 8 gtts	0.70 ml 16 gtts
Chlorination techniques	Amount for 5 PPM	Amount for 10 PPM
Household bleach 5% Sodium hypochlorie	0.1 Ml 2 gtts	0.2 ml 4 gtts
AquaClear Sodium dichloroisocyanurate		1 tablet
AquaCure, AquaPure, Chlor-floc Chlorine plus flocculating agent		8 PPM 1 tablet

*Providone-iodine solutions release free iodine in levels adequate for disinfecting, but scant data is available.

Measure with dropper (1 drop=0.05 ml) or tuberculin syringe

PPM-part per million gtts-drops ml-milliliter

Concentration of **Contact time in minutes at various water temperatures**

Halogen

	5 C / 40 F	15 C / 60 F	30 C / 85 F
2 PPM	240	180	60
4 PPM	180	50	45
8 PPM	60	30	15

NOTE: These contact times have been extended from the usual recommendations to account for recent data that prolonged contact time is needed in very cold water to kill *Giardia* cysts.

NOTE: Chemicals may not destroy Cryptosporidium.

d. Purification. Water purification is the removal of organic and inorganic chemicals and particulate matter, including radioactive particles. While purification can eliminate offensive color, taste, and odor, it may not remove or kill microorganisms.

(1) Filtration. Filtration purifying is a process by which commercial manufacturers build water filters. The water filter is a three-tier system. The first layer, or grass layer, removes the larger impurities. The second layer, or sand layer, removes the smaller impurities. The final layer, or charcoal layer (**not the ash but charcoal from a fire**), bonds and holds the toxins. All layers are placed on some type of straining device and the charcoal layer should be at least 5-6 inches thick. Layers should be changed frequently and straining material should be boiled. Remember that this is not a disinfecting method, cysts can possibly move through this system.

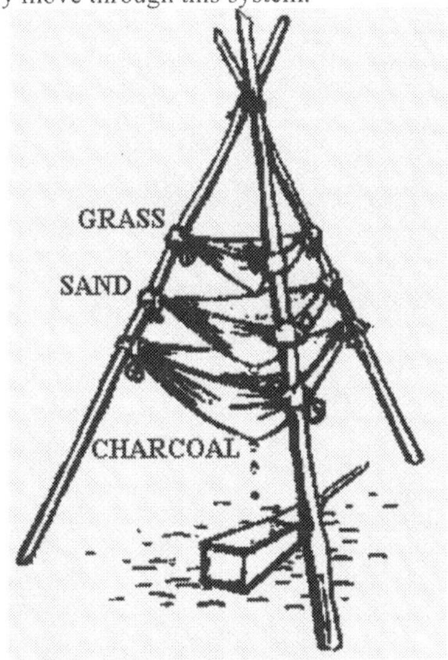

WATER FILTER

(2) Commercial Water Filters. Commercial water filters are generally available in most retail stores and may be with you. Understanding what the filter can do is the first step in safeguarding against future illnesses.

(a) A filter that has a .3 micron opening or larger will not stop Cryptosporidium.

(b) A filter system that does not release a chemical (i.e., iodine) may not kill all pathogens.

(c) A filter that has been overused may be clogged. Usage may result in excessive pumping pressure that can move harmful pathogens through the opening.

e. Potable. Potable indicates only that a water source, on average over a period of time, contains a "minimal microbial hazard," so the statistical likelihood of illness is acceptable.

(1) Sedimentation. Sedimentation is the separation of suspended particles large enough to settle rapidly by gravity. The time required depends on the size of the particle. Generally, 1 hour is adequate if the water is allowed to sit without agitation. After sediment has formed on the bottom of the container, the clear water is decanted or filtered from the top. Microorganisms, especially cysts, eventually settle, but this takes longer and the organisms are easily disturbed during pouring or filtering. Sedimentation should not be considered a means of disinfecting and should be used only as a last resort or in an extreme tactical situation.

5. **SOLAR STILLS.** (FMST.07.08k)

a. Solar stills are designed to supplement water reserves. Contrary to belief, they will not provide enough water to meet the daily requirement for water.

b. AboveGround Solar Still. This device allows the survivor to make water from vegetation. To make the aboveground solar still, locate a sunny slope on which to place the still, a clear plastic bag, green leafy vegetation, and a small rock.

(1) Construction.

(a) Fill the bag with air by turning the opening into the breeze or by "scooping" air into the bag.

(b) Fill the bag half to three-quarters full of green leafy vegetation. Be sure to remove all hard sticks or sharp spines that might puncture the bag.

CAUTION
Do not use poisonous vegetation. It will provide poisonous liquid.

(c) Place a small rock or similar item in the bag.

(d) Close the bag and tie the mouth securely as close to the end of the bag as possible to keep the maximum amount of air space. If you have a small piece of tubing, small straw, or hollow reed, inserts one end in the mouth of the bag before tying it securely. Tie off or plug the tubing so that air will not escape. This tubing will allow you to drain out condensed water without untying the bag.

(e) Place the bag, mouth downhill, on a slope in full sunlight. Position the mouth of the bag slightly higher than the low point in the bag.

(f) Settle the bag in place so that the rock works itself into the low point in the bag.

(g) To get the condensed water from the still, loosen the tie and tip the bag so that the collected water will drain out. Retie the mouth and reposition the still to allow further condensation.

(h) Change vegetation in the bag after extracting most of the water from it.

(i) Using 1-gallon zip-loc. bag instead of trash bags is a more efficient means of construction.

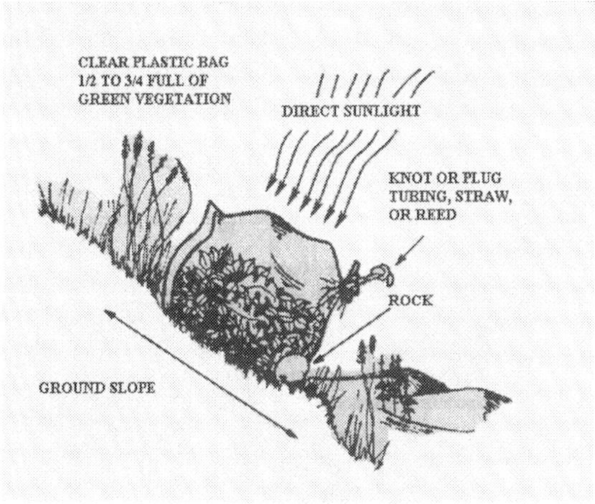

<u>ABOVE GROUND SOLAR STILL</u>

c. <u>BelowGround Solar Still</u>. Materials consist of a digging stick, clear plastic sheet, container, rock, and a drinking tube.

(1) Construction.

(a) Select a site where you believe the soil will contain moisture (such as a dry streambed or a low spot where rainwater has collected). The soil should be easy to dig, and will be exposed to sunlight.

(b) Dig a bowl-shaped hole about 1 meter across and 24 inches deep.

(c) Dig a sump in the center of the hole. The sump depth and perimeter will depend on the size of the container you have to place in it. The bottom of the sump should allow the container to stand upright.

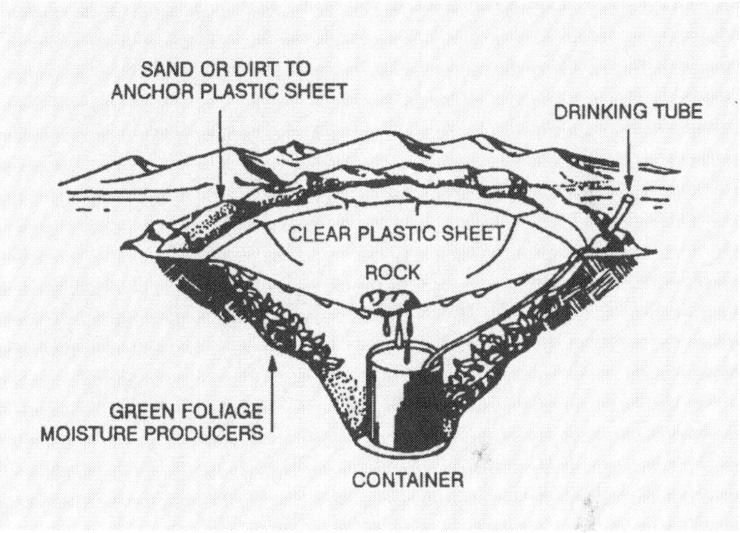

BELOW GROUND SOLAR STILL

(d) Anchor the tubing to the container's bottom by forming a loose overhand knot in the tubing. Extend the unanchored end of the tubing up, over, and beyond the lip of the hole.

(e) Place the plastic sheet over the hole, covering its edges with soil to hold in place. Place a rock in the center of the plastic sheet.

(f) Lower the plastic sheet into the hole until it is about 18 inches below ground level. Make sure the cone's apex is directly over the container. Ensure the plastic does not touch the sides of the hole because the earth will absorb the moisture.

(g) Put more soil on the edges of the plastic to hold it securely and prevent the loss of moisture.

(h) Plug the tube when not in use so that moisture will not evaporate.

(i) Plants can be placed in the hole as a moisture source. If so, dig out additional soil from the sides.

(j) If polluted water is the only moisture source, dig a small trough outside the hole about 10 inches from the still's lip. Dig the trough about 10 inches deep and 3 inches wide. Pour the polluted water in the trough. Ensure you do not spill any polluted water around the rim of the hole where the plastic touches the soil. The trough holds the polluted water and the soil filters it as the still draws it. This process works well when the only water source is salt water.

(k) Three stills will be needed to meet the individual daily water intake needs.

UNITED STATES MARINE CORPS
Mountain Warfare Training Center
Bridgeport, California 93517-5001

WMC
04/07/02

STUDENT HANDOUT

FORAGING PLANTS AND INSECTS FOR SURVIVAL USES

PURPOSE The purpose of this period of instruction is to familiarize the Wilderness Medicine Student with the potential resources for survival sustenance.

OUTLINE

1. GENERAL CONSIDERATIONS. There are few places without some type of edible vegetation. Plants contain vitamins, minerals, protein, carbohydrates, and dietary fiber. Some plants also contain fats. The following are general considerations:

a. Do not assume that because birds or animals have eaten a plant, it is edible by humans.

b. Poor plant recognition skills will seriously limit your ability to survive.

c. Plant dormancy and snowfall make foraging plants difficult during the winter months.

d. Plants generally poison by:

(1) Ingestion: When a person eats a part of a poisonous plant.

(2) Contact: When a person makes contact with a poisonous plant that cause any type of skin irritation or dermatitis.

(3) Absorption: When a person absorbs the poison through the skin, which can interrupt a bodily function.

(4) Inhalation: Poisoning can occur through the inhalation of smoke that contains poisonous plant residue.

A7

e. Plant properties can change throughout the growing season. Plants can be edible during certain periods while poisonous in others.

2. PLANT RECOGNITION. Plant recognition can take years of study to learn. The following is used as a general guideline.

a. Types of plants that should be avoided are:

(1) All mushrooms.

(2) Plants with a milky sap.

(3) White and yellow berries should be avoided, as they are almost always poisonous.

(4) Plants with shiny leaves should be considered poisonous.

(5) Plants that are irritants to the skin should not be eaten.

(6) All beans and peas should be avoided.

b. Edible Plants: There are several edible plants found at the Marine Corps Mountain Warfare Training Center. All are easy to recognize and many of the same plants are located throughout the United States and abroad.

(1) Wild Onions - Complete plant.

(2) Dandelion - Roots and leaves.

(3) Mallow - Leaves, stem and flowers.

(4) Watercress -Complete plant.

(5) Cattail - Root, stalk and stem.

(6) Bull Thistle - The flower.

(7) Juniper Tree - Berries and cambium layer.

(8) All Conifers - Cambium layer, needles, and nuts within the cones.

(9) Currants - Berries.

c. Preparing an edible plant. All edible plants should be boiled twice with a change of water to remove any bitterness, acids, or bacteria that the plant may contain.

3. PLANT TESTING PROCEDURE. Always adopt the following procedure when trying unknown plants as food.

a. Wait 8 hours without eating.

b. Select a plant that grows in sufficient quantity in the local area. Separate the part of the plant you wish to test root, stem, leaf, or flower. Certain parts of plants are poisonous while the, other parts may be edible.

c. Rub a portion of the plant you have selected on your inner forearm. Wait 15 minutes and look for any swelling, rash, or irritation.

d. Boil the plant or plant part in changes of water. The toxic properties of many plants are water-soluble or can be destroyed by heat. Cooking and discarding two changes of water can lessen the amount of poisonous material or remove it completely. These boiling periods should last at least 5 minutes each.

e. Place 1 teaspoon of the prepared plant food in the mouth and chew for 5 minutes, but do not swallow. If unpleasant effects occur (burning, bitter, or nauseating taste), remove the plant from the mouth at once and discard it as a food source. If no unpleasant effects occur, swallow the plant material and wait 8 hours.

f. If after 8 hours no unpleasant effects have occurred (nausea, cramps, and diarrhea), eat two tablespoonfuls and wait 8 hours.

g. In no unpleasant effects have occurred at the end of this 8-hour period, the plant may be considered edible.

h. Completely document and sketch the plant in a logbook to refer to for future use. This will aid in future procurement of this plant. If plant properties have changed, you will have to repeat the plant testing procedure.

NOTE: Have charcoal ready to eat. If any side effects occur, attempt to induce vomiting and immediately eat charcoal. Charcoal will absorb toxins and reduce the chance of Death.

4. MEDICINAL PLANTS One of the many things that hinder a survivor's ability to survive is medical problems. Injuries incurred will reduce survival expectancy and the ability to evade.

a. Coyote Willow: Coyote Willow is a thick forming shrub with clustered stems and very narrow leaves. It habits wet soils, especially riverbanks, sandbars, and silt flats.

(1) Drinking tea made from the inner bark can reduce fever.

(2) The dried and powdered inner bark can be used to stop bleeding.

(3) Administering a half-cup of willow charcoal dissolved in water can treat diarrhea.

(4) The Coyote Willow roots can be mashed and applied to tooth aches.

b. Yarrow: Yarrow is a plant (tall, usually not branched, with many white, slat topped group flowers). Do not confuse Yarrow with Water Hemlock.

(1) Rubbing a handful of crushed Yarrow flowers and leaves on any exposed can make insect repellent skin.

(2) Placing a Yarrow leaf poultice on the wound can stop bleeding.

(3) Relief from many rashes, including poison oak and ivy, can be achieved by applying a Yarrow leaf compress to the effected area.

(4) The Yarrow root can be chewed to relieve the pain of a toothache or break a fever.

(5) A potent anesthetic can be made by scrubbing fresh Yarrow roots in water to clean them. Once the roots are clean, crush them into a spongy mass and apply gently to the wounded area.

c. Tea To make a tea, place the plants in a container and pour in two cups of boiling water. Let the mixture steep for 20 minutes and drink.

d. Poultice: Place plant (Coyote Willow, Yarrow, etc.) in gauze or other similar material and fold it so the gauze will hold the plant in place. Put gauze in a larger cloth, about 6"x8", and roll the sides inward. Fold the cloth over without losing the plant. While boiling water, dip the bottom portion of the cloth containing the plant into the hot water by holding the edges. Keep the plant submerged in the boiling water until it becomes saturated. Bring the cloth straight up and with a twisting motion, wring the excess water back into the pot. Apply the poultice to the affected area as soon as it has cooled enough to place on the wound. To be effective, the poultice should be as hot as you can tolerate it.

e. Compress: A compress is made just like a poultice; except it is cold when applied to the wound. NOTE: Poultice or Compress should be applied for 1 to 24 hours, as needed. When applying a poultice, you may experience a throbbing pain as it draws out the infection and neutralizes toxins. When the pain subsides, the poultice has accomplished its task and should be removed. Apply a fresh poultice as needed until the desired level of healing has been reached.

5. POISONOUS PLANTS. Rarely will a survivor have an ideal means of killing large game, however there are certain plants that can aid the survivor. The two plants that we will talk about are the Water Hemlock and Monkshood.

a. Preparing Poisonous Plants: Once a poisonous plant has been located, dig up the root of the plant. The roots of the Water Hemlock and Monkshood generally grow 8 to 10 inches deep. Extreme caution must be used when handling the root. Do not handle the root without a barrier between your hands and the root. A barrier can be gloves, socks, T-shirt, or even moss. Split the root lengthwise to expose the inside of the root where the toxin is located. With the root split, rub the tip of your spear/arrow inside of the root opening. In a slow controlled manner, work from the bottom of the tip to the top. Once the tip is thoroughly coated, allow the toxin to dry and apply another coat to the tip. Continue to apply coat after coat of toxin, until the root is completely drained of its' toxin. You are now finished and the tip is ready for use.

NOTE: Do not eat the meat in the vicinity of the poisoned wound.

CAUTION: Do not apply poison to the actual point and edges of tip. This avoids accidental puncture to you. Wash and boil all gloves and clothing used as a barrier before wearing

6. INSECTS. Insects are the most abundant life form on earth and are an excellent survival food. They are easy to catch and provide 65-80% protein; compared to 20% for beef. They aren't too appetizing, but personal bias has no place in a survival situation. The focus must remain on maintaining your health.

a. Insects to avoid.

(1) All adults that sting or bite.

(2) Hairy or brightly colored insects.

(3) All Caterpillars.

(4) Insects that have a pungent odor.

(5) All spiders.

(6) Disease carriers like ticks, flies, or mosquitoes.

b. Edible Insects.

(1) Insect Larvae.

(2) Grasshoppers.

(3) Beetles.

(4) Grubs.

(5) Ants.

(6) Termites.

(7) Worms.

c. Foraging for Insects. One must be careful not to expend more energy harvesting food than can be replaced. For example, catching insects such as grasshoppers can become exasperating and tiring.

(1) At night grasshoppers climb tall plants and clings to the stalks near the top. They can be picked from the plants in the early morning while they are chilled and dormant.

(2) Dig for worms in damp humus soil, under rocks/logs or look for them on the ground after it has rained.

(3) Carpenter ants are found in dead trees and stumps, which can be gathered by hand.

(4) Most other insects can be found in rotten logs, under rocks, and in open grassy areas.

d. Preparing Edible Insects.

(1) Insects with a hard outer shell have parasites. Remove the wings and barbed legs before cooking.

(2) Drop worms into potable water for at least a half-hour. They will naturally purge themselves. You can either cook or eat them raw.

(3) Most other insects can be eaten raw. Cooking insects will improve their taste. If the thought of eating insects is unbearable, grind them into a paste and mix with other foods.

UNITED STATES MARINE CORPS
Mountain Warfare Training Center
Bridgeport, California 93517-5001

WMC.A8.0
05/02/01

STUDENT HANDOUT

SURVIVAL TRAPS AND SNARES

PURPOSE: The purpose of this class is to familiarize the student with some of the techniques used to take game. Game can be useful for both food and clothing.

OUTLINE

1. GENERAL TIPS AND TECHNIQUES FOR TRAPPING

a. General Tips. Knowing a few general hints and tips will make the trapping of animals much easier and considerably more effective. The eight general tips for trapping are:

(1) Know your game: Knowing the habits of the animal you want to trap will help lure it into your trap. Such things as when and where they move, feed, and water will help you determine where the trap can be most effectively placed.

A8

(2) Keep things simple: You don't have time in a survival situation to construct elaborate traps and they do not necessarily do a better job.

(3) Set traps in the right place: Animals will travel and stop in certain locations. That is where to best employ traps.

(4) Cover up your scent: Animals will avoid an area, which smells strange to them. Smoke from your fire is the best cover to use.

(5) Use the right type of trap: Some traps work better than others do (i.e. a deadfall works better on squirrels than a spring pole).

(6) Use the right size trap: Adjust the size of your traps to the size of the animal.

(7) Check traps: Check your traps twice daily: morning and evening. Checking your traps less than twice a day can allow your game to escape, rot, or be taken by other predators.

(8) Bait your traps: Bait of any type will add to your chances of success.

b. General Techniques. A general technique is the method in which the trap is intended to kill or hold the animal.

(1) Strangulation: This method strangles the animal, such as a snare.

(2) Mangling: This method crushes the animal, such as a deadfall.

(3) Entanglement: This method entangles the animal, such as a net.

(4) Live: This method holds the animal, such as a box trap.

c. Types of Triggers. There are three basic triggers used for all traps and path guards. Depending on the situation, variations of these triggers can be used.

(1) Puite Figure 4.

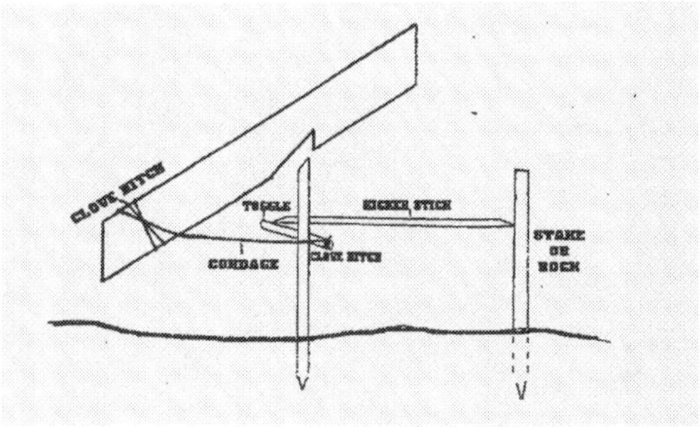

(2) <u>Toggle</u>.

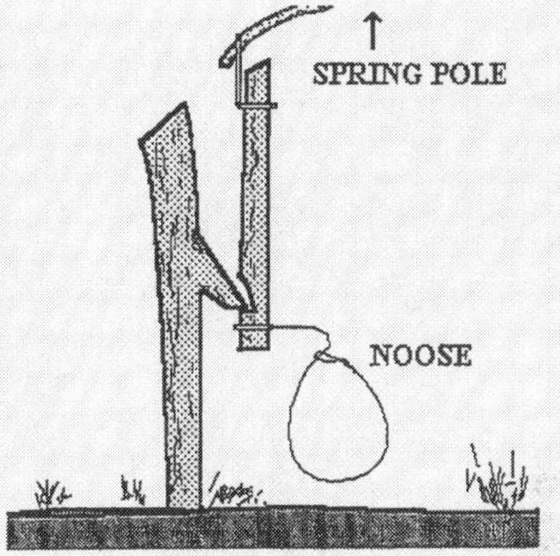

TOGGLE

(c) <u>Universal</u>.

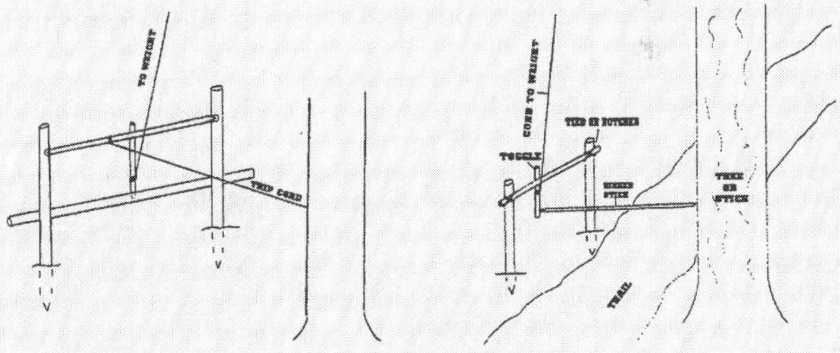

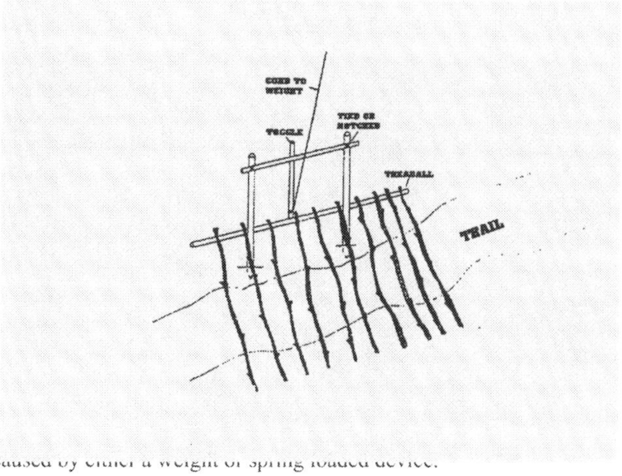

2. TRAPS. Traps are c
action is generally caused by either a weight or spring loaded device.

a. Puite Deadfall: The Puite deadfall requires a knife and piece of cord to construct. It is designed to mangle small rodents. The trigger for the deadfall is the Puite figure 4.

b. Spring Pole: The spring pole requires a small sapling and cordage to construct. The trigger for the spring pole is the toggle. It is designed to lift the animal off the ground; not allowing predatory animals to take your game. Remember, the trigger can not be so tight that the intended game can not set it off.

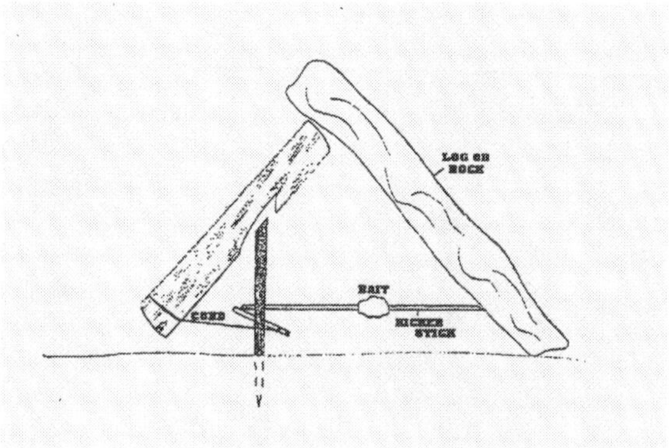

SPRING POLE

c. Box Trap: The box trap requires limited cordage to construct. It is designed to hold live small rodents and birds. The box trap trigger is the Puite figure 4.

3. PATH GUARDS Path guards is designed to protect and provide security for your shelter area against the enemy and predatory animals. They are classified into noise and casualty producing path guards.

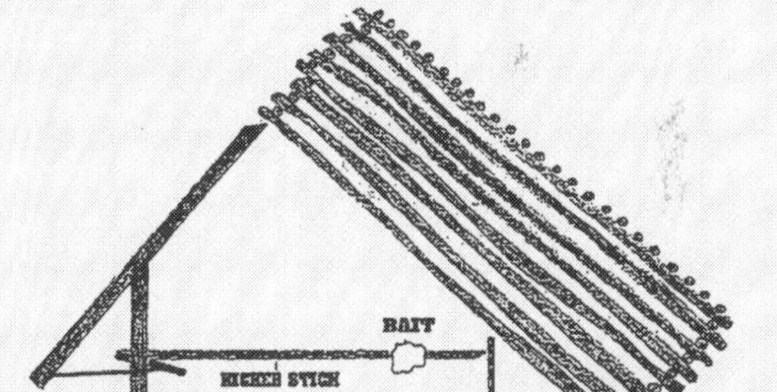

BOX TRAP

a. Noise producing path wards. Noise producing pathguards serve as an alarm for your shelter area. When triggered, it should produce some type of loud noise. Although construction can vary, depending on materials available, one example is as follows:

(1) Secure a young sapling to a universal trigger.

(2) At the end of the sapling, tie several pieces of metal to the sapling. Use what ever is available for metal.

(3) Camouflage the metal on the ground.

(4) When triggered, the sapling should swing back and forth, causing the metal rattle.

b. Casualty producing path wards. Casualty producing path guards, when triggered, should cause death or injury to the enemy or predatory animal. Tips should be poisoned as discussed in Survival Plant Uses class. Triggers for this type of path guard should be the universal triggers.

(1) Log Jerk

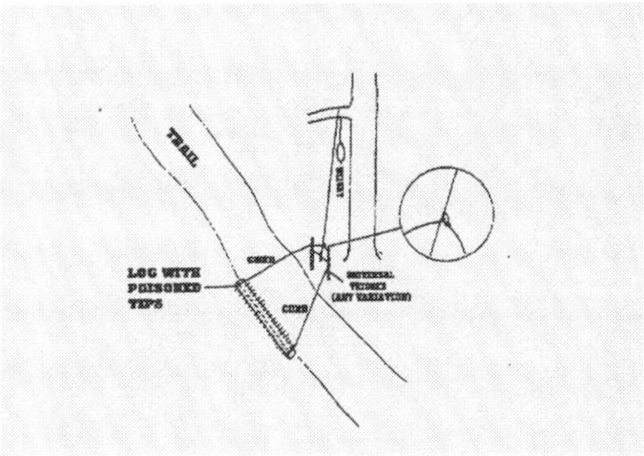

(2) Fish Hook Nightmare

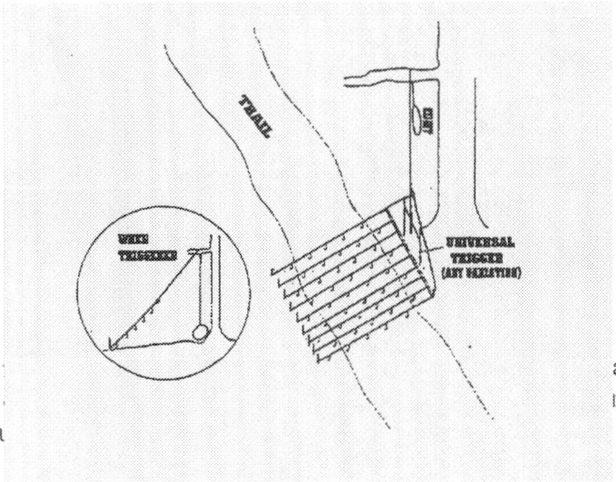

4. SNARING. A snare a loop at one end, which tightens, nd less time consuming to constr

a. Requirements for snaring there are five requirements too effectively employ snares. They are as follows:

(1) Wire: Although snares can be used with rope or cord, they are less effective than wire. Wire should have memory and resist kinking. Aircraft cable type in size 1/32 to 3/16 inch should be used. This type of wire prevents animal chew out and resists breakage. Remember, you want the smallest diameter cable capable of holding the animal.

(2) Locking device: A locking device is imperative for a snare to work properly. Locking devices secure the snare around the animal's neck. There are several methods available for a locking device.

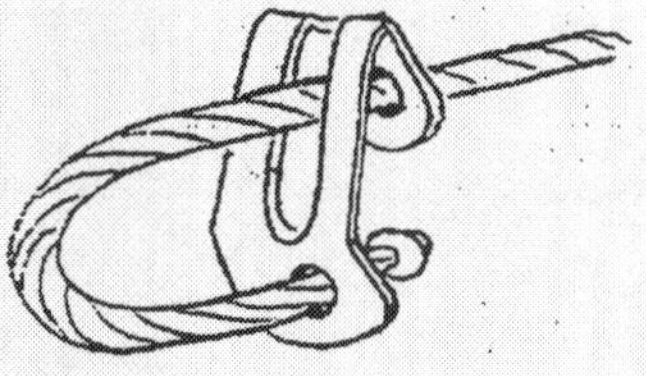

COMMERCIAL **BENT WASHER**

(3) Channelization: The objective of channelization is to force the animal to move through the snare. Remember, animals will take the path of least resistance.

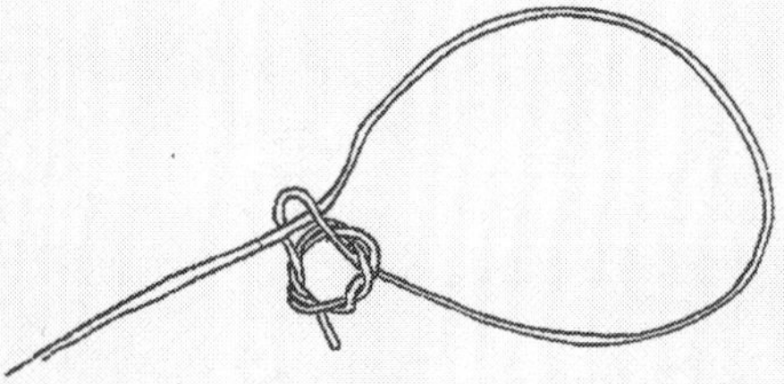

KNOT

(4) Loop size: Probably the single most important requirement in snaring. A correctly employed snare will have the snare holding the animal around the neck. Loop size is placed on the snare according to the intended animal. Too large will result in a body or leg catch, resulting in possible chew out or breakage. Too small will enable the animal to force the snare to the side, resulting in a miss. Additionally, the loop must be placed with specific ground clearance. The following chart will give a general guideline.

ANIMAL	NOOSE SIZE	GROUND
CLEARANCE		
SQUIRREL	2 1/2 TO 3 INCHES	1/2 TO 1 1/2 INCHES
RABBIT	4 TO 5 1/2 INCHES	1 1/2 TO 3 INCHES
RACCOON	6 INCHES	3 TO 4 INCHES
FOX	7 TO 10 INCHES	8 TO 10 INCHES
COYOTES	12 TO 14 INCHES	12 INCHES
BOBCAT	9 INCHES	8 INCHES

Note: (1) Noose size is the diameter of the snare loop.

(2) Ground clearance is measured from the bottom of the loop to the ground.

(5) Bait. In a survival situation, you will not be able to employ numerous snares. Baiting all snares increases your chances of success.

b. Snares although there are numerous ideas to employ snares, here are a few ideas.

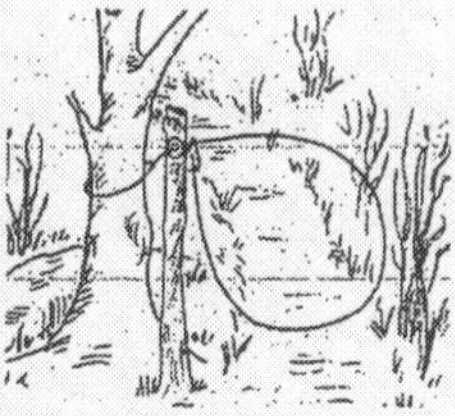

TRAIL SET

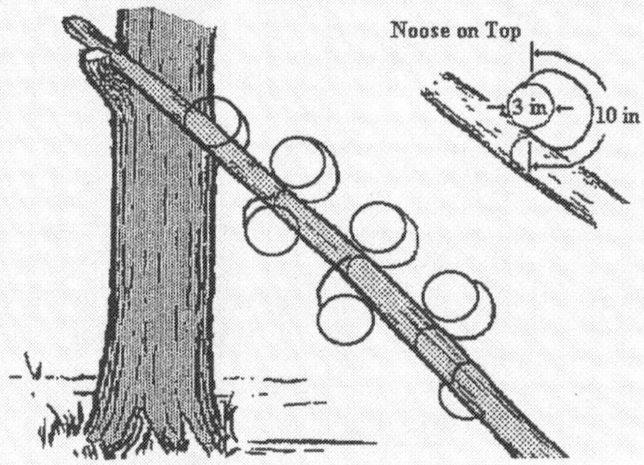

SQUIRREL POLE

DEN SET

CUBBIE SET

5. FISH. In a mountainous region, fish are normally an abundant resource. Not only are they a food source, the "left-overs" provide an excellent bait for traps and snares.

a. Fishing locations: Fishing in mountain streams is generally best done with a hand line. When fishing these streams, always look for these places to fish.

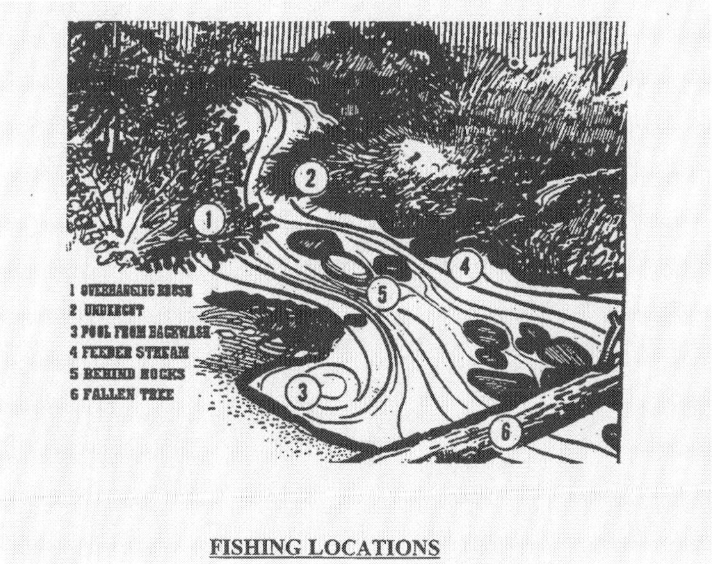

FISHING LOCATIONS

b. Expedient hooks: Although hooks should be carried in a survival kit, the survivor should be able to construct additional hooks if the situation arises. Expedient hooks are made to become lodged in the throat of the fish. Below are a few examples.

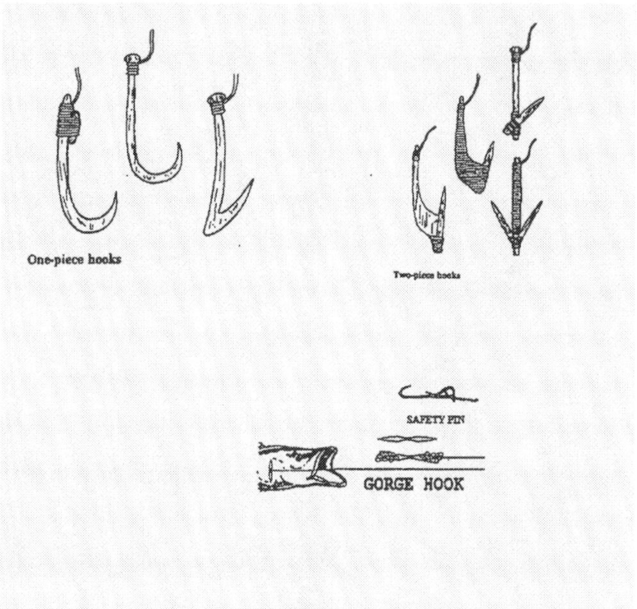

EXPEDIENT HOOKS

c. Fish traps: A fish trap can be effective if you have a shallow stream and time to construct it. A basic fish trap in nothing more than a barricade of rocks or sticks across a stream with another barricade using a funnel-type entrance which fish can be driven into but have a difficult time finding their way out. Once fish are trapped between these two barricades they may be speared, clubbed, or grabbed. This can be very effective when certain types of fish are moving in large groups to spawn. This type of trap is very effective in catching fish. A door can be constructed at the mouth of the trap so that excess fish can be kept live until needed.

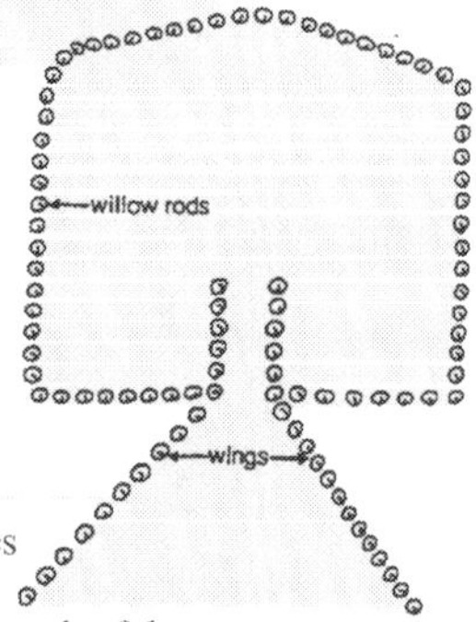

FISH TRAP

d. Set Lines: Set lines are an effective method of fishing which conserves energy. Put them out over night with several boated hooks attached. Place them with the hooks either on the bottom or suspended off the bottom, until you have determined where the fish are feeding.

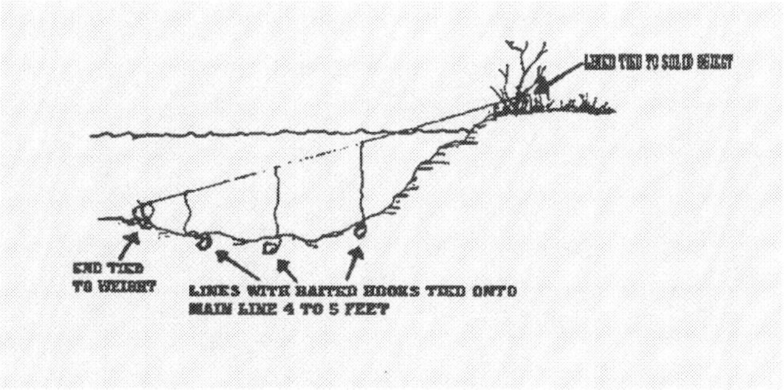

SET LINES

e. Netting: A gill net is most effective in still water, e.g., a lake (near the inlet and outlet are good locations) or back water in a large stream (for survival doesn't hesitate to block the stream). Nets can be constructed using the inner cords of parachute shroud lines. Place floats on top and weights on the bottom to keep the net vertical in the water. When the ice is on a lake, the fish are inclined to stay deeper. The smaller the mesh, the smaller the fish you can catch, but a small mest will still entangle a large fish. A mesh of 3 inches is a good standard. A gill net is very time consuming and requires allot of material to construct.

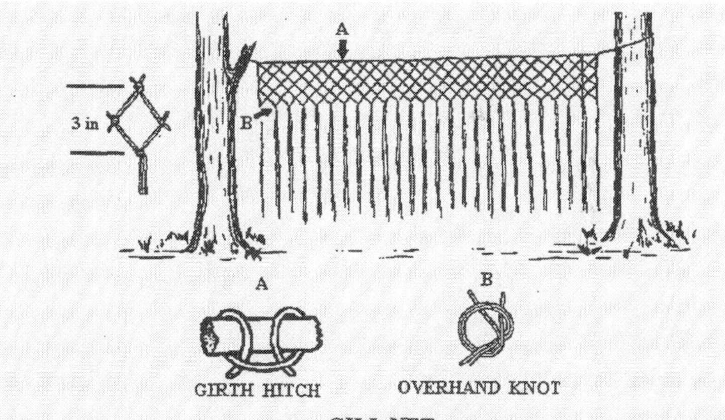

GILL NET

UNITED STATES MARINE CORPS
Mountain Warfare Training Center
Bridgeport, California 93517-5001

WMC
04/07/02

STUDENT HANDOUT

MOUNTAIN WEATHER

LESSON PURPOSE: The purpose of this period of instruction is to introduce you to weather patterns, the invisible aspects of weather, and some visible clues to help you forecast the weather. This lesson relates to all operations in the mountains at MWTC.

OUTLINE

1. GENERAL

a. The earth is surrounded by the atmosphere, which is divided, into several layers. The world's weather systems are in the troposphere, the lower of these layers. This layer reaches as high as 40,000 feet.

b. Dust and clouds in the atmosphere absorb or bounce back much of the energy that the sun beams down upon the earth. Less than one half of the sun's energy actually warms the earth's surface and lower atmosphere.

c. Warmed air, combined with the spinning (rotation) of the earth, produces winds that spread heat and moisture more evenly around the world. This is very important because the sun heats the Equator much more than the poles and without winds to help restore the balance, much of the earth would be impossible to live on. Where the air-cools, you can get clouds, rain, snow, hail, fog, frost, etc.

d. The weather that you find in any place depends on many things, i.e. how hot the air is, how moist the air is, how it is being moved by the wind, and especially, is it being lifted or not?

2. PRESSURE

a. All of these factors are related to air pressure, which is the weight of the atmosphere at any given place. The lower the pressure, the more likely are rain and strong winds.

b. In order to understand this we can say that the air in our atmosphere acts very much like a liquid.

c. Areas with a high level of this liquid would exert more pressure on the Earth and be called a "high pressure area".

d. Areas with a lower level would be called a "low pressure area".

e. In order to equalize the areas of high pressure it would have to push out to the areas of low pressure.

f. The characteristics of these two pressure areas are as follows:

(1) High-pressure area. Flows out to equalize pressure.

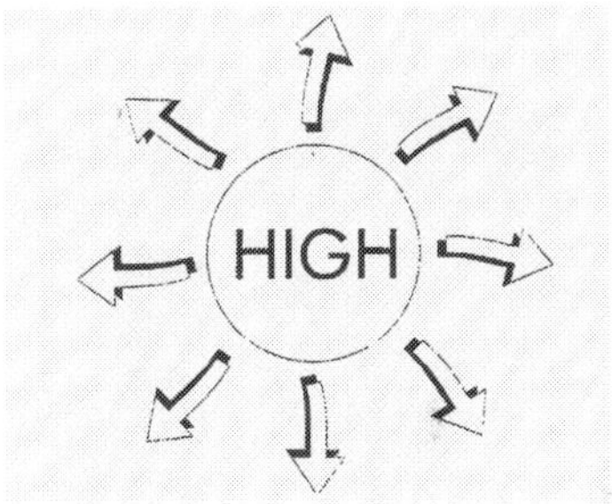

(2) Low pressure area. Flows in to equalize pressure.

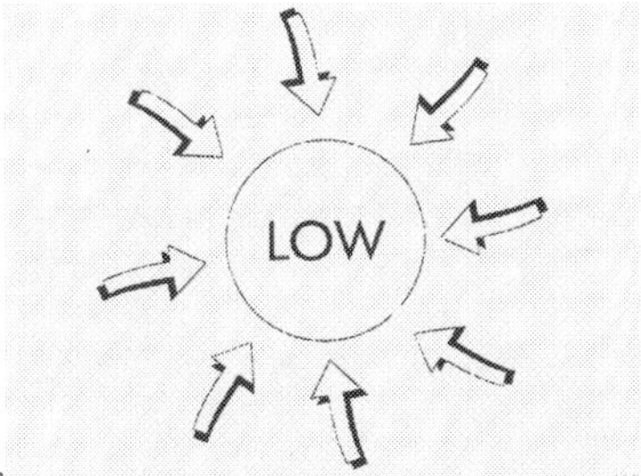

g. The air from the high-pressure area is basically just trying to gradually flow out to equalize its pressure with the surrounding air, while the low pressure is beginning to build vertically. Once the low has achieved equal pressure, it can't stop and continues to build vertically; causing turbulence, which results in bad weather.

NOTE: When looking on the weather map, you will notice that these differences in pressure resemble contour lines. These contour lines are called isobars and are translated to mean equal pressure area.

h. Isobars: Pressure is measured in millibars or another more common measurement is inch mercury.

i. Fitting enough, areas of high pressure are called ridges and areas of low pressure are called troughs.

NOTE: The average air pressure at sea level is:

29.92 inches mercury.
1,013 millibars.

j. As we go up in altitude, the pressure (or weight) of the atmosphere decreases.

EXAMPLE: At 18,000 feet in elevation it would be 500 millibars vice 1,013 millibars at sea level.

3. HUMIDITY Humidity is the amount of moisture in the air. All air holds water vapor, although it is quite invisible.

a. Air can hold only so much water vapors, but the warmer the air, the more moisture it can hold. When the air has all the water vapor that it can hold, the air is said to be saturated (100% relative humidity).

b. If the air is then cooled, any excess water vapor condenses; that is, it's molecules join to build the water droplets we can see.

c. The temperature at which this happens is called the "condensation point". The condensation point varies depending on the amount of water vapor and the temperature of the air.

d. If the air contains a great deal of water vapor, condensation will form at a temperature of 20°C (68°F). But if the air is rather dry and does not hold much moisture, condensation may not form until the temperature drops to 0°C (32°F) or even below freezing.

e. Adiabatic Lapse Rate: The adiabatic lapse rate is the rate that air will cool on ascent and warm on descent. The rate also varies depending on the moisture content of the air.

(1) Saturated Air = 2.2°F per 1,000 feet.

(2) Dry Air = 5.5°F per 1,000 feet.

4. WINDS. As we stated earlier, the uneven heating of the air by the sun and rotation of the earth causes winds. Much of the world's weather depends on a system of winds that blow in a set direction. This pattern depends on the different amounts of sun (heat) that the different regions get and also on the rotation of the earth.

a. Above hot surfaces, air expends (air molecules spread out), and move to colder areas where it cools and becomes denser, and sinks to the earth's surface. This forms a circulation of air from the poles along the surface or the earth to the Equator, where it rises and moves towards the poles again.

b. Once the rotation of the earth is added to this, the pattern of the circulation becomes confusing.

c. Because of the heating and cooling, along with the rotation of the earth, we have these surface winds. All winds are named from the direction they originated from:

(1) Polar Easterlies: These are winds from the polar region moving from the east. This is air that has cooled and settled at the poles.

(2) Prevailing Westerlies: These winds originate from approximately 30 degrees North Latitude from the west. This is an area where prematurely cooled air, due to the earth's rotation, has settled back to the surface.

(3) Northeast Tradewinds: These are winds that originate from approximately 30 degrees North from the Northeast. Also prematurely cooled air.

d. Jet Stream: A jet stream can be defined as a long, meandering current of high speed winds near the tropopause (transition zone between the troposphere and the stratosphere) blowing from generally a westerly direction and often exceeding 250 miles per hour. The jet stream results from:

(1) Circulation of air around the poles and Equator.

(2) The direction of airflow above the mid latitudes.

(3) The actual path of the jet stream comes from the west, dipping down and picking up air masses from the tropical regions and going north and bringing down air masses from the Polar Regions.

NOTE: The average number of long waves in the jet stream is between three and five depending on the season. Temperature differences between polar and tropical regions influence this. The long waves influence day to week changes in the weather; there are also short waves that influence hourly changes in the weather.

5. AIR MASSES. As we know, all -of these patterns move air. This air comes in parcels known as "air masses". These air masses can vary in size from as small as a town too as large as a country. These air masses are named for where they originate:

a. Maritime: Over water.

b. Continental: Over land.

c. Polar: Above 60 degrees North.

d. Tropical: Below 60 degrees North.

e. Combining these give us the names and description of the four types of air masses:

(1) Continental Polar: Cold, dry air mass.

(2) Maritime Polar: Cold wet air mass.

(3) Continental Tropical: Dry, warm air mass.

(4) Maritime Tropical: Wet, warm air mass.

ABBREVIATED MEANINGS
CP = Continental Polar.
MP = Maritime Polar.
CT = Continental
Tropical. MT =
Maritime Tropical.

f. The thing to understand about air masses; they will not mix with another air mass of a different temperature and moisture content. When two different air masses collide, we have a front, which will be covered in more detail later in this period of instruction.

g. Here are some other types of winds that are peculiar to mountain environments but don't necessarily affect the weather:

(1) Anabolic wind: These are winds that blow up mountain valleys to replace warm rising air and are usually light winds.

(2) Catabolic wind: These are winds that blow down mountain valley slopes caused by the cooling of air and are occasionally strong winds.

6. LIFTING/COOLING. As we know, air can only hold so much moisture depending on its temperature. If we cool this air beyond its saturation point, it must release this moisture in one form or another, i.e. rain, snow, fog, dews, etc. There are three ways that air can be lifted and cooled beyond its saturation point.

a. Orographic uplift: This happens when an air mass is pushed up and over a mass of higher ground such as a mountain. Due to the adiabatic lapse rate, the air is cooled with altitude and if it reaches its saturation point we will receive precipitation.

OROGRAPHIC UPLIFT

b. Convection effects: This is normally a summer effect due to the sun's heat radiating off of the surface and causing the air currents to push straight up and lift air to a point of saturation.

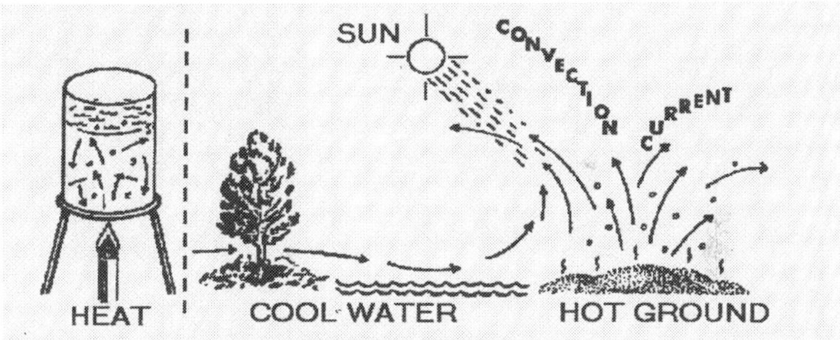

CONVECTION EFFECTS

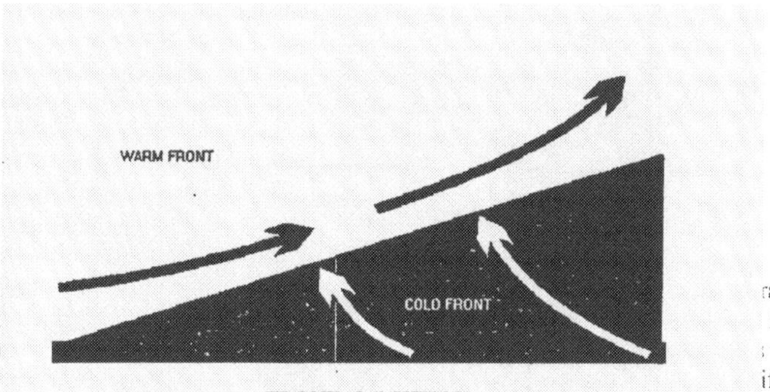

FRONTAL LIFTING

c. Frontal … rature content … is forced aloft; fi … where we receive … ifting is not uncom…

7. CLOUDS. Anytime air is lifted or cooled beyond its saturation point (100% relative humidity), clouds are formed. Clouds are one of our signposts to what is happening. Clouds can be described in many different ways; they can also be classified by height or appearance, or even by the amount of area covered, vertically or horizontally.

a. Cirrus: These clouds are formed of ice crystals at very high altitudes (usually 20,000 to 35,000 feet) in the mid-latitudes and are thin, feathery type clouds. These clouds can give you up to 24 hours warning of approaching bad weather, hundreds of miles in advance of a 'warm front. Frail, scattered types, such as "mare-tails" or dense cirrus layers, tufts are a sign of fair weather but predictive may be a prelude to approaching lower clouds, the arrival of precipitation and the front.

b. Cumulus: These clouds are formed due to rising air currents and are prevalent in unstable air that favors vertical development. These currents of air create cumiliform clouds that give them a piled or bunched up appearance, looking similar to cotton balls. Within the cumulus family there are three different types to help us to forecast the weather:

(1) Cotton puffs of cumulus are Fair Weather Clouds but should be observed for possible growth into towering cumulus and cumulonimbus.

(2) Towering cumulus are characterized by vertical development. Their vertical lifting is caused by some type of lifting action, such as connective currents found on hot summer afternoons or when wind is forced to rise up the slope of a mountain or possibly the lifting action that

may be present in a frontal system. The towering cumulus has a puffy and "cauliflower-shaped" appearance.

(3) Cumulonimbus clouds are characterized in the same manner as the towering cumulus, form the familiar "thunderhead" and produce thunderstorm activity. These clouds are characterized by violent updrafts, which carry the tops of the clouds to extreme elevations. Tornadoes, hail and severe rainstorms are all products of this type of cloud. At the top of the cloud, a flat anvil shaped form appears as the thunderstorm begins to dissipate.

c. Stratus: Stratus clouds are formed when a layer of moist air is cooled below its saturation point. Stratiform, clouds lie mostly in horizontal layers or sheets, resisting vertical development. The word "stratus" is derived from the Latin word "layer". The stratus cloud is quite uniform and resembles fog. It has a fairly uniform base and a dull, gray appearance. Stratus clouds make the sky appear heavy and will occasionally produce fine drizzle or Very light snow with fog. However, because there is little or no vertical movement in the stratus clouds, they usually do not produce precipitation in the form of heavy rain or snow.

d. As previously stated, clouds are formed when air is lifted to a point where it cools to its saturation point. We also know that frontal lifting affects our fronts, which produce the largest portion of our precipitation.

8. FRONTS. As we know, fronts often happen when two air masses of different moisture and temperature content interact. One of the ways we can identify that this is happening is by the progression of the clouds.

a. Warm Front: A warm front occurs when warm air moves into and over a slower (or stationary) cold air mass. Since warm air is less dense, it will rise naturally so that it will push the cooler air down and rise above it. The cloud you will see at this stage is cirrus. From the point where it actually starts rising, you will see stratus. As it continues to rise, this warm air-cools by the cold air and, this, receiving moisture at the same time. As it builds in moisture, it darkens becoming "nimbus-stratus", which means rain of thunderclouds. At that point some type of moisture will generally fall.

b. Cold Front. A cold front occurs when a cold air mass (colder than the ground that it is traveling over) overtakes a warm air mass that is stationary or moving slowly. This cold air, being denser, will go underneath the warm air, pushing it higher. Of course, no one can see this "air", but they can see clouds and the clouds themselves can tell us what is happening. The cloud progression to look for is cirrus to cirrocumulus to cumulus and, finally, to cumulonimbus.

c. Occluded Front. Cold fronts move faster than warm ones so that eventually a cold front overtakes a warm one and the warm air becomes progressively lifted from the surface. The zone of division between cold air ahead and cold air behind is called a "cold occlusion". If the air behind the front is warmer than ahead, it is a warm occlusion. Most land areas experience more occlusions than other types of fronts. In the progression of clouds leading to fronts, orographic uplift can play part in deceiving you of the actual type of front, i.e. progression of clouds leading to a warm front with orographic cumulus clouds added to these. The progression of clouds in an occlusion is a combination of both progressions from a warm and cold front.

9. USING PRESSURE AS AN INDICATOR. A very important factor of telling us what might happen is the pressure. As we know, low pressure or dropping pressure normally indicates deteriorating weather whereas high pressure usually gives us more good weather or clearing of bad weather. There are a couple of ways to monitor our pressure and are as follows:

a. Barometer: A barometer could be described as a pan of mercury with a tube leading out of the pan. Pressure from the atmosphere causes the mercury to rise in the tube.

(1) The tube is marked in millibars and the station that's reading these millibars will know how much it should rise for that location. Once again, if it rises more than normal, it would be considered a high-pressure reading.

b. Altimeter: Another means that is used to measure pressure is an altimeter, which is commonly used by mountaineers. It works like this:

(1) As you rise in elevation the pressure becomes less, thus allowing the needle in the altimeter to rise. If the needle rises without you rising with it, there is less pressure in the atmosphere than before and thus, a low-pressure area.

c. Contrail Lines: A basic way of identifying a low-pressure area is to note the contrail lines from jet aircraft. If they don't dissipate within two hours, that indicates a low-pressure area in your area. This usually occurs about 24 hours prior to an oncoming front.

d. Lenticulars: These are optical, lens-shaped cumulus clouds that have been sculpted by the winds. This indicates moisture in the air and high winds aloft. When preceding a cold front, winds and clouds will begin to lower.

10. USING SIGNS FROM NATURE. These signs will give you a general prediction of the incoming weather conditions. Try to utilize as many signs together as possible, which will improve your prediction. All of these signs have been tested with relative accuracy, but shouldn't be depended on 100%. But in any case you will be right more times than wrong in predicting the weather. From this we can gather as much information as needed and compile it along with our own experience of the area we are working in to help us form a prediction of incoming weather. The signs are as follows:

a. A spider's habits are very good indicators of what weather conditions will be within the next few hours. When the day is to be fair and relatively windless, they will spin long filaments over which they scout persistently. When precipitation is imminent, they shorten and tighten their snares and drowse dully in their centers.

b. Insects are especially annoying two to four hours before a storm.

c. If bees are swarming, fair weather will continue for at least the next half day.

d. Large game such as deer, elk, etc., will be feeding unusually heavy four to six hours before a storm.

e. When the smoke from a campfire, after lifting a short distance with the heated air, beats downward, a storm is approaching. Steadily rising smoke indicates fair weather.

f. A red sun or sky in the morning indicates nearing rain due to the excessive amount of moisture in the air.

g. A red evening sky indicates that the air contains so little moisture that rain within the next 24 hours is highly improbable.

h. A gray, overcast evening sky indicates that moisture carrying dust particles in the atmosphere have become overloaded with water; this condition favors rain.

i. A gray morning sky indicates dry air above the haze caused by the collecting of moisture on the dust in the lower atmosphere; you can reasonably a fair day.

j. When the setting sun shows a green tint at the top as it sinks behind clear horizon, fair weather is probable for most of the next 24 hours.

k. A rainbow in the late afternoon indicates fair weather ahead. However, a rainbow in the morning is a sign of prolonged bad weather.

l. A corona is the circle that appears around the sun or the moon. When this circle grows larger and larger, it indicates that the drops of water in the atmosphere are evaporating and that the weather will probably be clear. When this circle shrinks by the hour, it indicates that the water drops in the atmosphere are becoming larger, forming into clouds, rain is almost sure to fall.

m. In fair weather, air currents flow down streams and hillsides in the early morning and start drifting back up towards sunset. Any reversal of these directions warns of a nearing storm.

n. When the breeze is such that the leaves show their undersides, a storm is likely on the way.

o. It is so quiet before a storm, that distant noises can be heard more clearly. This is due to the inactivity of wildlife a couple of hours before a storm.

p. When in the mountains, the sight of morning mist rising from ravines is a good sign of clear weather the rest of the day.

q. A heavy dew or frost in the morning is a sign of fair weather for the rest of the day. This is due to the moisture in the atmosphere settling on the ground vice in the form of precipitation such as rain, snow, etc.

UNITED STATES MARINE CORPS
Mountain Warfare Training Center
Bridgeport, California 93517-5001

FMST.07.42
03/14/02

STUDENT HANDOUT

RESCUE PHILOSOPHY FOR SWIFTWATER RESCUE TEAMS

TERMIANL LEARNING OBJECTIVE In a swift water rescue scenario, organize a swift water rescue, in accordance with the reference. (FMST.07.42)

ENABLING LEARNING OBJECTIVES.

(1) With out the aid of reference, from a given list select the correct components of rescue preparation, in accordance with the reference. (FMST.07.42a)

(2) With out the aid of reference, from a given list select the correct role description of the rescue team members, in accordance with the reference. (FMST.07.42b)

(3) With out the aid of reference, from a given list select the correct methods of rescue, in accordance with the reference. (FMST.07.42c)

OUTLINE.

1. **COMPONENTS OF RESCUE PREPARATION**. (FMST.07.42a) These components comprise the backbone of both technical skills and the proper mental attitude required of a rescue professional. Remember that this course only provides you with the first component.

a. Training. Provides the students with the basic technical skills required to perform a safe rescue, these can be gained by attending an approved SWIFTWATER RESCUE COURSE.

b. Practice. Use of the skills learned in training.

c. Experience. Gained through advanced training, practical exercises and real-world situations.

B1

d. Judgment. The combination and application of skills for a given situation. Remember that every rescue situation is different. Don't get into a SOP mindset.

2. **PRIORITIES OF SAFETY**. Most of you have heard the saying that "It is better to be judged by twelve then carried by six". With that in mind you must realize that, at times, there is no way to affect a safe rescue. If you place yourself or your team in a hostile environment, from which you cannot escape, you are adding to the problem, not assisting in the solution. For this reason we set priorities of safety.

a. Self-aid. Be able to rescue yourself.

b. Buddy-aid. Ensure all safety measures are in force and that team members can rescue each other in an emergency.

c. Victim-aid. Only after self and buddy-aid are ensured should you attempt to rescue the victim. This will enable you to concentrate and focus your entire effort on the rescue.

3. **ROLES OF THE TEAM MEMBERS**. (FMST.07.24b) Often when dealing with rescues, you will find yourself working with outside agencies and organizations. In order to successfully coordinate the entire effort the Incident Command System (ICS) was developed. In swiftwater rescue situations there are four specific roles that must be filled.

a. Incident Commander. Is in overall command of the rescue efforts. Does not necessarily have the technical knowledge to conduct the rescue, but must be able to coordinate the rescue effort.

b. Rigger. Personnel responsible for equipment and rope systems, i.e. establishing the tension diagonal or the tyrolean high line.

c. Gofer. This group of individuals fills all other supporting roles, i.e. Communications, logistics, medevac, accounting, equipment accountability etc.

d. Rescuer. The individuals who are actually conducting the rescue.

4. **COMPONENTS OF SEARCH AND RESCUE**. In order to successfully complete a rescue there is four objectives that must be met. In order to remember them we use the acronym LAST.

a. Locate. This is the search in the search and rescue business.

b. Access. The determination of the assets, equipment, technical skills, and personnel required safely reaching and extracting the victim.

c. Stabilization. Providing the on scene medical attention required in order to extract the victim without causing further injury or aggravating an existing condition.

d. Transport. Removing the victim from the hostile environment to a safe one where further treatment can be provided.

5. **METHODS OF RESCUE**. (FMST.07.42c) In any rescue, we must access the situation to determine the best techniques to apply. Bearing in mind the self-aid, buddy-aid, victim-aid concept as well as time, weather, victim's condition, etc. In all cases we will attempt the lowest risk methods first, while preparing for a higher risk option. In all cases remember the KISS principle.

a. Reach. Use a ladder, fire hose, stick, paddle, rifle or any other sturdy object to extend your reach in order to reach the victim.

b. Throw. Get some type of floatation device to the victim. Or throw a line to the victim in order to pendulum or pull him/her to the shore.

c. Row. Putting a boat in the water to reach the victim.

d. Go and Tow. These are the contact rescues that, while look good on film, put you at the extreme risk of becoming a second victim.

e. Helo. Most civilian helicopter crews are not trained in SAR evolutions. Military aircraft and crews are a better bet. Several uses for helos are in the support role. Locating the victim, inserting and extracting teams and moving equipment are some of these roles. If using the helo in a direct role it is better to use a qualified SAR swimmer/diver to hook-up or pick-up victims.

UNITED STATES MARINE CORPS
Mountain Warfare Training Center
Bridgeport, California 93517-5001

FMST.07.25
03/14/02

STUDENT HANDOUT

PERSONAL AND TEAM EQUIPMENT

TERMINAL LEARNING OBJECTIVE As a member of a swiftwater rescue team, maintain swiftwater rescue equipment, in accordance with the reference. (FMST.07.25)

ENABLING LEARNING OBJECTIVES.

(1) Without the aid of references, from a given identify the correct items of personal equipment, in accordance with the reference. (FMST.07.25a)

(2) Without the aid of references, from a given identify the correct items of team equipment, in accordance with the reference. (FMST.07.25b)

OUTLINE.

1. **PERSONAL EQUIPMENT**. (FMST.07.25a) The following items of personal equipment should be in each man's possession for a swiftwater rescue.

   a. Dry Suit. Warmest, most versatile, full body protection available. Extra thermal layers can be added for varying water temperatures. Provides wind protection out of the water and provides protection against contaminated water.

   b. Wet Suit. Excellent protection from flotsam. Variety of thick nesses for thermal protection. Insulating quality is degraded in moving water due to the flushing effect.

   c. Footwear. Primary is full sole scuba booties. You can use tennis shoes or jungle boots with heavy wool socks.

   d. Gloves. Neoprene is recommended.

   e. Hoods. Wetsuit hood or pile hood for insulation. Remember 30-50% of your body heat losses can occur through the head and neck region of your body if left unprotected.

B2

f. Helmet. Protects you from flotsam and impact injuries. Should be padded and ventilated to allow water to flow through.

g. Knife. Fixed or locking blade attached to the vest, handle down.

h. Whistle. Required for communication. Best without "pea".

i. Eye Protection. Protects subject from debris.

j. Personal Floatation Device. This is the single most important piece of equipment. Jackets should provide at least 15 lbs. of floatation. Proper fit is a priority. Rescue vests with a "Blow-Out" belt should only be used by trained professionals.

2. **TEAM EQUIPMENT**. (FMST.07.25b) The following is a list of some of the team equipment for a typical four-man team. Of course the situation will dictate which equipment will be required for the mission on hand, but all hands should be familiar with this equipment. This is only a list. The uses and application of this gear, along with practical application exercises later in this course will provide you with the hands-on training and working knowledge to properly and safely use this equipment.

QTY	ITEM
1	300 ft, l Static Rope
4	150 ft, 11 mm Static Rope
25	85 Locking Stubai Carabiners
4	2" Pulleys
2	Rescue Figure 8
6	3 ft, 7mm Prussic
20	15 ft, 1" Tubular Nylon Runners
2	Ascenders
3	Throw Bags
2 pr	Fins
1	Rescue PFD
1	Self Bailing Raft (13 or 14 foot)
1	Oar Frame w/ Oars, or Seven Paddles
3	Radios (waterproofed)
1	Medical Kit
1	Litter

NOTE: This is not a complete list of rescue equipment. However, it is a good basic load.

UNITED STATES MARINE CORPS
Mountain Warfare Training Center
Bridgeport, California 93517-5001

FMST.07.41
03/14/02

STUDENT HANDOUT

SWIFTWATER TERMINOLOGY AND DYNAMICS

TERMINAL LEARNING OBJECTIVES In a swift water rescue scenario, evaluate swift water river conditions in accordance with the reference. (FMST.07.41)

ENABLING LEARNING OBJECTIVES.

1) Without the aid of reference, from a given list correctly identify the swiftwater terminology, in accordance with the reference. (FMST.07.41a)

2) Without the aid of reference, from a given list correctly identify the classifications of rivers, in accordance with the reference. (FMST.07.41b)

OUTLINE.

1. **SWIFTWATER TERMINOLOGY**. (FMST.07.41a)

a. Hydraulic or Hydraulic effect: Movement of water caused by pressure.

b. Eddy: Horizontal reversal. Current is blocked by an obstacle creating a vacuum which, when filled, causes water to flow upstream.

c. Eddy Fence: Reversal line between downstream current and the eddy. This fence can be very strong and the water flow may vary in height by several inches.

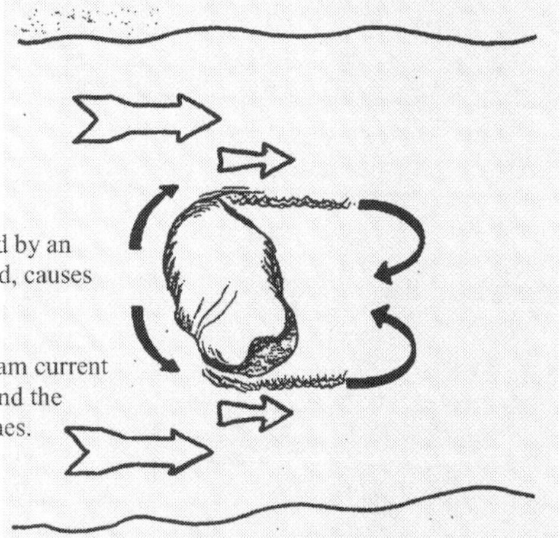

EDDY/EDDY FENCE

B3

d. Hole, Stopper, Keeper: Vertical reversal as water flows over an obstacle the main current follows the base of the river causing a surface vacuum. Downstream water flows back upstream to fill this void. This sets up a recirculating effect.

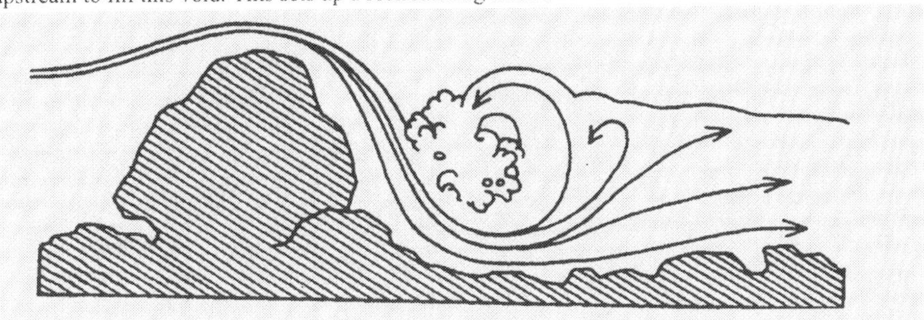

HOLE/STOPPER/KEEPER

e. Standing Wave. Haystack: Series of waves in the main channel caused by obstacles, ledges, or increases in the flow rate, gradient, or volume.

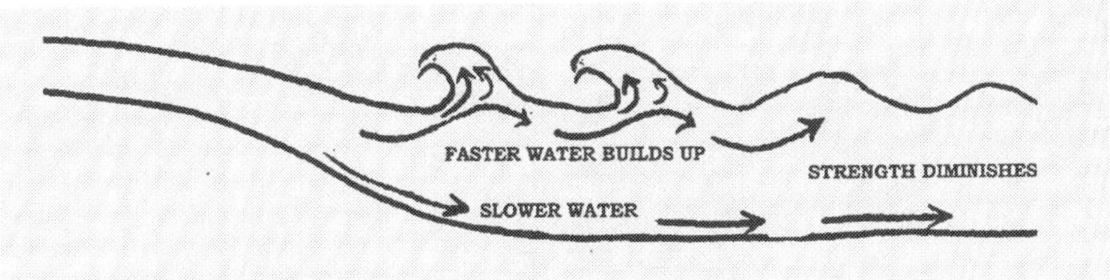

STANDING WAVES/HAYSTACKS

f. Downstream V. Tonal: Represent largest volume of water or main channel.

g. Upstream V: Caused by obstacles submerged just below the surface.

h. Strainer: Debris, logs, small trees on flooded banks, which allow water to flow through but will not allow a body to pass.

STRAINER / GUARD RAIL

STRAINER / TREE

i. Cushion, Pillow: Build-up of water on the upstream side of obstacles or the outside of curves.

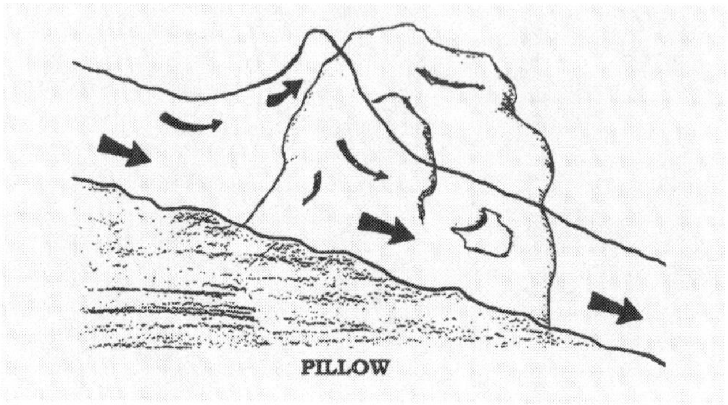

j. Confluence: Merging of streams.

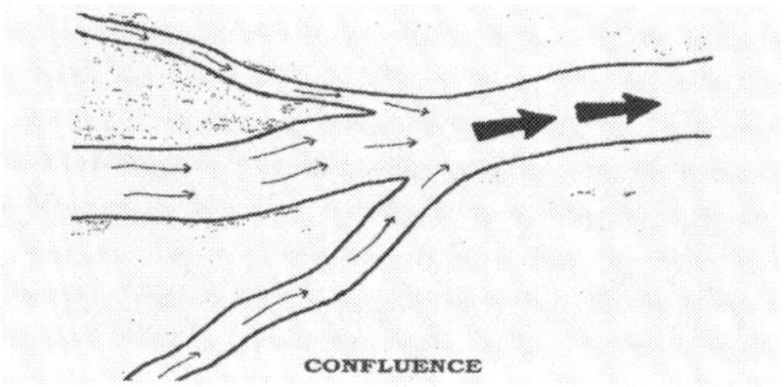

k. River Right. On the right side of the river looking downstream.

l. River Left. On the left side of the river looking downstream.

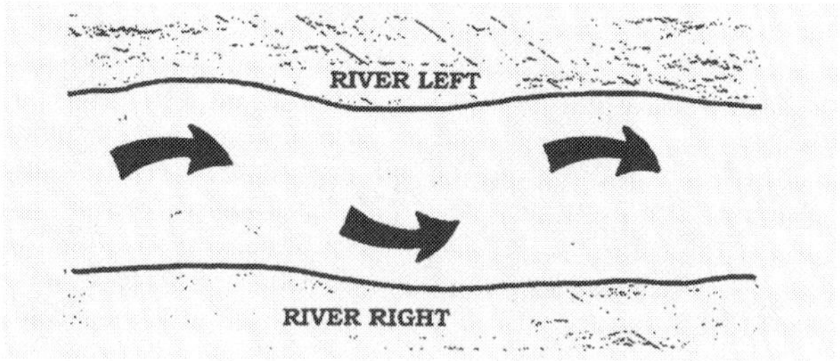

2. **CLASSIFICATIONS OF RIVERS**: (FMST.07.41b)

a. Class 1. Moving water, few obstacles, small ripples or waves.

b. Class 2. Moving water, clear channels, wide waves to 3', no scouting required.

c. Class 3. Rapids, narrow channels that require complex maneuvering, waves capable of flipping rafts or swamping canoes, scouting required.

d. Class 4. Long hazardous rapids, precise boat handling in rough water, scouting required, rescues hazardous.

e. Class 5. Severe, extended, extremely violent rapids, professional boat handlers and scouting required. Rescues extremely hazardous, situations are life threatening.

f. Class 6. Only Expert/Professional River Runners. Rescue may be impossible.

g. CLASSES U DON'T EVEN THINK ABOUT IT! Waters are not navigable.

3. **RIVER STRENGTH**.

a. Volume of Water. The volume of water in a given stretch of the river between two points is based on a simple formula and is measured in Cubic Feet Per Second (CFS). Simply take the river depth x river width x river velocity = volume of water in CFS.

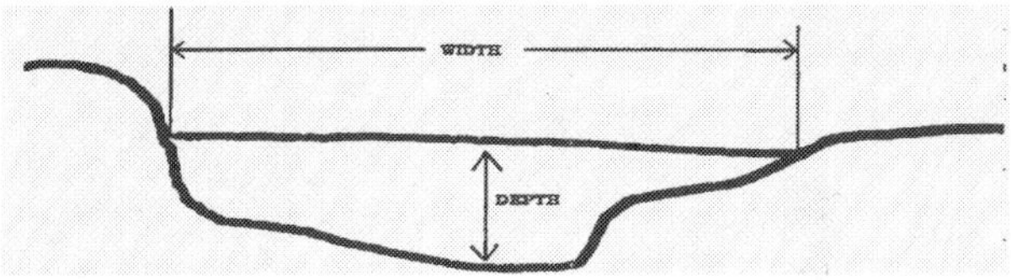

For example: 6 feet deep x 100 feet wide x 2 feet per second = 1,200 CFS

b. River Velocity. In order to determine the velocity or current speed measure a 100-foot section of the river and drop a float at the upstream end. Then, time its movement to the lower end and use the chart below.

TIME IN SECONDS	VELOCITY FT PER SEC	VELOCITY MPH
5	20	13
10	10	7
15	6.7	5
17	5.9	4
20	5	3.5
25	4	3
29	3.5	2.5
37	2.7	2
50	2	1.5
80	1.3	1
110	0.7	0.5

NOTE: Should the shape of the river change by becoming wider or narrower, once you have the flow in CFS, you can determine the new velocity by using the following formula:

$$\text{Velocity} = \frac{\text{Volume in CFS}}{\text{Width x Depth}}$$

c. Force Exerted. The velocity of the water will determine the force exerted X's the amount of exposed area under pressure. For simplicity we will use the following chart:

CURRENT VELOCITY	FORCE EXERTED ON OBJECT		
MPH	LEGS RAFT	BODY	WRAPPED
3	17	34	168
6	67	134	672
9	151	302	1,512
12	269	538	2,688

4. **LOW HEAD DAMS**. Low head dams are used throughout the world for irrigation, flood control, at hydroelectric plants and for many other uses. Basically they are pouring over obstacles that create hydraulics, keeper holes, and recirculating currents at their bases. Unlike natural obstacles, that rarely span the entire river and that allow you to be "flushed" out of the bottom or sides, low head dams create a hydraulic that extends from shore to shore.

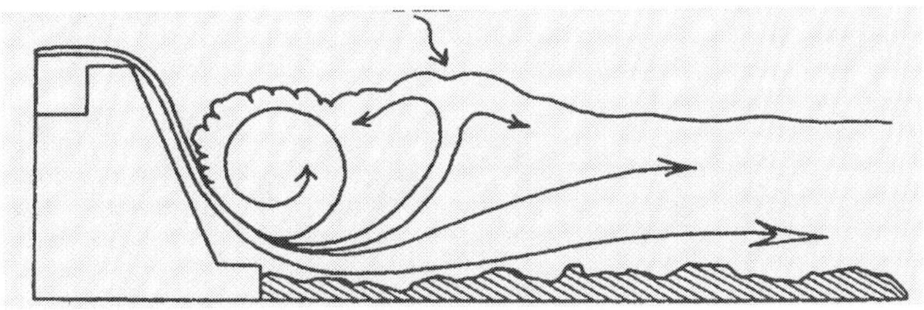

a. The size of the hydraulic created is based on several denominators:

(1) Height of drop.

(2) Angle of drop.

(3) Volume of water CFS.

(4) Velocity of water MPH.

(5) Bottom composition of runout.

b. The degree of difficulty for the rescuer is also compounded by:

(1) Water depth below dam.

(2) Debris build-up.

(3) Dam construction/conformation.

(4) Lack of access.

(5) Lack of natural anchors.

(6) Distance between riverbanks.

c. The basic hydrology is the same as any other pour over obstacle. As the river flows over the dam, the water follows the face of the dam to the river bottom. When sufficient speed diminishes the water rebounds to the surface and continues downstream. This flow has caused a vacuum or pocket directly below the dam that must be filled. Therefore, when the water rebounds to the surface, some of it flows back upstream to form a vertical reversal or pour-over eddy. The line of demarcation for the upstream/downstream flow is called the boil line. The shape and angle of the dam face and the volume and velocity of the water flowing over the dam create the distance and width of the boil line.

5. **REVIEW**.

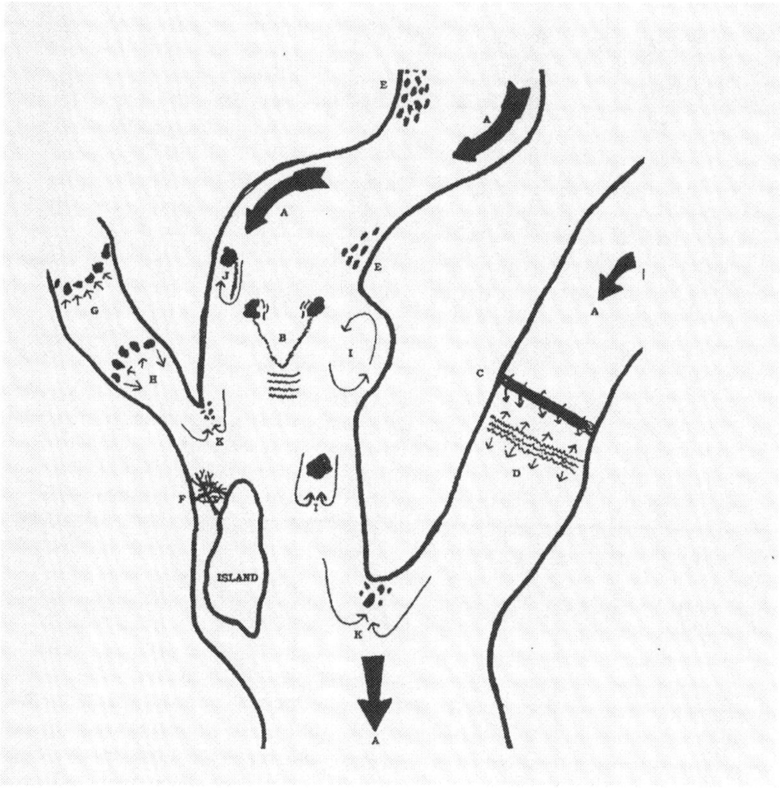

A. Main Current	G. Frowning Hole
B. Downstream V	H. Smiling Hole
C. Low head Dam	I. Eddy
D. Boil Line	J. Eddy Fence
E. Rock Garden	K. Confluence
F. Debris Pile/Strainer	

UNITED STATES MARINE CORPS
Mountain Warfare Training Center
Bridgeport, California 93517-5001

FMST.07.24
04/14/02

STUDENT HANDOUT

COMMUNICATIONS

TERMINAL LEARNING OBJECTIVES Given a casualty in a swift water environment, assignment as a member of a rescue team, and a whistle, employ swift water rescue communication signals to support mission requirements per the reference. (FMST.07.24)

ENABLING LEARNING OBJECTIVES.

(1) In a summer mountainous environment, prior to swift water rescue operations, when the instructor verbally states a hand and arm signal, the student will demonstrate that signal, in accordance with the references. (FMST.07.24a)

(2) In a summer mountainous environment, standing at the edge of a swift water stream, when the instructor blasts a whistle a varied amount of times, the student will focus their attention in the proper direction, in accordance with the references. (FMST.07.24b)

OUTLINE.

1. **RELIABLE COMMUNICATION**. In many cases time is critical in a SWR scenario. Therefore immediate, understandable communication is a must. Radios, bull horns, and voice work well in some situations, but wet equipment, dead batteries, and ambient noise levels can render these methods ineffective. Therefore, rescue professionals have developed a series of simple hand and arm, and whistle signals. Any rescue team can develop their own signals. The only requirement is that all hands understand the signals and that they are briefed to any other agency you are working with.

B4

2. **HAND AND ARM SIGNALS**. (FMST.07.24a)

ACTION	MEANING
1 hand extended over head	Require Assistance
Both hands extended forming an "O" with one hand touching The head	Okay
Extend both hands above head, then point left/right	Move in direction indicated
Extend both hands above head, wave, then point left/right	Eddy out in direction indicated
Cross both arms over chest	Require medical assistance or Corpsman

3. **WHISTLE BLASTS**. (FMST.07.24b)

ACTION	MEANING
One Blast	Stop/Attention
Two Blasts	Upstream
Three Blasts	Downstream
Three Blasts Repeated	Emergency

UNITED STATES MARINE CORPS
Mountain Warfare Training Center
Bridgeport, California 93517-5001

FMST.0
7.27
04/14/02

STUDENT HANDOUT

THROW BAGS AND FOOT ENTANGLEMENTS

TERMINAL LEARNING OBJECTIVES In a swiftwater rescue scenario employ a throw bag, in accordance with the reference. (FMST.07.27)

ENABLING LEARNING OBJECTIVES.

1) Without the aid of reference, provided with a throw bag and swimming victim, rescue the victim, in accordance with the reference. (FMST.07.27a)

OUTLINE.

1. **THE THROW BAG**: The throw bag is very simply a cloth bag with a small float insert, packed with 50 to 70 feet of 1/4" to 3/8" floating polypropylene or spectra/poly line. The concept is to hold the exposed end of the line and throw the packed bag to a swimmer. As the bag flies, the rope deploys enabling you to reach a victim 50 to 70 feet from shore. It is used either from a standing position, or a moving position on open shores.

   a. Standing Procedures. (FMST.07.27a)

1. Assume a stable belay position and ensure your footing is solid. Avoid loose, sloped or wet rock.

2. Attract victim by blowing your whistle (1 blast). Show him/her the throw bag and yell, "grab the rope". NOTE: If the victim grabs the bag rather than the rope, the rope will continue to deploy to its full length.

3. Grasping the exposed end of the rope, throw the bag underhand in a smooth motion directly at the victim. You may need to lead them in order to score a direct hit.

4. Tell the victim to place the rope over his/her upstream shoulder. This will establish a good ferry angle.

B5

5. Do not wrap the rope around yourself or tie it off at anytime. You may have to release the victim to avoid an obstacle or strainer. Use a dynamic belay and pull the victim to shore or pendulum the victim to other rescuers downstream.
6. If available have other rescuers downstream to recover the victim or repeat the procedure if you miss!

b. Moving Procedures. The moving procedures are the same as the standing procedures except you are on a smooth, open shoreline, so you move slightly downstream of the victim as you pull them gently ashore.

2. **FOOT ENTRAPMENTS AND VERTICAL PINNINGS**. These scenarios are caused by a variety of actions. Swimmers put their feet down or fishermen, in waders, get their feet stuck between rocks and are pushed over by the current. Swimmers are stuck in a strainer. Rafters get wrapped on a rock with victims pinned between the raft and rock or kayakers drop down a vertical shoot and get lodged in the rocks at the bottom. Whatever way it occurs you must respond quickly in order to save the victim. Initially take the following precautions:

   1. Position upstream spotters, to provide warning of boats or debris.

   2. Plan alternate methods if initial try doesn't work.

   3. Have back-up rescuers downstream in order to retrieve the victim.

   a. Once these actions are complete conduct the rescue using one of the five following methods.

      1. Bechel-Ray One Line Method. The simplest and safest method. Get a rescuer on both sides of the river. Establish a rope between the two rescuers and downstream of the victim. Walk upstream until the rope hooks the victim. If you can position the line near the victims ankles you may be able to pop him/her loose and the downstream safety can pick him/her up. If not you can haul the victim upstream until he/she is clear of the obstacle, then by one rescuer playing out line and the other retrieving it you can bring the victim ashore. Remember to bring the victim to the shore with the medevac capability.

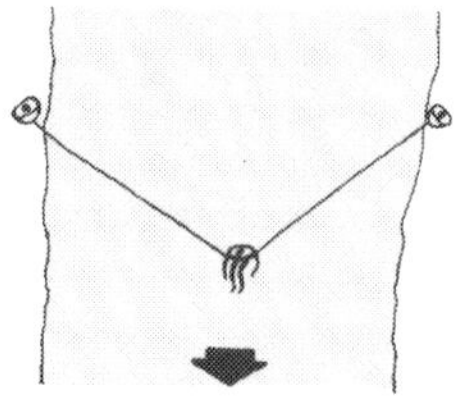

      2. The Strong Swimmer-variation. If the previous method doesn't work, leave the line in place and anchor it to stabilize the victim above the surface. Then have a "strong swimmer" go down one side of the rope, as would a tension diagonal. This rescuer is not tied in. Upon arrival the rescuer can readjust the line on the victim, attempt to pull them free, give them a PFD and helmet, or if a debris pile or rock is near, get on it and assist in further efforts.

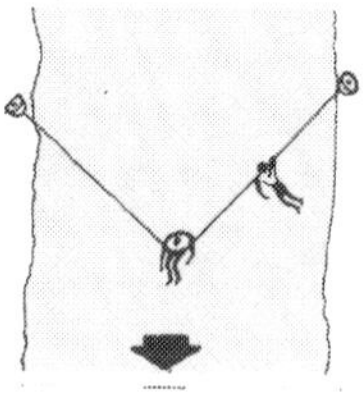

3. Two Line Loop. There are two variations on this method both of which are applied using the initial rope used in the Bechel-Ray attempt.

(1) Deploy a second rope across the river and hook it onto the first line. Allow it to go one-half way down the rope to the victim. Pull at a 90° angle to rope #1. This will establish a mechanical advantage and pull the victim loose.

(2) Once the victim is free walk the line downstream trapping the victim in a loop. Then by releasing slack on line one, you can bring him/her ashore.

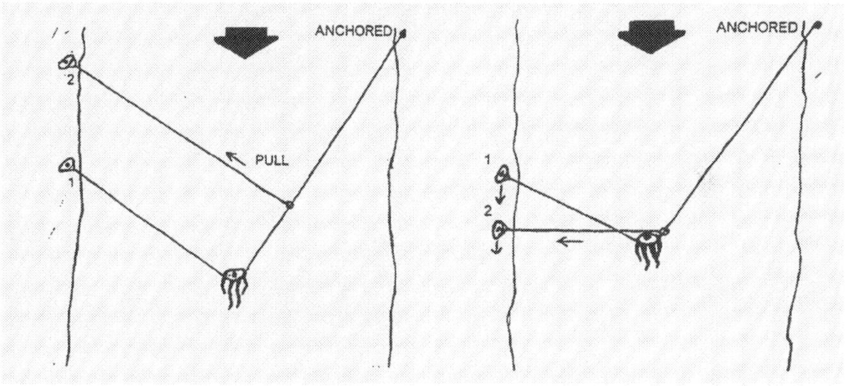

4. European Tethered Rescuer Lower. Using a rescue vest with a "blow-out" belt establish an upstream belay and lower the tethered rescuer to the victim. The rescuer's body will block the water forming an eddy around the victim enabling him to be pulled free. Ensure that the rescuer has a working rescue vest, as this will enable him to cut away if the situation dictates.

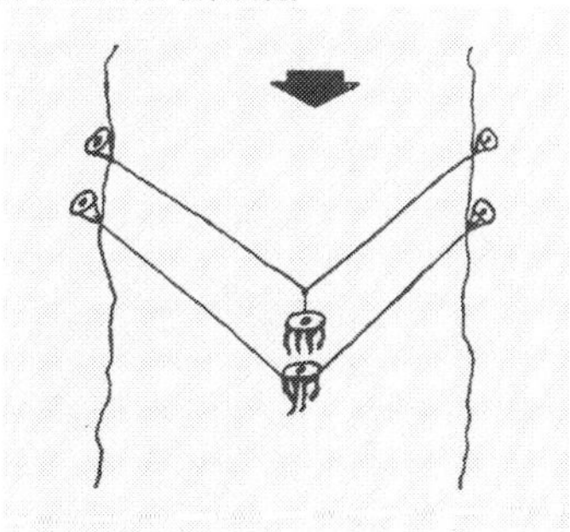

5. Mechanical Advantage Systems. As a last resort you can use a mechanical advantage system. This method is usually used only for body recoveries, as the amount of force exerted on the victim is tremendous and can cause severe injuries.

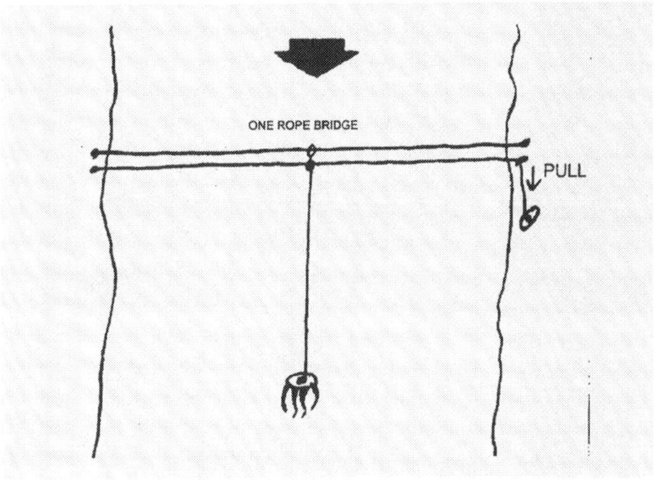
ONE ROPE BRIDGE
PULL

UNITED STATES MARINE CORPS
Mountain Warfare Training Center
Bridgeport, California 93517-5001

FMST.07.28
04/14/02

STUDENT OUTLINE

DEFENSIVE SWIMMING AND FERRY ANGLE

TERMINAL LERNING OBJECTIVES Given a swift water river and throw bags, perform aggressive swift water swimming to meet mission requirements per the reference. (FMST.07.28)

ENABLING LEARNING OBJECTIVES.

(1) Without the aid of reference, demonstrate the defensive swimming position, in accordance with the reference. (FMST.07.28a)

(2) Without the aid of reference, demonstrate use of the ferry angle to cross the river, in accordance with the reference. (FMST.07.28b)

OUTLINE

1. **THE DEFENSIVE SWIMMING POSITION (DSP)**. (FMST.07.28a) The DSP is assumed by lying flat on your back, chin on chest (to read the river & select a route), with your feet up, heels just below the level of your buttocks. Holding your feet in this position will enable you to fend off of shallow rocks and will reduce the likelihood of a foot entrapment or pinning. Your arms and hands should be kept close to the body but may be used like oars on a boat to maneuver. Your head will remain upstream except to cross a strainer or to over take a victim in a swimming rescue.

2. **FERRY ANGLE**. (FMST.07.28b) A ferry angle is used to maneuver diagonally across a river. By placing your body at a 45° angle to the current, the river will push you across the river. This is because the force of the current will travel down the exposed, upstream, side of the body moving it laterally more rapidly then the current pushing straight. This is like traversing a slope with edged skis vice sideslipping down on flat skis. In order to cross faster or to reach an eddy you may roll to your belly and execute the rescue crawl stroke or breaststroke. Remember to maintain your ferry angle with your head upstream this will enable you to travel diagonally across the river.

B6

3. **STRAINER CROSSING**. The first rule is to avoid the strainer or shallow submerged obstacle. Read the river and maneuver to the main channel by using downstream "V's" or tongues. If the obstacle cannot be avoided, roll on your belly, head downstream and use the rescue crawl stroke. You must get your body moving faster than the current. Upon arrival at the obstacle use both hands to "vault" yourself over. A well-timed dolphin kick will assist in keeping your legs clear of entanglements. Once clear of the obstacle, resume the defensive swimming position. Bear in mind that it is usually safer to remain in the main current or to eddy hop downstream than it is to attempt to get out on a flooded riverbank, the outside of a bend or on slippery rocks. Also, use the dolphin kick rather than placing your feet on the bottom and kicking off. This will reduce the likelihood of foot entrapments.

UNITED STATES MARINE CORPS
Mountain Warfare Training Center
Bridgeport, California 93517-5001

FMST.07.25
03/14/02

STUDENT HANDOUT

FEET-WET RESCUES AND C-SPINE ROLLS

TERMINAL LEARNING OBJECTIVES In a swift water rescue scenario, conduct a swimming rescue, in accordance with the reference. (FMST.07.25)

ENABLING LEARNING OBJECTIVES.

(1) Without the aid of reference, while in a river and provided with a victim, perform a contact rescue, in accordance with the reference. (FMST.07.25a)

(2) Without the aid of reference, while in a river and provided with a simulated unconscious victim, perform a C-spine roll, in accordance with the references. (FMST.07.25b)

OUTLINE.

1. **FEET WET RESCUE GENERAL CONSIDERATIONS**. (FMST.07.25a)

a. Always wear the proper personal equipment, i.e. PFD, helmet, drysuit, etc.

b. Don't make physical contact if it can be avoided. Make him/her "chase" you ashore. Use a life ring, can, or boogie board, and tow it along with the victim. Toss him a throw bag and swim/tow him/her until you reach the shore, then pendulum or haul the victim in. Remember, to a panicked victim YOU are a floatation device.

2. **CONTACT RESCUE**.

a. Approach the victim from upstream. Use the rescue crawl stroke to catch-up to him/her. Talk to the victim and attempt to get him/her to swim to shore or "chase" you to the shore.

b. Explain to the victim the defensive swimming position and attempt to get them in this position. Assume this position yourself, in contact rescues this is called the "Reverse and Ready" position. Tell the victim what you are going to do prior to grabbing them.

c. Gain contact by grabbing the shoulders of the life jacket or utilize a cross-chest carry. If the victim is unconscious and face down use a C-Spine roll down and maintain stability as much as possible.

d. Assume a good ferry angle with the victim downstream of you and use his/her body to protect you from obstacles.

e. If the victim panics and attempts to climb on top of you, break away or lock the victim up and go for an alligator roll. Reestablish control and calm the victim as soon as possible.

f. Once contact is made remember that you may be held accountable under the "abandonment" laws if you quit.

3. **TOWED VICTIM SWIM**. When towing a victim it is essential that you establish a ferry angle. If the victim is on a boogie board or float, explain to him/her what you are doing and have them help by kicking their feet. If you are towing by the PFD grasp the shoulder towards the far bank and remain in the DSP (defensive swimming position) in case your victim panics.

4. **LIVE BAIT RESCUES**. The final go and tow method is the "live bait" rescue. Using a rescue vest with a "blow out" belt and a mule team on the shore to belay you, swim to the victim as explained in contact rescues. Once you establish contact you are, in effect, a throw bag with hands. The belayer will then either pendulum you ashore or hauls you in. In an emergency you can cut away and conduct a normal contact rescue.

5. **C-SPINE ROLLS AND STABILITY**.

a. Basic Rules: (FMST.07.25b)

(1) Suspect C-Spine involvement in any head injury.

(2) Stabilize and immobilize, use C-Collar, Sam Splints, or a field expedient method.

(3) Keep the victim in the water until a spine board or equivalent is available, or there are enough rescuers to move the victim safely.

(4) Observe for other signs, i.e. cranial bleeding, bleeding from ears, eyes, nose (Do not attempt to stop this bleeding) or unequal/unresponsive pupil response.

b. Methods:

(1) First, approach the victim from behind. Reach under the arms and grasp the victim's head with both hands, then roll the victim. This is the least desirable method as you cannot swim properly or evaluate the ABC's of the victim.

(2) Second, approach the victim from the side, extend his/her arms above and alongside his/her head, reach across and roll the victim onto your arm maintaining the victims arms in place to ensure stabilization. Evaluate the ABC's, provide rescue breathing and use your free arm to swim.

(3) Finally, approach a victim wearing a PFD from the head. If face down, cross your arms thumbs inboard and use your forearms to immobilize the neck. Then roll the victim over. If face up, simply approach as if gaining contact on a contact rescue and maintain control of the C-Spine.

UNITED STATES MARINE CORPS
Mountain Warfare Training Center
Bridgeport, California 93517-5001

WMC

04/14/02

STUDENT HANDOUT

BOATS AND BOAT HANDLING

LESSON PURPOSE. The purpose of this period of instruction is to introduce the student to the boats available at MWTC for river rescue evolutions.

OUTLINE

1. **GENERAL**. There are a wide variety of boats and hybrid river craft available to rescue crews. Whitewater rafts, rigid raiders, zodiacs, airboats, and regular pleasure boats can all be used in the proper situations. All have been used both safely and have caused deaths and injuries when used improperly. Wherever you are, learn the equipment available, both its assets and its limitations.

2. **WHITEWATER RAFTS**. The whitewater raft of today is an inflatable, self-bailing, multi-chambered vessel with several tie rings and foot cups. Additionally, the materials used in construction resist rips, tears, and punctures much better than their predecessors. The size most often found on local rivers is 12 to 14 feet long due to the size of the rivers. In the MWTC SAR locker, we have a 13-foot self-bailer. This will be our primary rescue boat.

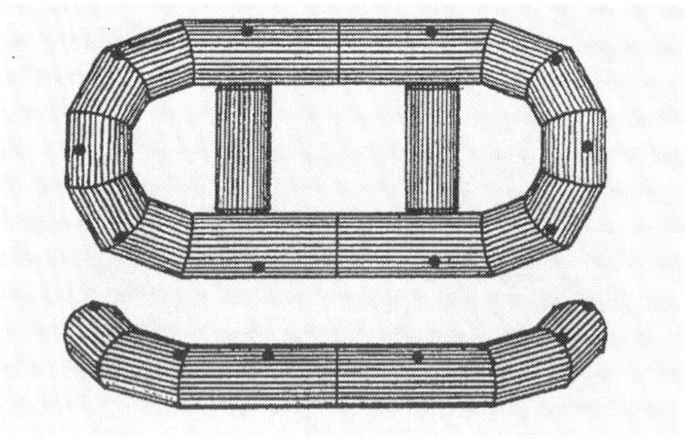

B8

3. **RAFTING COMMANDS**. In order to control the direction of a peddler powered boat the coxswain or guide must use a series of simple, concise, and understandable commands.

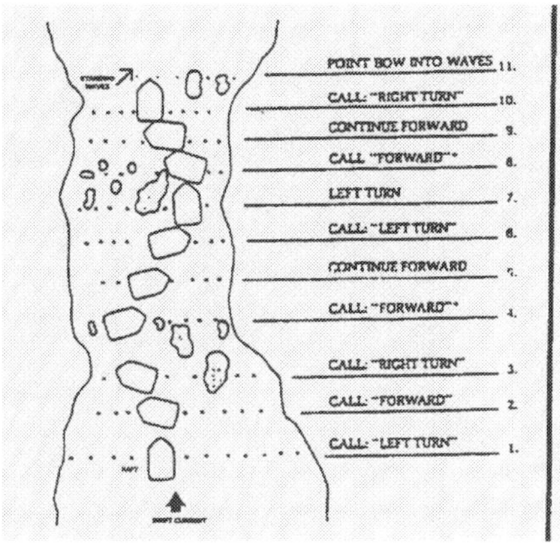

NORMAL TACTICS APPROACH

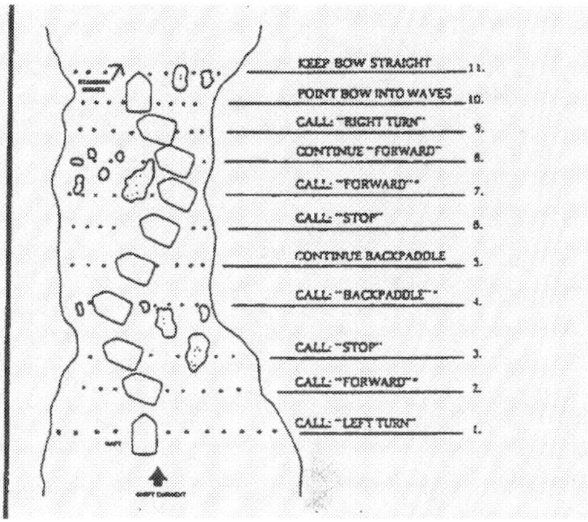

CONSERVATIVE/FAST CURRENT

a. Forward - All hands paddle forward together.

b. Back Paddle - All hands back paddle together.

c. Stop - Stop paddling - Remove blades from water.

d. Left Turn - Left side back paddles, right side forward.

e. Right Turn - Right side back paddles, left side forward.

f. High Side - Move quickly to the down streamside of the boat.

4. **PADDLE STROKES**. In order to control the raft from the coxswain position, either with a crew or alone on a tether system, you will need to apply the following strokes:

a. Sweep - Stroke in a wide arc around the fantail of the boat.

b. Draw - Reach out 90° away from the boat and "draw" it straight back to you.

c. Pry - Put paddle blade beside raft and "pry" the water away from the boat.

d. "J" - Used by lone guide to keep boat straight. Regular straight stroke finished with an outboard stroke or tail.

UNITED STATES MARINE CORPS
Mountain Warfare Training Center
Bridgeport, California 93517-5001

WMC
04/14/02

STUDENT HANDOUT

ROPE SYSTEMS AND BOAT ANCHORS

LESSON PURPOSE: The purpose of this period of instruction is to introduce the student to the rope systems and boat anchors used in swiftwater rescue situations.

OUTLINE

1. **TWO AND FOUR POINT TETHERING SYSTEMS**.

a. With all rope systems the first order of business is to get a rope across the river. Best case is to throw a rope bag across to someone already on the other side. Next best is to use another boat or raft to ferry the rope across. Last we use a swimmer to drag the rope across. Remember to use a smaller line on the swimmer and not to tie it to him/her. When swimming, start high and pendulum across using a good ferry angle.

b. Clip in on the side "D" rings forward for a two point or on the two forward and two aft "D" rings for a four-point tether system. For a two point you now need to simply load in a rescuer (if used) and move the boat into the river by taking slack on the far side and giving slack from the near side. For a four point man the two nearside lines and load a far side aft lineman and a rescuer (if used) into the boat and ferry the boat across. The far side aft belayer will get out and the boat is on a four point system ready for action.

c. This system is good for Class II Rivers with clear banks, few mid stream obstacles, and little floating debris. Use more belayers as required by river speed and boatload.

d. In all rope systems the whistle and hand and arm signals used to communicate are those taught in the communications class earlier in this course.

e. To break down the system simply reverse the procedures. The only difference is that the last man rides in the boat instead of swimming.

2. **THREE POINT AND BOATMEN'S SELF-EQUALIZING ANCHORS.**

a. These anchors have in common the fact that they are self equalizing. This means that, regardless of the direction of pull, all anchor points ("D" rings, thwarts, etc) will remain under equal tension.

b. Using ropes the knots required are the double figure 8 on a bight, a directional figure 8, and a figure 8 on a bight.

c. Using webbing (one inch tubular nylon) a water tape knot is required.

d. If you have an abundance of carabineers, simply clip into the anchor points and back into the small loop on the double figure 8 on a bight, or into the directional figure 8. If not run the rope through the small loop. Remember that this is not preferred as it causes rope on rope stress and rubbing that can melt the rope.

e. When using webbing the same concept applies. Ensure you put a twist in the webbing at the contact point to ensure that a single blow-out will not collapse the entire system.

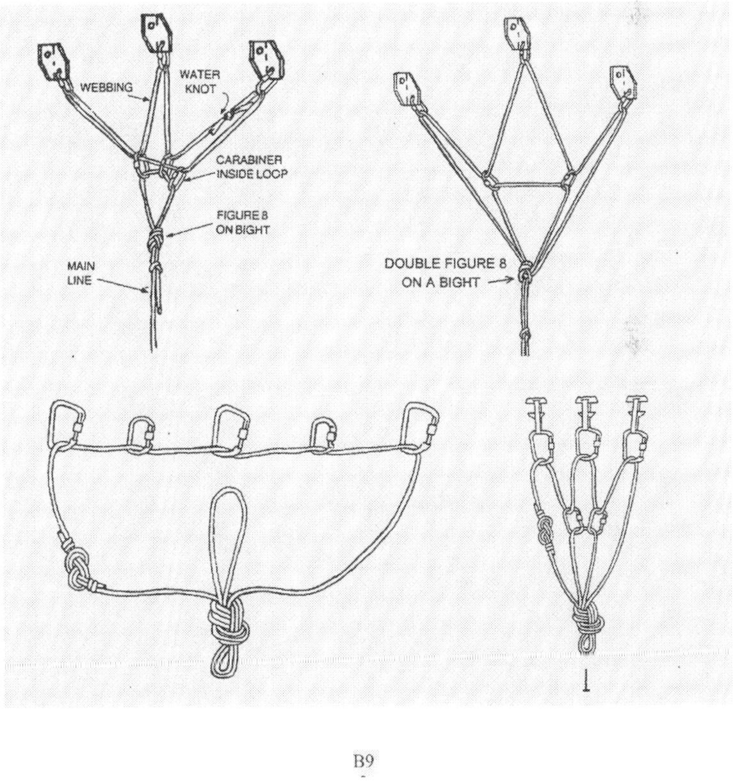

3. **PIG-RIG FOR MECHANICAL ADVANTAGE.** At times you will find yourself with ropes that are long enough to span the river, but not long enough to establish a mechanical advantage system. In these situations, you can use a piggy-back line or "pig-rig" by following the procedures below:

a. Tie off the mainline on either ends, or tie off the far side and set up a belay device on the near side using a munter hitch or rescue 8.

b. Establish an anchor for the pig-rig below the near side anchor.

c. Tie a prussic on the mainline and clip in a carabiner. Tie a figure 8 on a bight into the end of the pig-rig line and clip it into the prussic's carabineer.

d. Run the pig-rig line back to the anchor carabineer and then back through the prussic carabiner.

e. Pull on the running end towards the anchor. This will give you a 4:1 mechanical advantage on the main line.

f. If more tension is required you can either retie the nearside anchor (if tied), or use your belaying device to tension the mainline before moving the prussic.

g. If no prussic cord is available, you can accomplish this with a directional figure 8 on the mainline. However, this will only allow you to tighten the system once. A better method is to use the pig-rig line to tie a prussic directly on to the main line and run your system off of a directional figure 8 tied below the prussic on the pig-rig line. Remember that if the lines are of the same diameter this system will not work as the prussic will slip, therefore, you should use a kragur knot with ropes of the same diameter.

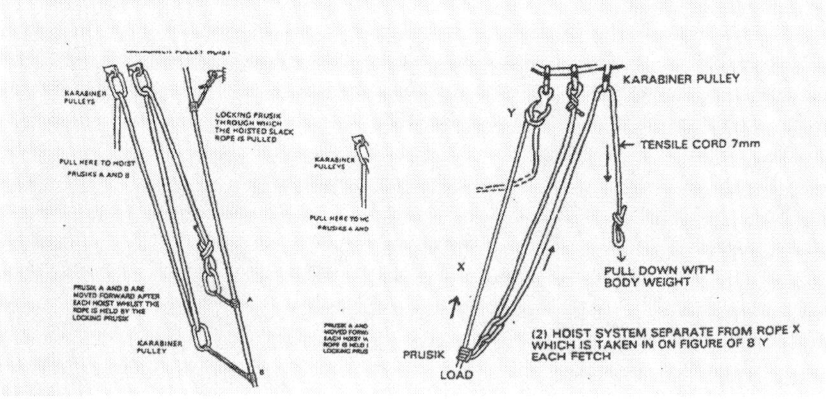

4. **TYROLEAN HIGHLINE AND TELFER LOWERING SYSTEMS** Both of these systems are used to maintain a boat as a stable rescue platform. The lateral movement (side to side/shore to shore) in both systems is controlled by mule teams on the shore. The main difference is that in the tyrolean highline the movement up and down-stream is also controlled from the shore, while in the telfer lowering system, the rescuer in the boat controls that aspect of movement. Both systems have good and bad points and both should be mastered. The situation will dictate which system you will use. Follow the steps below to establish these systems:

a. First you must establish TWO lines across the river. You can use any method taught to accomplish this.

b. Establish one line as the anchor line and tighten it as taught in the "one rope bridge" class or use a pig-rig, if required. If you have uneven mule teams set it at a slight down-stream angle towards the weak team's shore (no more then 10 to 15 degrees) to assist the weak team.

c. Rig the boat with a self-equalizing anchor system.

d. Rig the lowering systems as shown in the diagrams. Either tyrolean or telfer.

e. Load a rescuer or rescue team into the boat and execute. Remember the rescuer can act as coxswain and assist in maneuvering the boat by establishing a proper ferry angle.

f. Communications will be as taught earlier. Ensure you have a shore commander who maintains eye contact with the rescuers and mule teams on both the near and far shores.

g. When using the telfer system, a counter hitch, stitch plate or rescue 8 can be used as a belay device. This enables you to tie-off when you reach the victim so that you can affect the rescue without worrying about controlling the boat.

h. When using either of these systems it is a good idea to set-up a mechanical advantage system between the control (central) pulleys aid the raft/rescue platform.

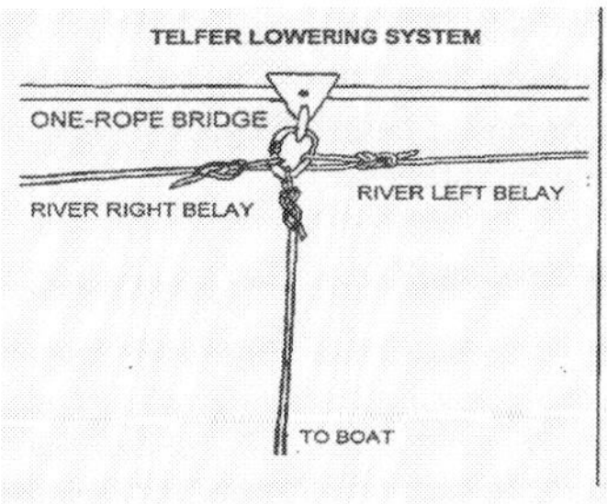

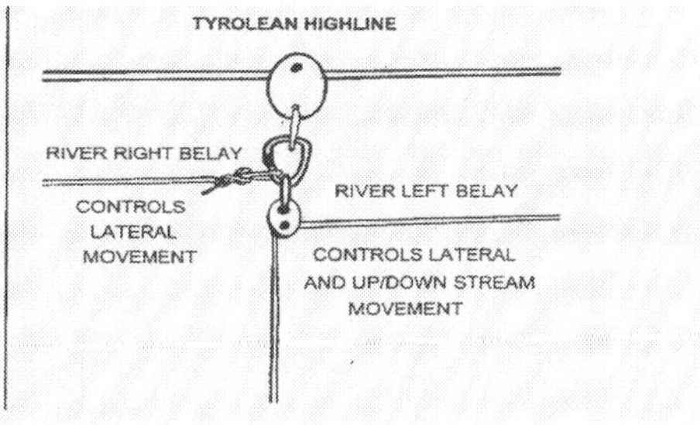

d. Equip the victims with PFD's and remove to the boat or onto the rope system. Consider the following points:

(1) The victims weight might be the only thing stopping it from rolling. You might need to transfer a rescuer in a one for one exchange.

(2) On hard pack or asphalt the vehicle will shift more easily.

(3) If the vehicle is sideways keep the victims on the upstream side. The high side rule is reversed.

(4) Vehicles roll slowly.

(5) Helo's are a last resort.

UNITED STATES MARINE CORPS
Mountain Warfare Training Center
Bridgeport, California 93517-5001

FMST.07.3
0
04/14/02

STUDENT HANDOUT

STREAM CROSSING

TERMINAL LEARNING OBJECTIVE: Given a mountain swift water environment, and a staff, conduct a mountain stream crossing to meet mission requirements per the reference. (FMST.07.30)

ENABLING LEARNING OBJECTIVES:

(1) Without the aid of references, choose from a given list the considerations in site selection for a stream crossing, in accordance with the references. (FMST.07.30a)

(2) Without the aid of the reference, choose from a given list the safety precautions taken while conducting a stream crossing, in accordance with the reference. (FMST.07.30b)

(3) In a summer mountainous environment, execute the individual preparations taken prior to crossing a stream, in accordance with the reference. (FMST.07.30c)

(4) In a summer mountainous environment and given a staff, cross a mountain stream using the staff method, in accordance with the reference. (FMST.07.30d)

(5) In a summer mountainous environment, cross a mountain stream using team methods, in accordance with the reference. (FMST.07.30e)

1. **SITE SELECTION** (FMST.07.30a)

   a. Mountain streams and rivers are military obstacles and therefore are danger areas for units crossing them. In order to reduce the time in the vicinity of the danger area, a recon team should precede the main body and select the best crossing site. The site selection for a stream crossing should include these eight considerations:

B11

(1) Look for logjams, rocks or fallen trees that will provide a dry crossing if possible.

(2) If a dry crossing is not possible, select a crossing point at a wide and shallow point where the current is slower.

(3) Avoid sharp bends. They can be deep with a strong current on the outside of the bend.

(4) Look for a firm, smooth bottom. This is because large rocks and boulders provide poor footing and cause a great deal of turbulence in the water.

(5) It may be easier to cross several small channels of water rather than one large one.

(6) Do not cross just above rapids, falls, or logjams, taking a fall or slipping could have serious consequences.

(7) Cross in the early morning. The water level will be lower since there has been less daylight for the snow to melt. Also, on sunny days, you will have more time to dry clothing and equipment.

(8) There should be a suitable spot downstream for safety swimmers.

2. **SAFETY PRECAUTIONS** (FMST.07.30b). The following two safety precautions must be taken while conducting stream crossings.

   a. There must be a safety line at a 45-degree angle downstream across the stream. This is for anyone who slips and is swept downstream to grab so that he is capable to stopping himself.

   b. There must be safety swimmers downstream. These are strong swimmers who are positioned downstream to help anyone who is swept downstream. The safety swimmers will use throw bags.

NOTE: The safety line, as well as any other lines that must be taken across the stream, will be taken across by using the lead swimmer method taught in ONE-ROPE BRIDGE.

3. **INDIVIDUAL CROSSING PREPARATIONS.** (FMST.07.30c) Prior to beginning to cross a stream, there are certain preparations that each individual should take. The five preparations are as follows:

   a. Wear pack with shoulder straps fastened snugly. Waterproof the pack for buoyancy if possible.

   b. Weapons will be slung diagonally over the shoulder with the weapon itself being between the pack and the individual's back.

c. Button all pockets and remove blousing garters. This prevents the water from creating added drag against the individual, which it would do if it could flow into open pockets.

d. Wear boots to protect the feet, but remove socks and insoles to keep them dry:

e. Wear the minimum amount of clothing. This reduces the amount of clothing that must be dried after the crossing.

f. Wear helmet in slow moving, shallow water.

NOTE: If you are in a tactical situation, the actual situation will dictate which of the above precautions will be taken.

4. **INDIVIDUAL CROSSING METHODS**. There are three individual methods that may be used.

a. Staff Method. (FMST.07.30d) A strong staff or pole about 6 feet long is used as a crossing aid. It should be strong enough to support the Marine's weight and trimmed clean of any branches. Placing both hands on the pole, the Marine should place the staff just upstream of his intended path. He should use the staff as the third leg of a tripod and should move only one leg or the staff at a time. He should face upstream using the staff to retain his balance. The staff is also used as a probe to discover bottom irregularities that could trip the Marine. The Marine should drag his feet instead of picking them up.

STAFF CROSSING METHOD

b. Swimming. This is an obvious method, if your Marines are good swimmers. This is not always the case, so usually this method is not a preferred one.

(1) In fast, shallow water, the Marine should angle across on his back with his feet downstream and his head up. He should use his hands to tread water and his feet to fend off obstructions.

(2) In fast, deep water, the Marine should angle across the stream on his stomach with his head upstream, to establish a proper ferry angle.

c. Belayed Method. In chest deep or water with a strong current, a rope can assist greatly. A rope will be secured from bank to bank, with the far anchor slightly downstream from the near anchor. The rope will be anchored off so that it lays at a minimum of 45 degrees. The Marine attaches himself to the rope by using a sling rope as safety line, tying a bowline around his waist and a figure-of-eight loop with a steel locking carabiner inserted. He then attaches the carabiner to the crossing line and crosses using the current to assist him.

5. **TEAM CROSSING METHODS** (FMST.07.30e)

a. Line Abreast. Small units (squad to platoon) can cross in moderate currents up to chest deep, by lining arms in a line abreast or chain method. The largest man of the chain is placed on the upstream side of the group. The group will enter the stream parallel to the flow of the stream. The middleman of the chain will control the group's movement and give the command when to step.

LINE ABREAST

b. Line Astern Method. Three or more men can line up facing the current. The upstream man, who should be the largest man in the group, breaks the current while the downstream men hold him steady. The upstream man may use a staff, similar to the individual staff method, to steady himself. All men side step at the same time with one man calling the cadence.

LINE ASTERN

c. Huddle Method. Between three and eight men can face inboard as in a football huddle. They will wrap their arms around each other's shoulders and cross the stream in this formation. The upstream man should change position as they cross because the entire formation will rotate. This prevents one man from becoming exhausted in the upstream position.

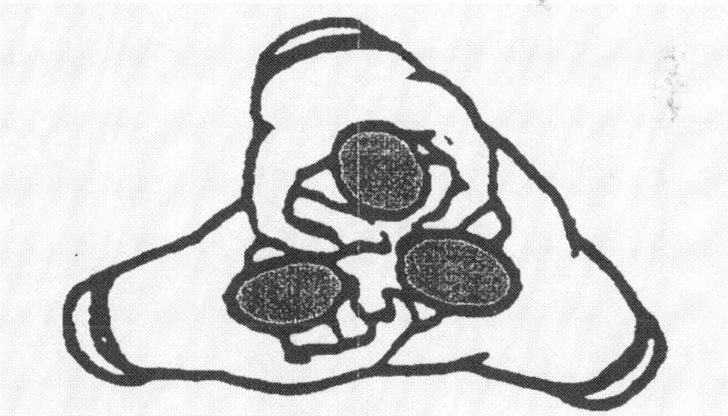

HUDDLE METHOD

UNITED STATES MARINE CORPS
Mountain Warfare Training Center
Bridgeport, California 93517-5001

WMC
04/09/02

STUDENT OUTLINE

ROPE MANAGEMENT

LESSON PURPOSE: To familiarize the student with some of the terms and procedures for working with a rope.

BODY

1. **TERMS USED IN ROPE WORK**

a. Bight. A simple bend in the rope in which the rope does not cross itself.

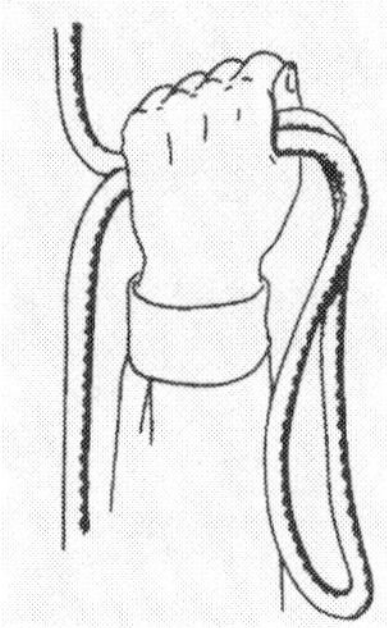

BIGHT OF ROPE

b. Loop. A simple bend in the rope in which the rope does cross itself.

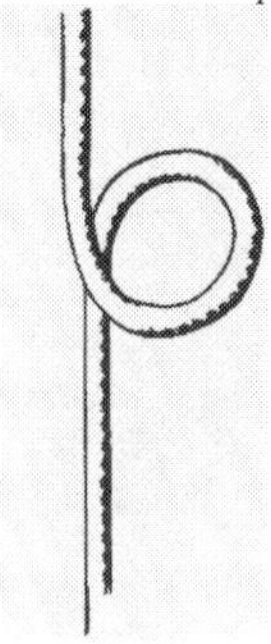

LOOP

c. Half Hitch. A loop which runs around an object in such a manner as to bind on itself.

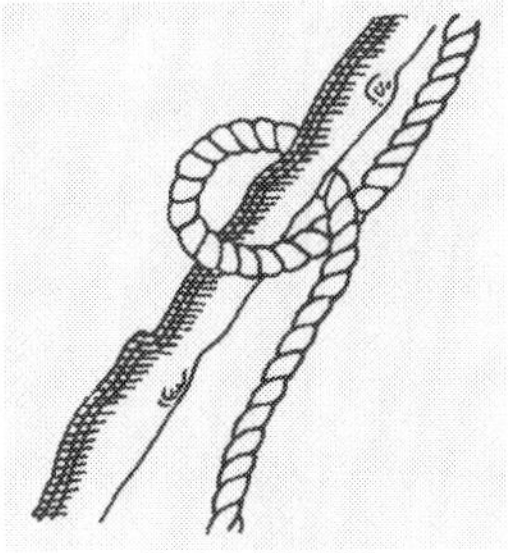

HALF HITCH

d. Standing End. The part of the rope which is anchored and cannot be used, also called the static end.

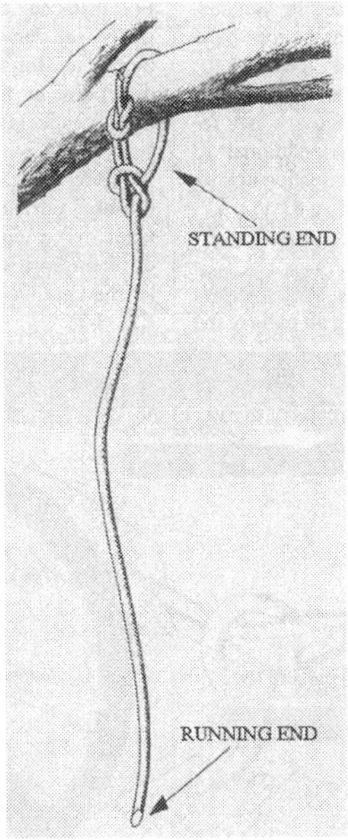

e. Running End. The free end of the rope which can be used.

f. Lay. The same as the twist of the rope. (Applies only to hawser laid ropes, such as manila.)

g. Pigtail. The short length left at the end of a rope after tying a knot or coiling a rope. It may or my not be tied off with a secondary knot, depending on the circumstance.

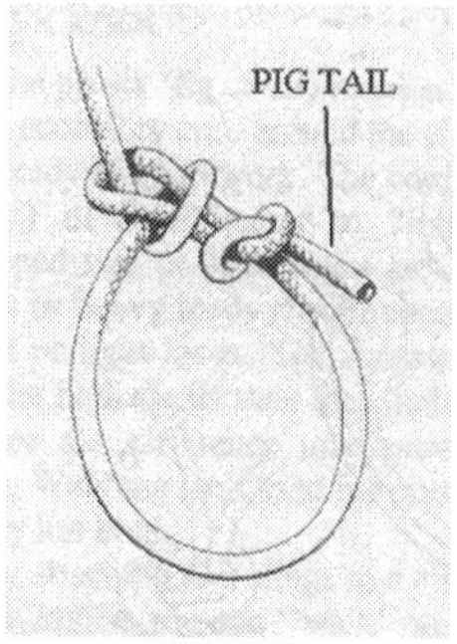

PIG TAIL

h. Stacking (or Flaking). Taking off one wrap at a time from a coil, and letting it fall naturally to the ground.

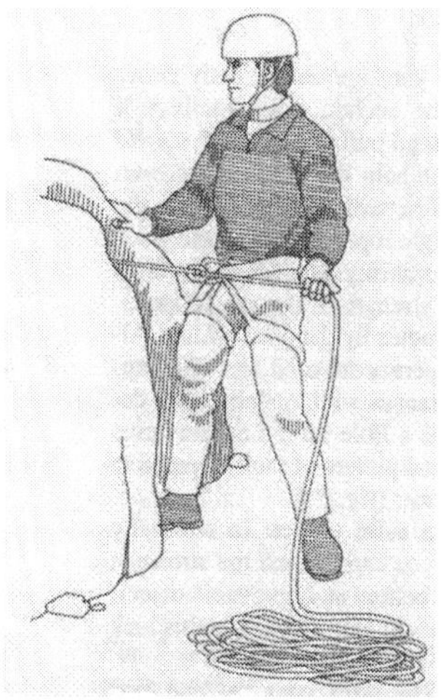

FLAKED OUT ROPEON THE GROUND

i. Dressing the knot. This involves the orientation of all of the knot parts so that they are properly aligned, straightened, or bundled, and so the parts of the knot look like the accompanying pictures. Neglecting this can result in an additional 50% reduction in knot strength.

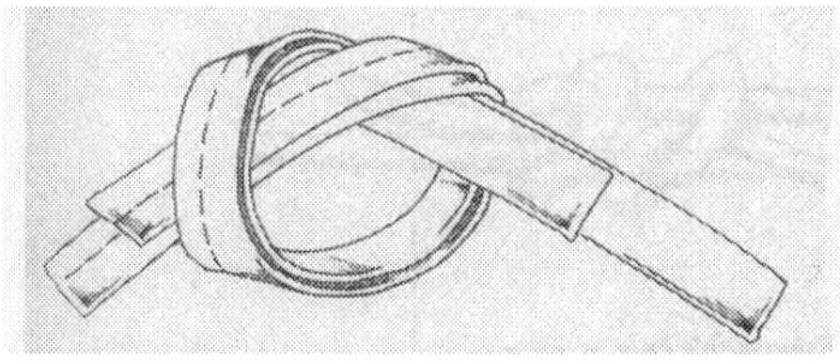

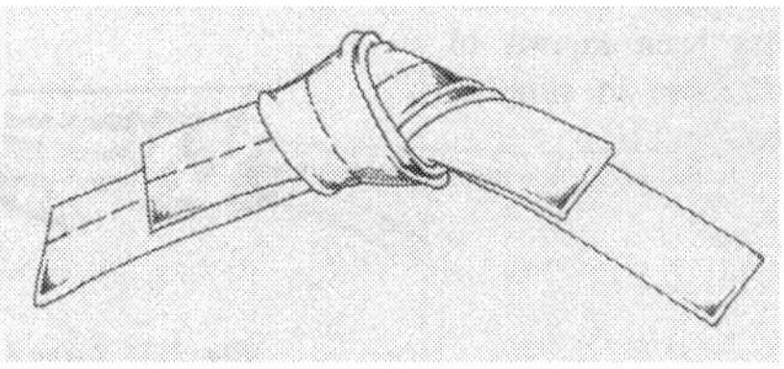

NON-DRESSED DOWN WATER TAPE KNOT   DRESSED DOWN WATER TAPE KNOT

j. Setting the knot. This involves tightening all parts of the knot so that all of the rope parts bind upon other parts of the knot so as to render it operational. A loosely tied knot can easily deform under strain and change character.

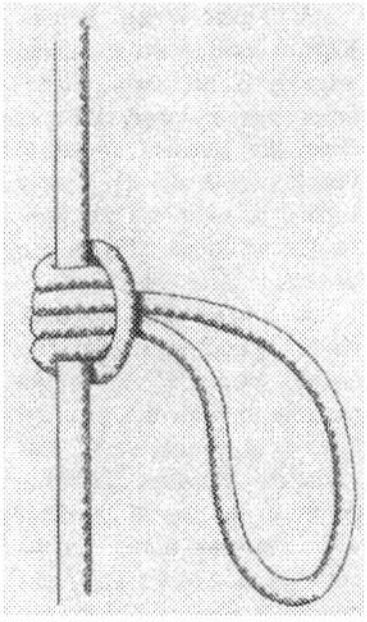

A FULLY SET PRUSIK KNOT

2. CONSIDERATIONS FOR THE CARE OF ROPE

a. The rope should not be stepped on or dragged on the ground unnecessarily. Small particles of dirt will get into and through the sheath causing unnecessary wear to the rope within.

b. The rope should never come in contact with sharp edges of any type. Nylon rope is easily cut, particularly when under tension. If a rope must be used around an edge which could cut it, then that edge must be padded or buffed using fire hose if available, or several small sticks.

c. Keep the rope as dry as possible. If it should become wet, hang it in large loops, above the ground, and allow it to dry. A rope should never be dried out by an open flame, or be hung to dry on metal pegs, as this will cause rust to get in the rope thus rendering it unserviceable.

d. Never leave a rope knotted or tightly stretched longer than necessary.

e. When using rope installations, never allow one rope to rub continually against another.

NOTE: With manila ropes this will cause the rope to fray, whereas nylon ropes can melt under the friction that this causes.

f. The rope should be inspected prior to each use for frayed or cut spots, mildew, rot or defects in construction.

g. Mark all climbing ropes at their midpoints to facilitate establishing the midpoint for a procedure requiring you to use the middle of the rope. The rope should be marked with a bright colored adhesive tape.

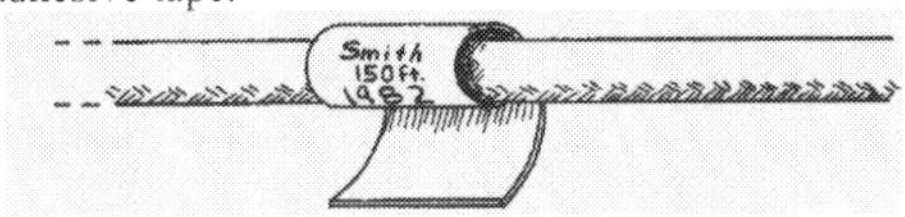

MARKING THE MIDDLE OF THE ROPE

h. The rope should not be marked with paints or allowed to come in contact with oils or petroleum products for these products will weaken it.

i. A climbing rope should NEVER be used for any other purpose except for mountaineering, i.e., towing vehicles.

j. The ends of a new rope or ends caused by a cut should be cut with the rope cutter contained in the MACK and marked with a serial number.

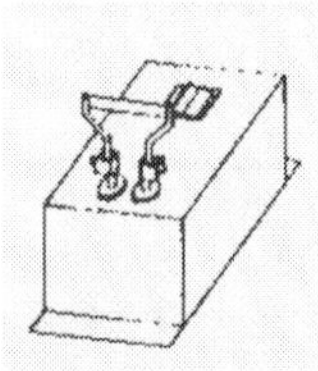

ROPE CUTTER

k. The rope should never be subjected to high heat or flame as this can significantly weaken it.

l. To clean rope use mild soap and rinse thoroughly with water. A rope washer can be used to clean or rinse the rope.

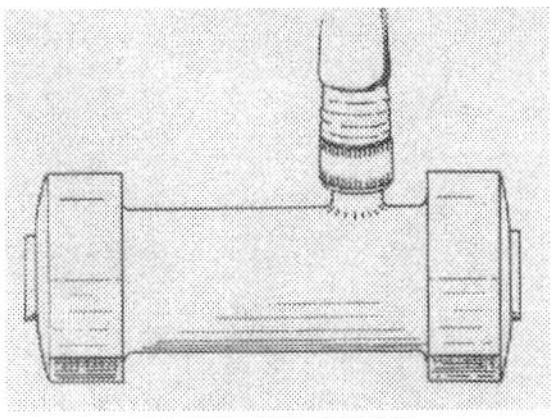

ROPE WASHER

m. When not in use, ropes should be coiled and hung on wooden pegs rather than on nails or any other metal object. They should be stored in a cool place out of the direct rays of the sun.

n. When in areas of loose rock, the rope must be inspected frequently for cuts and abrasions.

o. Always maintain an accurate Rope Log whenever using a rope.

p. Ropes 300-600 foot in length should be Mountain Coiled.

3. **INSPECTION OF ROPE**

a. All ropes have to be inspected before, during, and after all operations. Kernmantle rope is harder to inspect than a laid rope. I.e. green line. The Assault Climber must know what to look and feel for when inspecting a rope. Any of the below listed deficiencies can warrant the retirement of a rope.

(1) Excessive Fraying. Indicates broken sheath bundles or PIC breakage.

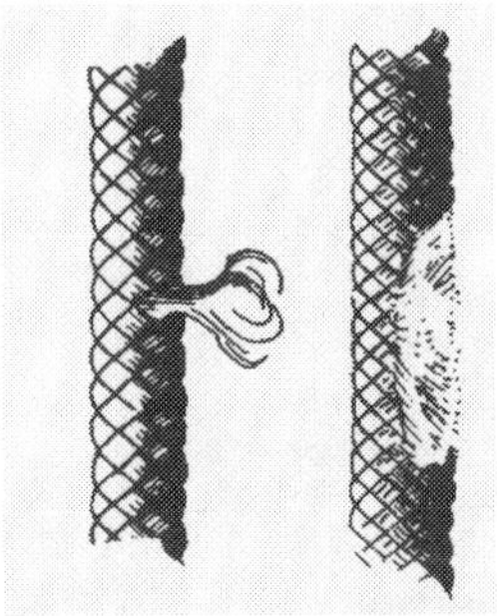

DAMAGED AND EXPOSED ROPE

(2) Exposed Core Fibers. Indicates severe sheath damage. (When you can see the inner core fibers)

(3) Glossy Marks. Signify heat fusion damage, also called a booger.

(4) Uniformity of Diameter / Size. May indicate core damage, noted by an obvious depression (hour glass) or exposure of white core fibers protruding from the sheath (puff).

(5) Discoloration. A drastic change from the ropes original color may indicate chemical change or damage.

(6) Stiffness or Soft Spots. Could signify core damage.

NOTE: Dynamic ropes measuring between 10mm and 12mm are marked at each end of its pigtails with a number "1" indicating that the rope is UIAA approved for single rope lead climbing. Dynamic ropes measuring between 8mm and 9mm are marked at each end of its pigtails with a "1/2" number indicating that two of these ropes are required to conduct a lead climb.

4. **ROPE LOG.** The purpose of the Rope Log is to maintain an accurate record for the use of each rope contained within the Marine Assault Climbers Kit (MACK). Due to the turnover of personnel, and the fact that no one person may have the same rope twice, the rope log is used to ensure the safe use, serviceability, and account of each rope.

a. Serial number. By assigning each rope a serial number, responsible units can determine information about a rope. As soon as the ropes are cut to the desired length for their intended purpose, each rope will be assigned a serial number by the responsible unit. That rope should then be labeled with that serial number in some permanent manner. The best method for this is as soon as the ends of the new rope have been whipped and fused, mark both ends of the rope with the serial number and then dip the ends of the rope, in effect, laminating the serial number to the rope ends.

(1) A rope serial number has five parts. The proper format for a rope serial number is as follows:

1. Type of rope: (S) - Static or (D) - Dynamic.

2. Last two digits of the year the rope was manufactured. Each rope has a shelf life of two years, after that it must not be used for any mountaineering purpose.

3. Four-digit number for that individual rope which is assigned by the responsible unit, and should be assigned sequentially as new ropes are issued out.

4. The length and diameter of the rope. The length may be recorded in feet, and the diameter in millimeters.

5. Responsible unit code. Example: S-96-0001-150/11mm-F2/8

Example meaning: This rope is a 150 foot, 11mm, static rope, manufactured in 1996. It is the first rope issued by Fox Co 2/8.

b. Recording Information in a Rope Log. Once a new rope has been serialized, the rope log for that rope should be started up. At a minimum, it should contain the following information:

(1) The rope serial number.

(2) Manufacturer. Depending on the manufacturer, ropes of the same type and diameter may vary in tensile strength, stretch factor, and durability.

(3) Date of manufacture. Five years from this shelf life date, ropes are considered to have reached their normal expiration date, and should be destroyed.

(4) Date in service. This is to be recorded for tracking purposes to establish how long a rope is in service. Two years is considered maximum rope life in service.

(5) Each time a rope is used, the using unit is responsible to record how that rope was used and how much use the rope received. Additionally, before checking out a rope and prior to turning it back in, the rope must be inspected by qualified personnel, and initial in the "Inspected By" block of the rope log.

NOTE: ANY TIME A DYNAMIC ROPE IS SUBJECTED TO A FALL FACTOR 2, THAT ROPE SHOULD NOT BE USED AGAIN FOR MOUNTAINEERING.

DYNAMIC					
SERIAL #					
MANUFACTURER					
DATE OF MANUFACTURE					
DATE IN SERVICE					
INDICATE THE DATE AND NUMBER OF EACH SPECIFIC USE					
CLIMBS	FALLS (FACTOR)	TOP ROPES	RAPPELS	REMARKS	INSPECTED BY

ROPE LOG

5. **COILING A ROPE**. There are two types of rope coils frequently used at MWTC. The Mountain Coil and the Butterfly Coil.

a. Mountain Coil. This coil is useful for carrying the rope over a pack or over a climber's shoulder and neck. It can be used for short time storage. The mountain coil can be tied in the following manner:

(1) Sit down with your leg bent at a 90-degree angle, heel on the deck. Starting at one end, the rope is looped around the leg in a clockwise fashion, going over the knee and under the boot sole until the entire rope is coiled.

(2) If coiling a 150-foot rope, use only one leg and offset the other, when coiling a 300-foot rope, use two legs and keep them together.

(3) With the starting end of the rope, form a 12-inch bight on the top of the coils.

(4) Uncoil the last loop and along the top of the coils, wrap 4-6 times towards the closed end of the bight.

(5) The end of the rope being wrapped is then placed through the closed end of the bight.

(6) The running end of the bight is then pulled snuggly to secure the coil.

(7) To prevent the coil from unraveling, the two pigtails are tied together with a square knot.

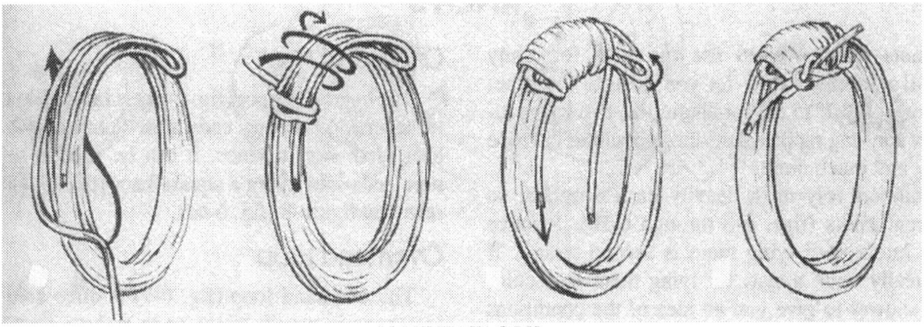

MOUNTAIN COIL

b. Butterfly Coil. This method is used for carrying a rope when the individual needs to have maximum use of his upper body, (i.e. while climbing), without the encumbrance of a large rope coil hanging across his chest.

(1) Coiling the Butterfly Coil

(a) Step 1: Find the middle of the rope, then form a three foot bight laying both ropes in the upraised palm at the two foot point.

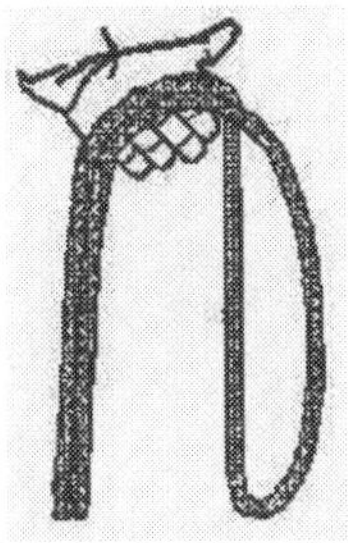

STARTING THE BUTTERLFY COIL

(b) Step 2: Form another two-foot bight with the running end. Place the rope at the two-foot bight along side on top of the original bight ensuring the running end is on the same side as the original bight.

(c) Step 3: Continue making two foot bights, laying them alternately into your palm until there is only six to eight feet remaining. At that point, begin wrapping the two pigtails horizontally four to six times at the mid way point of the ropes in a bight from bottom to top.

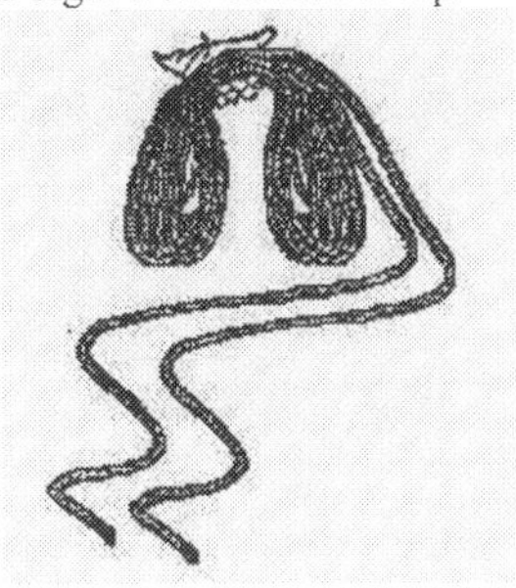

TWO FOOT BIGHTS ON BOTHS SIDES WITH 6 TO 8 FEET REMAINING

(d) Step 4: After completing your wraps, form a bight with the remaining pigtail and then thread it underneath your palm and upwards to one-foot above the coiled rope.

(e) Step 5: With the remaining pigtail, thread it through the one-foot bight in step four.

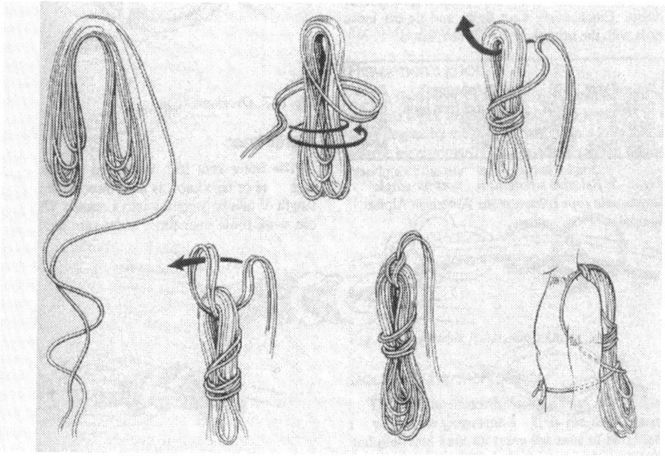

TYING AND CARRYING THE BUTTERFLY COIL

(2) Carrying the Butterfly coil. Separate the running ends, placing the coil in the center of the back of the carrier, then fun the two ends over his shoulders so as to form shoulder straps. The running ends are then brought under the arms, crossed in the back over the coil, brought around the body of the carrier and tied off with a square knot at his stomach.

6. **ROPE THROWING**. To insure that the rope will not get tangled when deployed, certain steps must be taken.

a. With a stacked rope, anchor off the standing end.

b. Take the opposite end of the rope and make 6-8 coils and place them in your strong arm. These wraps will serve as a throwing weight that you can aim.

c. 10-15 feet from the strong-arm coils create a second set of 6-8 wraps and place them in your weak arm.

d. From the edge of the cliff, sound off with the command "STAND-BY FOR ROPE". Just before you release the coils, sound off with the command "ROPE". At that time drop the weak arm coils from the cliff.

e. While taking aim, throw your strong-arm coils overhand or sidearm hard enough to hit your intended target.

f. If the throw was misdirected due to wind, tree, etc., reorganize and attempt to re-deploy the rope.

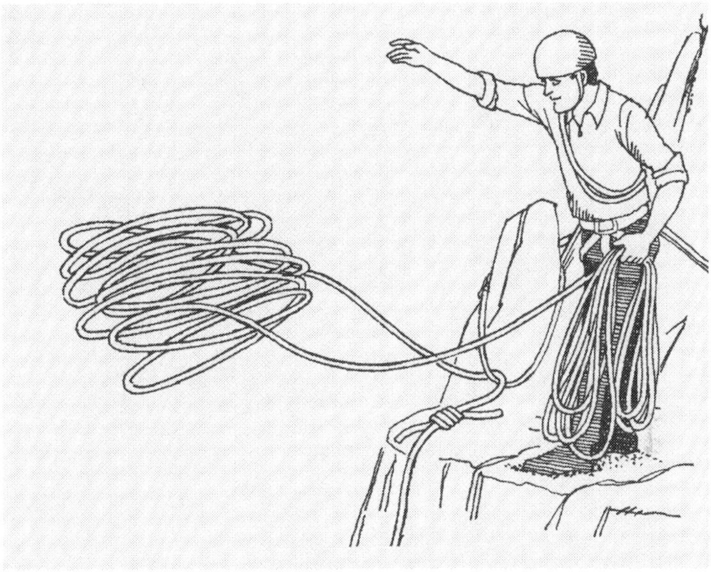

THROWING THE ROPE

6. **MOUNTAINEERING KNOTS.**

a. **Class I** – *End of the Rope Knots*

(1) Square Knot. Used to tie ends of two ropes of equal diameter together. It should be secured by overhand knots on both sides of the square knot.

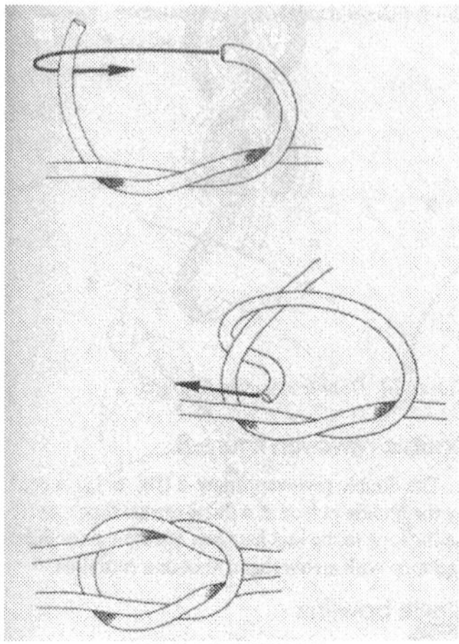

SQUARE KNOT

(2) Double Fisherman's Knot. It is a self-locking knot used for tying two ropes of equal diameter together. It can be tightened beyond untying.

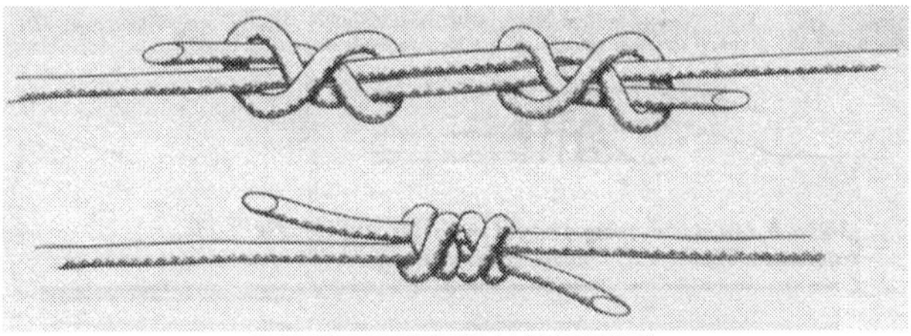

DOUBLE FISHERMAN'S KNOT

(3) Water/Tape Knot. Used to secure webbing or tape runners. It is constructed by tying an overhand knot (without twists) in one end of the tape, and threading the other end of the tape through the knot from the opposite direction. After the knot is dressed down, each pigtail should be a minimum of two inches long.

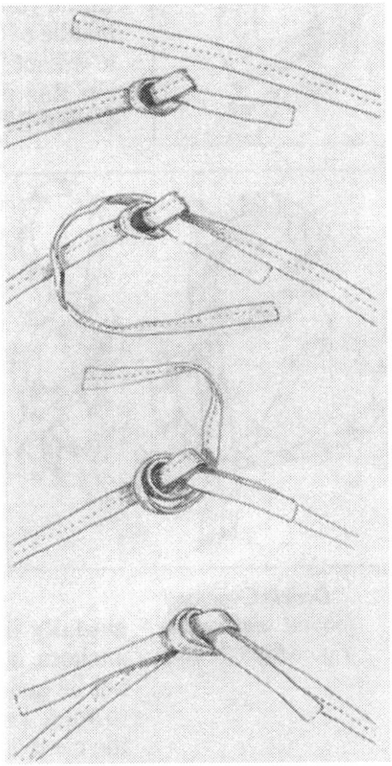

WATER TAPE KNOT

(4) Double Sheet Bend. Used to tie the ends of two or more ropes of equal or unequal diameter together.

SHEET BEND

b. **Class II** *Anchor Knots*

(1) Bowline. Used to tie a fixed loop in the end of a rope. This knot is always tied with the pigtail on the inside and secured with an overhand knot.

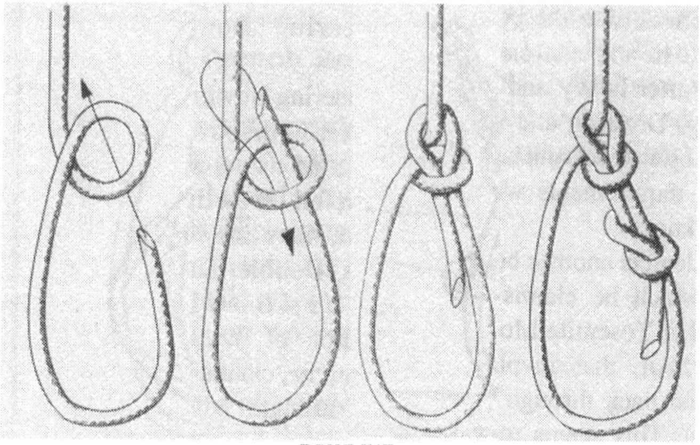

BOWLINE

(2) Round Turn with Two Half Hitches or a Bowline. A loop which runs around an object in such a manner as to provide 360 degree contact and may be used to distribute the load over a small diameter anchor. It will be secured with two half hitches or a bowline.

ROUND TURN AND TWO HALF HITCHES AND

ROUND TURN AND A BOWLINE

(3) Clove Hitch. This knot is an adjustable hitch. It could be considered a middle-of-the-rope anchor knot at the end-of-the-rope when used in conjunction with a bowline or round turn and two half hitches.

(1) Around-the-object Clove Hitch.

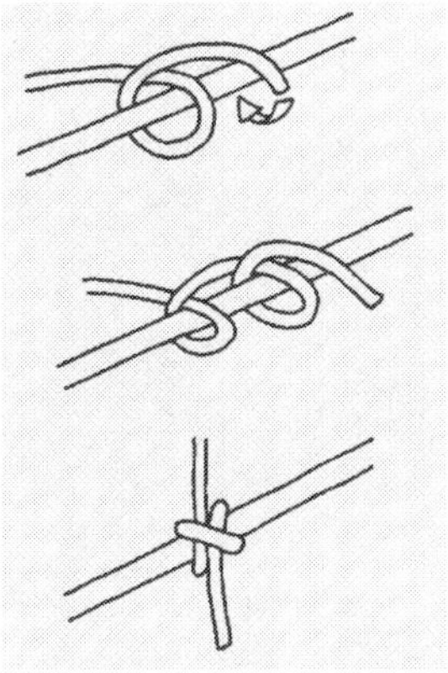

AROUND THE OBJECT CLOVE HITCH

(2) Over-the-object Clove Hitch.

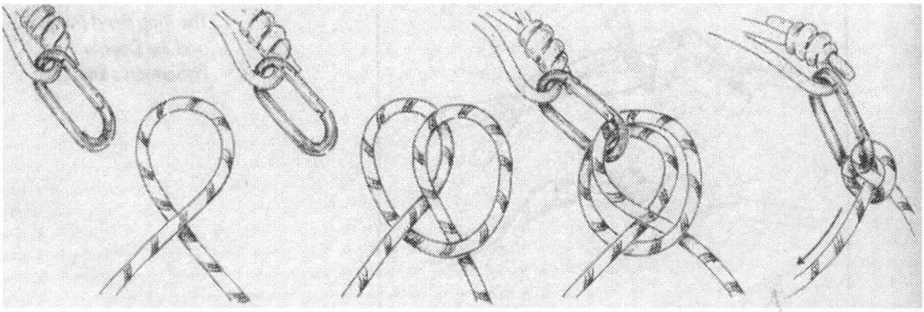

OVER THE OBJECT CLOVE HITCH

c. **Class III -** *Middle-of-the-rope*

(1) Figure-of-Eight-Loop. This is a strong knot that can be readily untied after being under load.

FIGURE 8 LOOP

(2) Double Figure of Eight Loop. The double Figure of Eight Loop is a strong knot and the double loop reduces the wear and strength loss from the rope bending around the carabiner by splitting the load between the two loops.

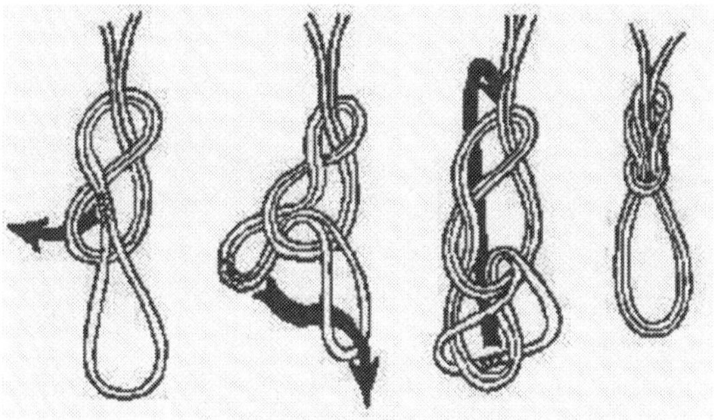

DOUBLE FIGURE 8 LOOP

(3) Two Loop Bowline. This knot can be used to construct a self-equalizing belay or to tie the middle person in on three people on a rope.

TWO LOOP BOWLINE

d. **Class IV** - *Special Knots*

(1) Prusik Knot. This knot functions by introducing friction that can be alternately set and released. For best results, tie the knot with a smaller diameter cord on a larger diameter cord. If slippage occurs, more wraps maybe used.

(a) Middle-of-the-Rope Prusik. This knot is created with an endless loop also known as a Prusik Cord. Do not tie this knot with tape due to less friction.

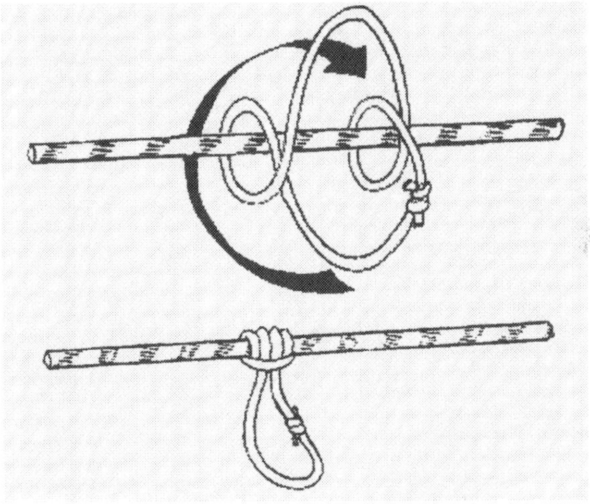

MIDDLE OF THE LINE PRUSIK

(b) End-of-the-Rope Prusik. This knot is always secured with a bowline. Do not tie this knot with tape due to less friction.

END OF THE LINE PRUSIK

(c) French Prusik. This can be constructed with a Prusik Cord or a single strand of cord with figure-of-eight loops on each end.

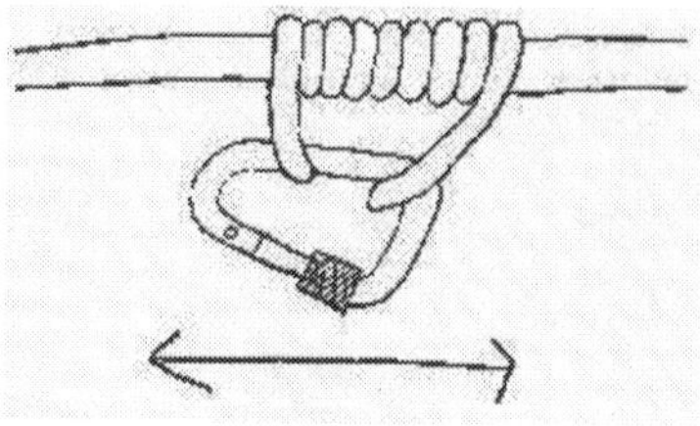

FRENCH PRUSIK

(2) Retraced Figure of Eight. Used to tie the end of the climbing rope into a harness or swammi wrap. The pigtail may or may not be secured with an overhand.

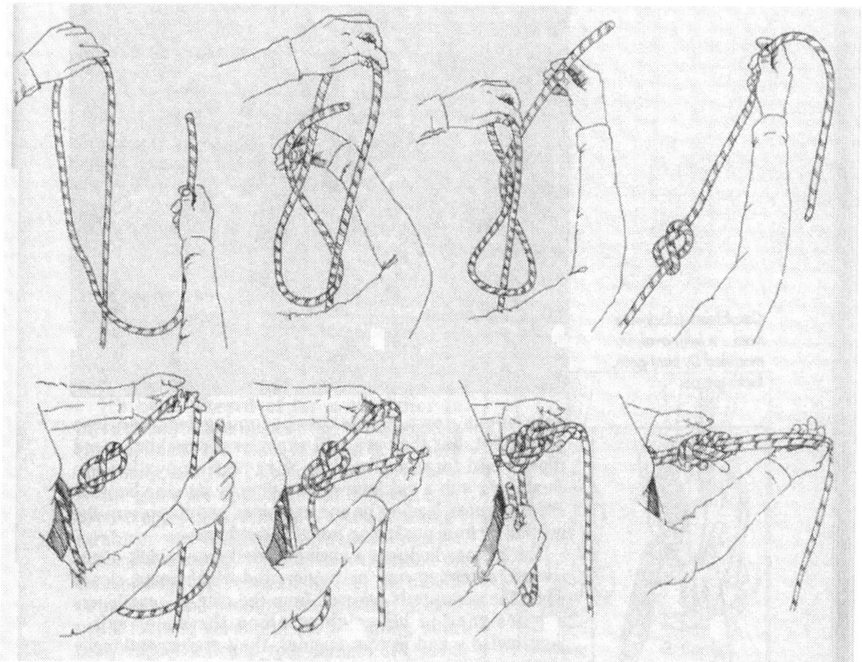

RETRACED FIGURE 8 IN HARNESS

(3) Directional Figure-Of-Eight. When tied and tensioned is applied to the standing and running ends of the rope, the knot will not pull apart. The loop will point toward the direction of pull.

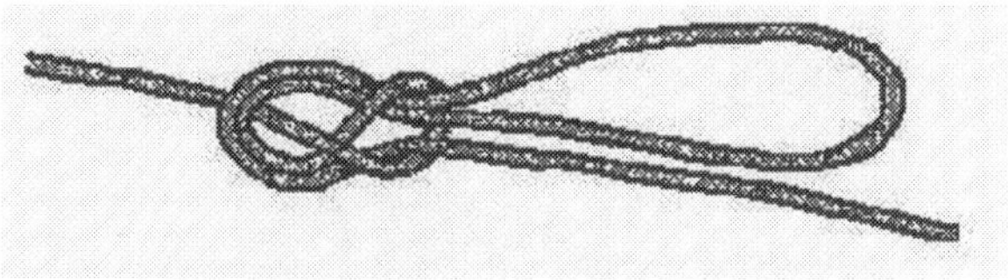

DIRECTIONAL FIGURE 8

(4) Slip Figure 8. This knot is used for retrievable anchors and fixed rope installations for the ease of untying the knot.

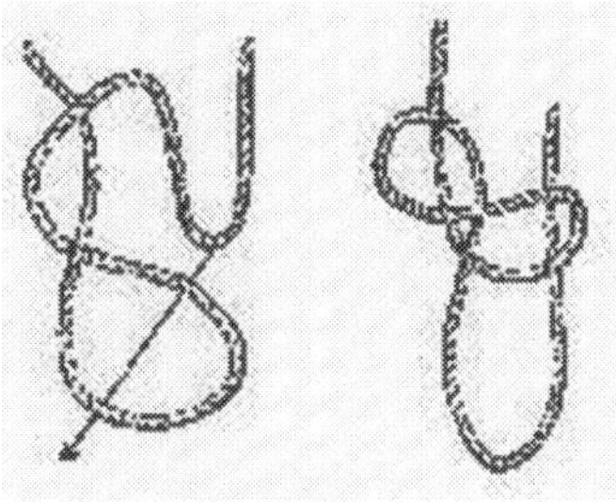

SLIP FIGURE 8

(5) Kliemheist. This is a friction knot.

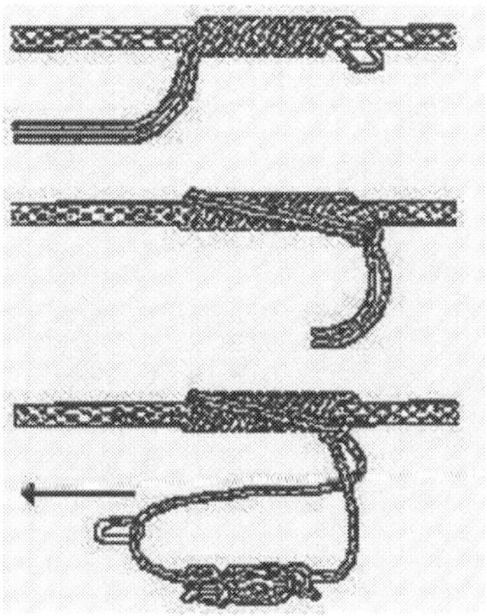

KLEIMHEIST

(6) Overhand Knot. Can be used to secure primary knots on itself.

OVERHAND

(7) M nter Hitch. This is a simple hitch in the rope that is clipped into a carabiner to put friction on the line.

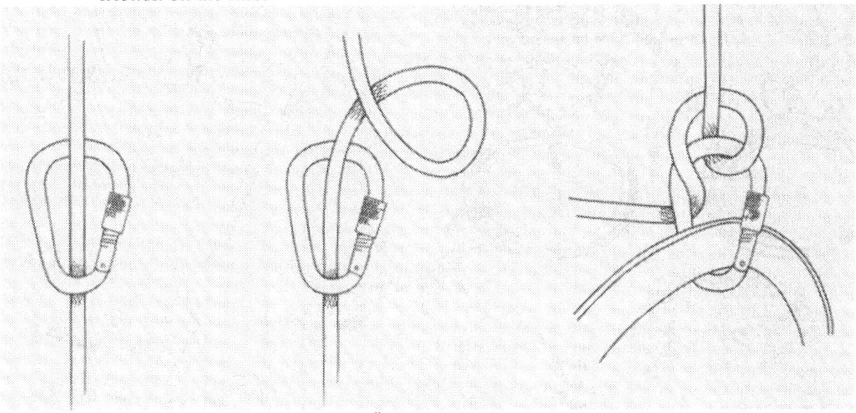

MÜNTER HITCH

(8) Timber Hitch. A Timber Hitch is used to fix a rope to a pole or equivalent for hoisting or towing purposes. It has the capability of casting off easily.

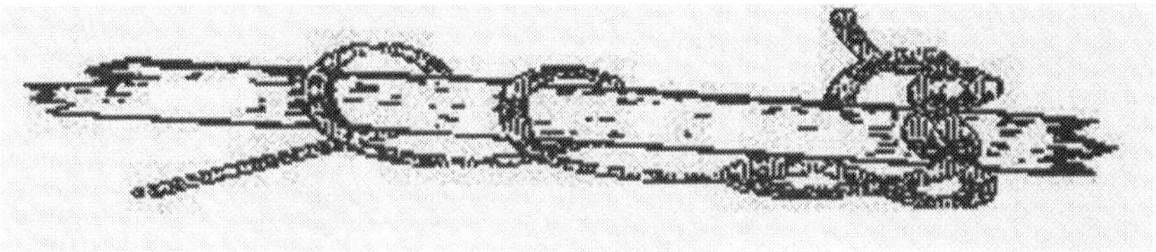

TIMBER HITCH

(9) Mariner's Hitch. This knot's advantage over the prusik is that it can be released under load.

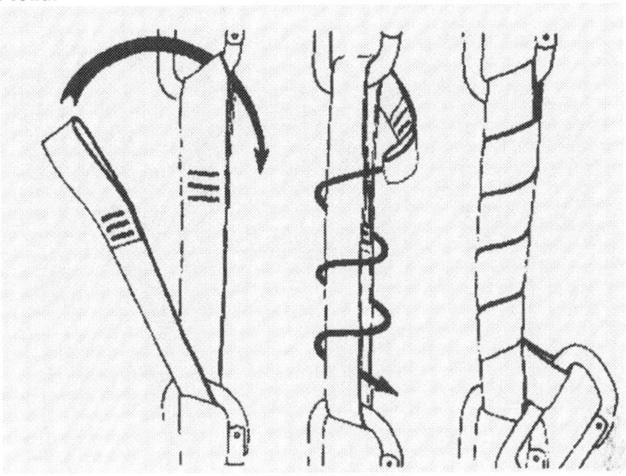

MARINER'S HITCH

(10) Krägur Knot. This friction knot is used to prevent unnecessary slippage when tied onto a rope of the same diameter.

KRÄGUR KNOT

(11) Rappel Seat. This is used as an expedient support harness for rappelling, crossing rope bridges, etc. It is constructed as follows:

(a) Center the sling rope on the left hip.

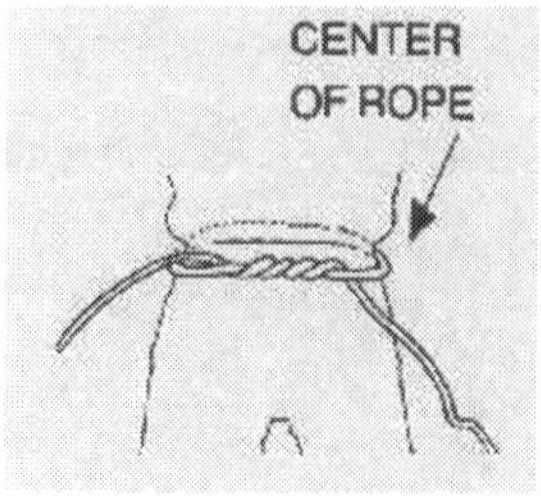

(b) Wrap the sling rope around the waist and tie at least one half hitch around itself (preferably two) in the front.

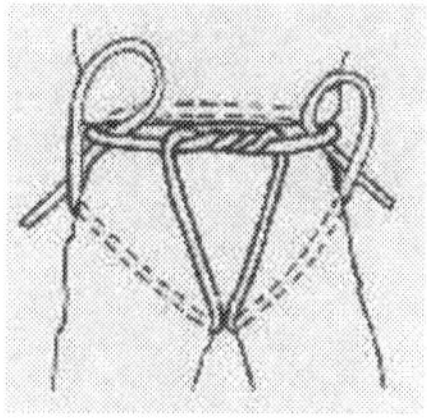

(c) Bring the running ends down through the legs, up over the buttocks, and over the original waist wrap, down between the waist wrap and the waist, and over itself, forming a bight. Cinch this up tightly.

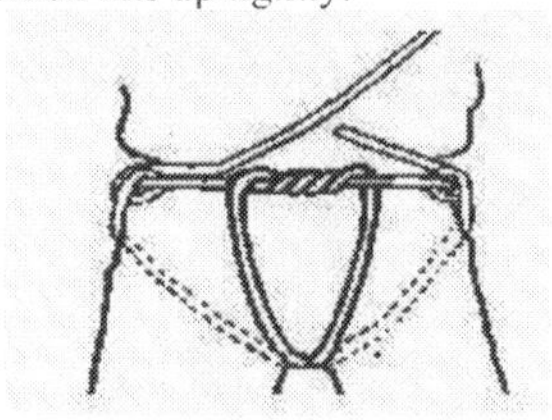

(d) Now tie a square knot with two overhand knots on the left hip ensuring that the overhand knots encompass all the wraps.

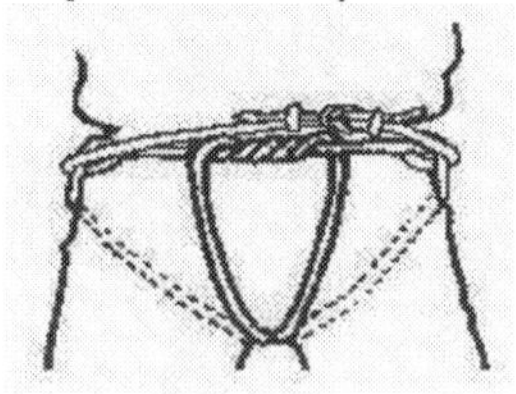

(e) Tuck any excess rope into a pocket.

(12) Bowline on a coil. Used by the first and last men on a climbing rope to tie into the rope. An overhand knot is used behind the knot. It distributes the force of a fall over a larger area of a climber's waist and is preferable to a single bowline around the waist. The bowline on a coil can also be used to take up excess rope. The bowline on a coil should have 4-6 wraps around the waist.

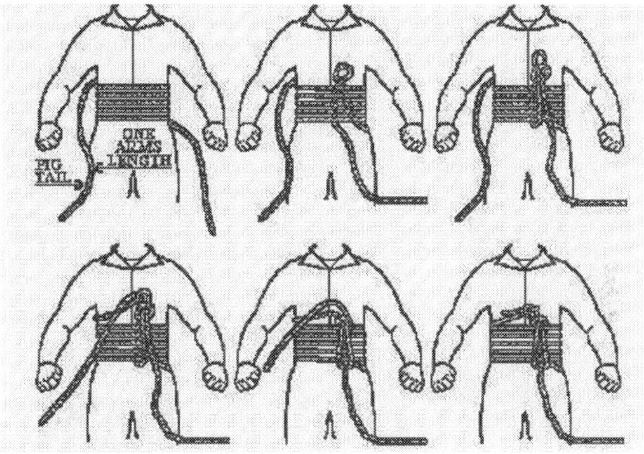

BOWLINE ON A COIL

(13) Swami wrap (Swami belt). The swami wrap (four to six wraps of rope or nylon webbing tied around the waist) popular with rock climbers who wanted to increase the working length of their climbing ropes. It also offers some improvement over the bowline on the coil in distributing the forces sustained in a fall over a larger area of the midsection of the body, but does not eliminate the problem of suffocation when hanging. The Swami wrap is constructed as follows:

(a) Take a sling rope and find the middle.

(b) Place the middle of the sling rope in the small of your back.

(c) Bring both pigtails to the front and continue to wrap around your body until both pigtails are about 15-20 inches long.

(d) Tie the pigtails on the left side with a square knot and two overhand knots.

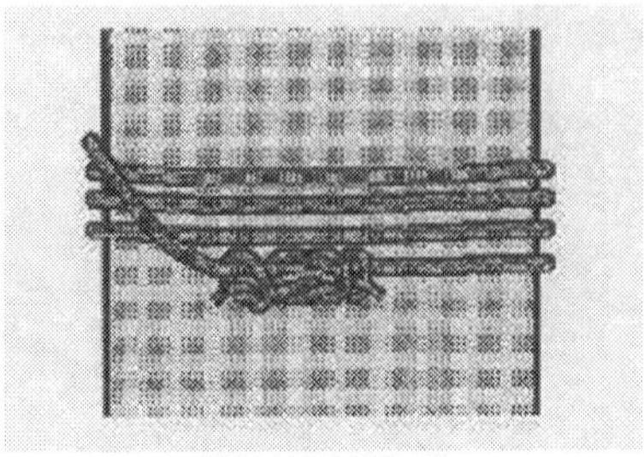

SWAMMI WRAP

(14) Three Loop Bowline. This knot will provide three bights, each of which can be adjusted against the others.

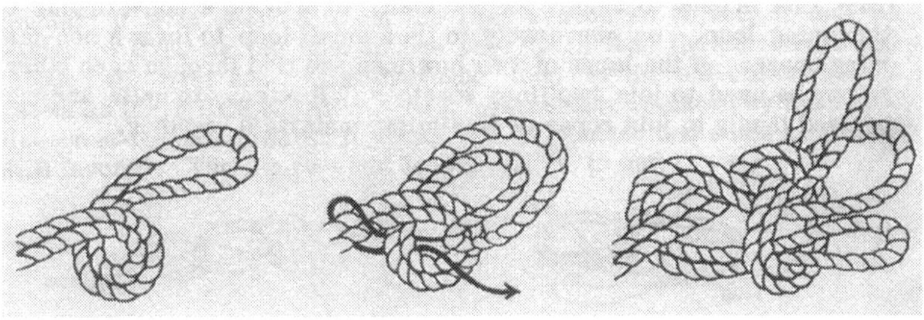

THREE LOOP BOWLINE

7 **KNOT STRENGTH.** Rope, cordage and webbing is strongest when loaded in a straight line. When bending a rope or web to create a knot, the strength of the rope is reduced. All knots should be dressed properly for maximum effective use.

KNOT	RELATIVE STRENGTH OF KNOT
No Knot	100%
Figure 8	75-80%
Bowline	70-75%
Double Bowline	70-75%

Double Fisherman	65-70%
Water Tape Knot	60-70%
Clove Hitch	60-65%
Overhand Knot	60-65%
Square Knot	45%

1. **KNOT TESTING TIME LIMIT.** The following times must be met to pass the knot tying portion of this course.

KNOT	UNIT OPERATIONS TIME LIMIT	SML, ACC TIME LIMIT (BLINDFOLDED)
Square knot	60 Seconds	30 Seconds
Double Fisherman's Knot	60 Seconds	30 Seconds
Water/Tape Knot	60 Seconds	30 Seconds
Round Turn and a Bowline	60 Seconds	30 Seconds
Round Turn and Two Half Hitches	60 Seconds	30 Seconds
Clove Hitch (around the object)	60 Seconds	30 Seconds
Munter Hitch	60 Seconds	30 Seconds
Slip Figure 8	60 Seconds	30 Seconds
Figure 8 Loop	60 Seconds	30 Seconds
Directional Figure 8	60 Seconds	30 Seconds
End of the Line Prusik	90 Seconds	45 Seconds
Retrace Figure 8	90 Seconds	45 Seconds
Rappel (Swiss) Seat	2 Minutes	60 Seconds

UNITED STATES MARINE CORPS
Mountain Warfare Training Center
Bridgeport, California 93517-5001

SMO
05/02/01

STUDENT OUTLINE

NOMENCLATURE AND CARE OF MOUNTAINEERING EQUIPMENT

LESSON PURPOSE. The purpose of this period of instruction is to introduce you to the types of equipment used here at MWTC and how to care for it so as to prevent its untimely failure. This lesson relates to all climbing and installation work that you perform here at MWTC.

OUTLINE

1. **ROPES.** All ropes used in the military must meet UIAA standards or U. S. Federal Test Standard 191A. Most ropes have a 5-year shelf life and maximum 2-year service life.

a. Static. Black in color.

(1) Construction. Kernmantle

(2) Minimum tensile strength. 7500 lbs.

(3) Maximum elongation. 1.5%

(4) Diameter. 11mm

(5) Sizes.

(a) 165 ft ± 5 ft

(b) 300 ft ± 10 ft

(6) Usage. Rescue operations and bridging where a low amount of elongation is desirable under a working load.

b. Dynamic. Olive Drab in color.

(1) Construction. Water-resistant treated Kernmantle to reduce friction.

(2) Minimum tensile strength. 6500 lbs.

(3) Maximum elongation. 6%

(4) Diameter. 10.5mm and 11mm

(5) Size. 165 ft $\pm$ 5 ft.

(6) Usage. For lead climbing/party climbing.

c. Maxim Dry Rope. Olive Drab or Multi-Colored.

(1) Construction. Water-repellent treated Kernmantle.

(2) Minimum tensile strength. 3472 lbs.

(3) Maximum elongation. 6%

(4) Diameter. 9mm

(5) Size. 150 ft $\pm$ 5 ft

(6) Usage. For glacier travel/ice climbing

d. Gold Line II.

(1) Construction. Eight strand braided nylon plymor.

(2) Minimum tensile strength. 4500 lbs.

(3) Maximum elongation. 20%

(4) Diameter. 11 mm

(5) Size. 300 ft or 600 ft spools

(6) Usage. Sling Ropes and litters only.

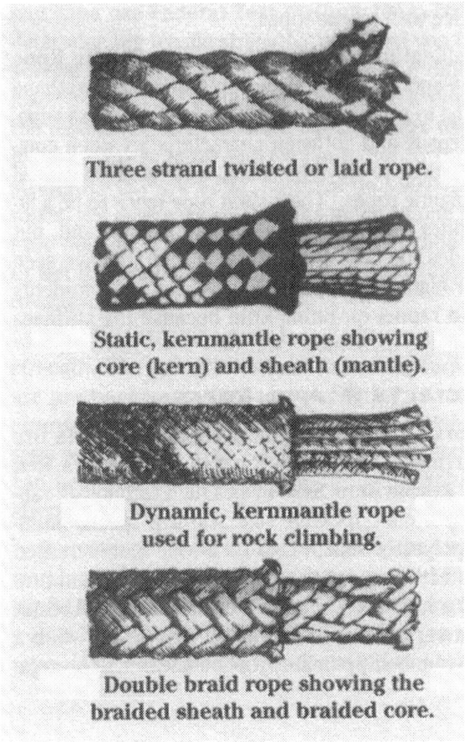

INSIDES AND OUTSIDE SHEATHS OF VARIOUS TYPES OF ROPES

NOTE: Sling ropes are made from 15 foot lengths of plymor or dynamic rope ONLY. Twenty-five foot practice coils should be constructed with static rope, but dynamic rope can be used.

2. ADVANTAGES/DISADVANTAGES

a. Advantages of Nylon Rope

(1) High strength to weight ratio.

(2) Good energy absorption in dynamic ropes.

(3) Flexible.

(4) Rot resistant, not affected by frost.

b. Disadvantages of Nylon Rope

(1) Low melting point. Nylon fuses at 400°F and melts at 480°F.

(2) Susceptible to abrasions and cuts.

(3) Affected by chemicals and light.

c. <u>Advantages of Manila Rope</u>

(1) Easily gripped.

(2) Hard wearing.

(3) Does not deteriorate in heat.

d. <u>Disadvantages of Manila Rode</u>

(1) Heavy, kinks, especially when wet. Absorbs water and swells.

(2) Burns at +300°F.

(3) Edible by rodents.

3. GENERAL INFORMATION

a. Nylon rope stretches under tension and will rupture at between 30% and 70% elongation, depending on construction.

b. Nylon rope loses as much as 30% strength when wet.

c. Temperatures as low as 250°F will damage a nylon rope.

4. NYLON WEBBING

a. The type of nylon webbing available is tubular. Tubular nylon webbing is very strong and flexible. All rules that apply to nylon rope apply to tubular nylon webbing. The size of nylon webbing used is:

(1) 1 inch tubular nylon. Tensile strength approximately 4,000 - 4,500 lbs., depending on the manufacturer.

b. Pre-sewn Spectra Runners. Tensile strength approximately 5,500 lbs.

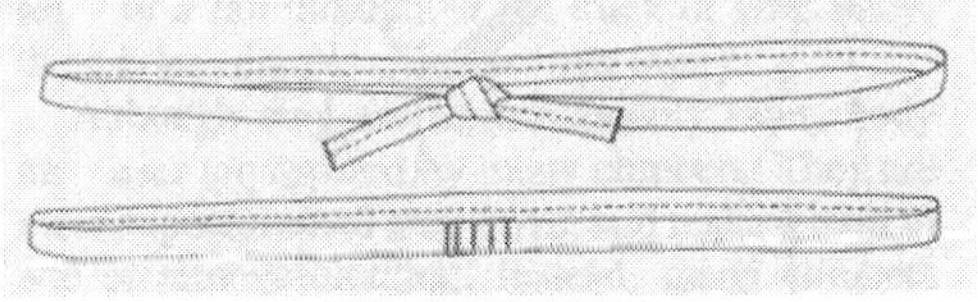

NYLON WEBBING

NOTE: These are minimum strengths. Some manufactures make even stronger webbing.

1. **CARABINERS**. Also commonly known as snap links. There are two different types of carabiners frequently used at MWTC, the steel locking carbine and the non-locking aluminum carabiners.

a. Steel Locking Carabiner. There are two different sizes of carabiners:

(1) Steel Locking Stubai 82.

(2) Large Locking "D" Stubai 85.

b. Aluminum Non-Locking Carabiner.

(1) The standard "D" non-locking carabiner with a tensile strength of 2000 lbs. or more.

2. **NOMENCLATURE OF A CARABINER**.

a. There are four parts to a non-locking carabiner. They are as follows:

(1) Gate.

(2) Gate pivot pin.

(3) Locking pin

(4) Body.

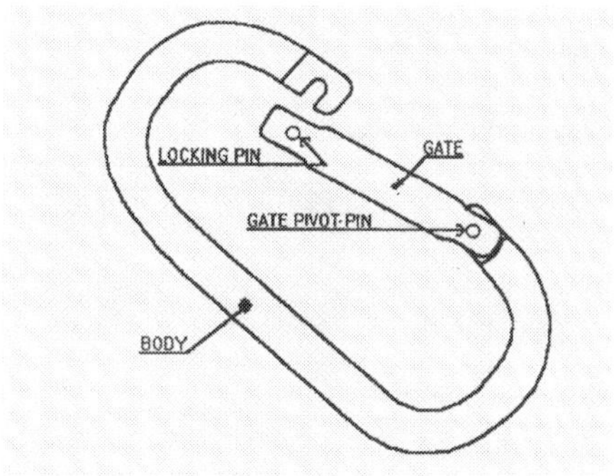

STEEL NON-LOCKING CARABINER

b. There are five parts to a locking carabiner. They are as follows:

(1) Gate

(2) Gate pivot pin

(3) Locking notch

(4) Locking nut

(5) Body

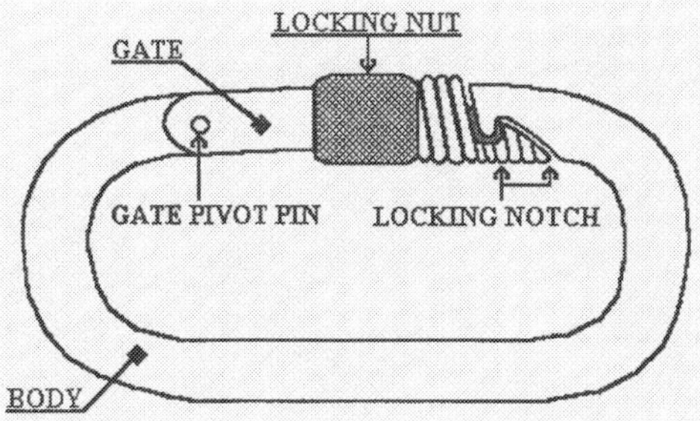

STEEL LOCKING CARABINER

3. **CARE, MAINTENANCE AND SERVICEABILITY OF A CARABINER**. The following inspections and maintenance should be performed on a routine basis:

a. Care.

(1) Avoid dropping of a carabiner onto a hard surface, hairline factures could occur.

(2) Keep the carabiners off of the bare ground because dirt and grit can get into the working parts.

(3) If the carabiner has fallen from a significant height, it should be retired.

b. Maintenance.

(1) Remove all dirt, moisture and grime after use.

(2) Lubricate with tri-flow graphite and wipe off with a clean cloth.

c. Serviceability Checks.

(1) The gate snaps shut with no friction and with no gap between the locking pin and locking notch.

(2) There is no excessive side to side movement of the gate

(3) The pivot pin is tight.

(4) That the spring-loaded gate functions properly.

(5) The locking pin is tight.

(6) The locking nut travels freely and locks securely.

(7) There are no cracks or flaws in the metal.

NOTE: The weakest part of a carabiner is the gate. If an engraver is used to mark a carabiner, it should be applied to the gate and not the load bearing side.

UNITED STATES MARINE CORPS
Mountain Warfare Training Center
Bridgeport, California 93517-5001

WMC
04/11/02

LESSON PLAN

BALANCE CLIMBING

1. LEARNING OBJECTIVES

a. TERMINAL LEARNING OBJECTIVE. In a summer mountainous environment, execute balance climbing, in accordance with the references. (WMC)

b. ENABLING LEARNING OBJECTIVES.

(1) With the aid of references, state orally the safety requirements for balance climbing, in accordance with the references. (WMC.C3a)

(2) With the aid of references, state orally the individual preparations for a balance climb, in accordance with the references. (WMC.C3b)

(3) In a summer mountainous environment, execute the duties of a spotter for a balance climb, in accordance with the references. (WMC.C3c)

(4) In a summer mountainous environment, execute the commands used between the climber and the spotter during a balance climb, in accordance with the references. (WMC.C3d)

(5) In a summer mountainous environment, execute each type of hold used in balance climbing, in accordance with the references. (WMC.C3e)

(6) Without the aid of references and given the acronym "CASHWORTH", describe orally the nine considerations for proper body position while climbing, in accordance with the references. (WMC.C3f)

OUTLINE

1. **SAFETY PRECAUTIONS**. There are two safety precautions that always apply to balance climbing. They are as follows:

C3

a. Never climb more than 10 feet above the ground. By this it is meant that the climber's feet are never more than 10 feet above the ground.

b. A spotter is required for all balance climbs.

SPOTTER WATCHING THE CLIMBER

2. **INDIVIDUAL PREPARATIONS**. Prior to beginning a balance climb there are seven things that the climber must do to prepare himself. They are as follows:

a. Helmet with a serviceable chinstrap must be worn.

b. Sleeves rolled down_to give hand and arm freedom of movement. Blouse tucked in to your trousers. In case of fall, it may catch on a rock and cause you to flip over sideways.

c. All watches, rings, and jewelry must be removed before climbing.

d. Gloves will not be worn, as they can slip, and also give a false feel for the rock.

e. Unblouse trousers, if they restrict movement.

f. Soles of boots clean and dry.

g. Select route where vegetation is minimal. Never use vegetation for hand or foot holds.

3. **DUTIES OF THE SPOTTER**. The spotter is the balance climber's partner who, rather than climbing himself, acts as the safety man for the climber during the climb. The five duties of the spotter are as follows:

a. Positions himself directly behind the climber before the climb starts.

b. Maintains his position facing the cliff, directly below the climber and approximately 3-4 feet away from the base of the cliff, for the duration of the climb. He will move diagonally as necessary to remain below the climber.

c. The spotter will stand with his feet shoulder width apart and arms ready to stop the climber if he falls.

d. If the climber falls, the spotter will not "catch" him; he will prevent the climber from falling further down the hill. He will do this by pushing the climber towards the base of the cliff, thereby preventing him from tumbling backwards.

e. At no time will the spotter allow anyone to come between himself and the face of the cliff while a balance climb is taking place. He will require anyone who wants to pass by his position to go behind him.

4. **SPOTTING AND CLIMBING COMMANDS**. The following are the commands used by both the spotter and climber.

COMMAND	GIVEN BY	MEANING
"Last Name, climbing."	Climber	I am ready to climb.
"Climb Climber."	Spotter	I am ready to spot you.
"Last Name, off Climb."	Climber	I am off the climb.
"ROCK!"	Climber or Spotter.	A rock has been knocked off the rock face and is falling.
"FALLING!"	Climber	I am Falling.

a. If the command "ROCK" is given, all personnel in the vicinity will take the following action:

(1) If close to the cliff face, move against the cliff face with your face against the cliff face and your hands between you and the cliff face.

(2) If not close to the cliff face, look up to locate the rock and avoid it.

5. **ACTIONS IF FALLING**. If, while making a balance climb, the climber feels himself slipping and beginning to fall, he will take the following action:

a. Sound the command "falling".

b. Push himself away from the rock face.

c. Maintain proper body position as follows:

(1) Head up.

(2) Hands out toward the rock.

(3) Body relaxed.

(4) Feet kept below the body, slightly apart.

(5) Ensure you face the cliff face as you fall.

6. **TYPES OF HOLDS**. There are five basic holds that are used in balance climbing. They are as follows:

a. Push Holds.

(1) Most effective when hands are kept low.

(2) Often used in combination with a pull hold.

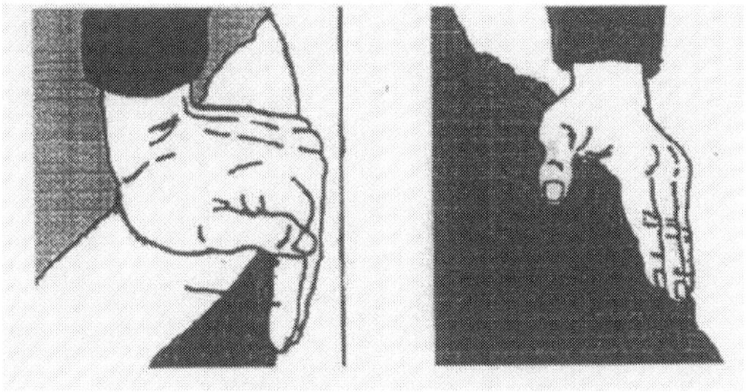

TYPES OF PUSH HOLDS

b. Pull Holds.

(1) The easiest hold to use and, consequently, often overused.

(2) Can be effective on small projections.

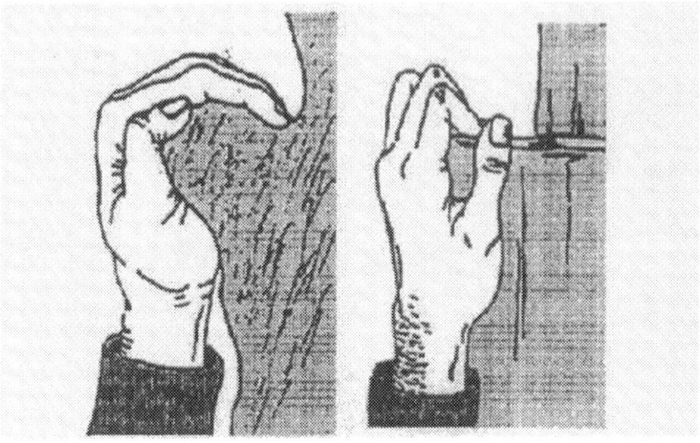

TYPES OF PULL HOLDS

c. Foot Holds

(1) Feet should be positioned with the inside of the foot to the rock.

(2) Use full sole contact as much as possible.

(3) Avoid crossing your feet. If you must cross your feet use a change step. A change step is a method of substituting one foot for the other foot on the same foothold.

(4) Making maximum use of footholds, climbing with your feet, is an effective means of conserving your body strength, since your leg muscles are stronger than your arm muscles.

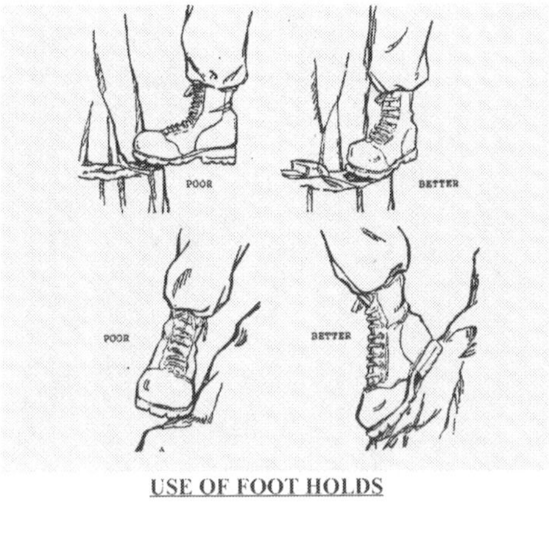

USE OF FOOT HOLDS

d. Friction Holds. A friction hold is anytime you are relying on the friction of your foot or hand against the face of the rock for traction, rather than pushing/pulling against a projection on the face of the rock.

(1) It is a type of hold that feels very insecure to an inexperienced climber.

(2) The effectiveness of this type of hold is dependent upon many things i.e., type, condition and angle of the rock face, type of boot soles, confidence, etc.

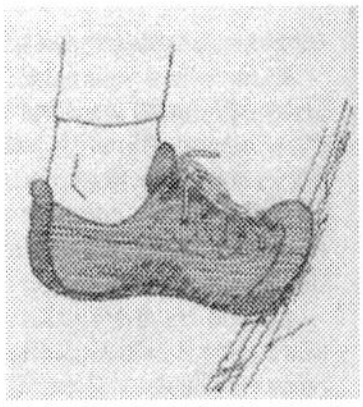

FRICTION USED IN SMEARING

e. Jam Holds. This type of hold involves jamming/wedging any part of your body or your entire body into a crack/opening in the rock.

(1) An important consideration is that you do not jam such that you cannot free that portion of your body after you complete the move. This sounds like a ridiculous statement; however, you must remember that after you complete your move, you may be withdrawing the portion of your body that you used from a different angle than you inserted it.

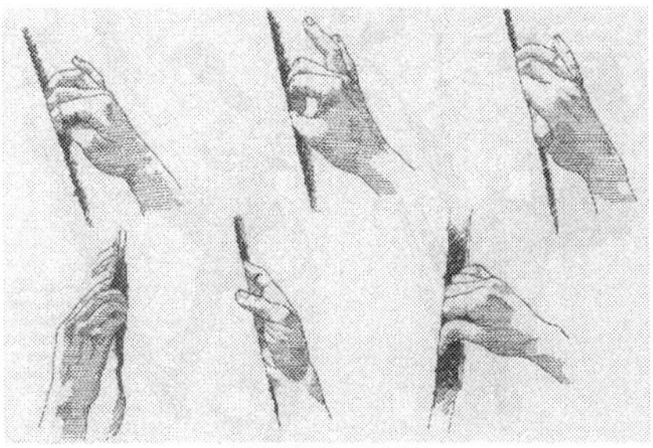

FINGER JAMS

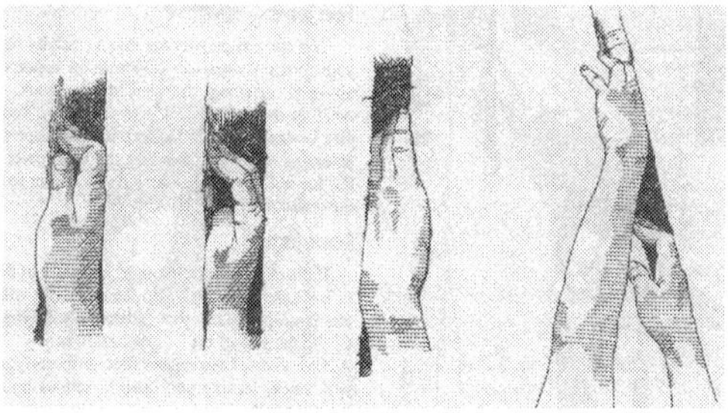

HAND JAMS

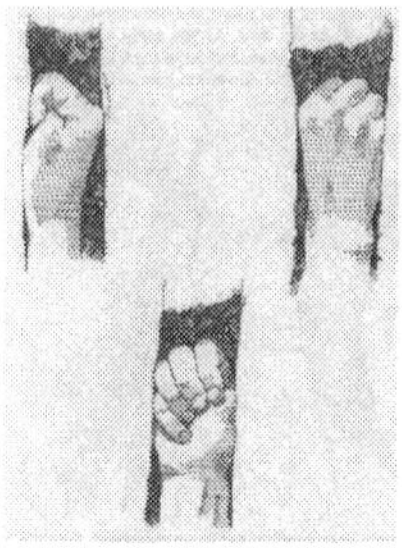

FIST JAMS

7.) **COMBINATION HOLDS.** The five types of holds just mentioned above are not just used individually. They are most often used in combinations with each other. Some examples are:

a. Chimney Climbing. This is when you insert your entire body into a crack in the rock and by using both sides of the opening, and possibly all five types of basic holds, move up' the crack.

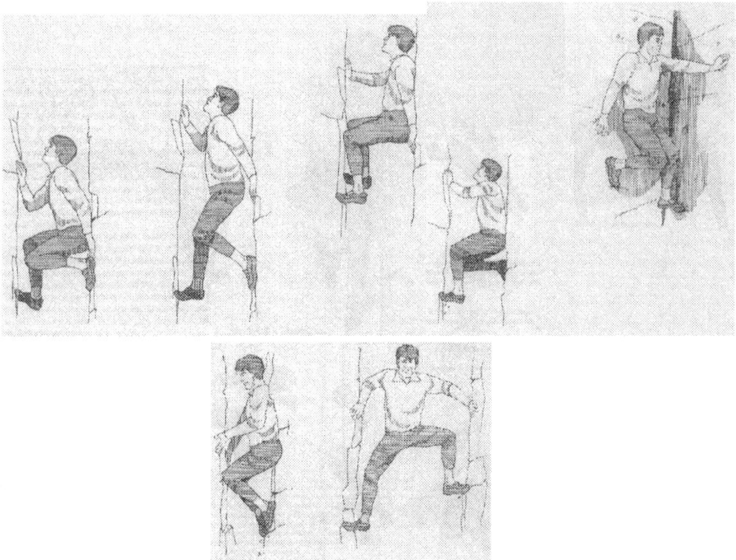

CHIMNEY CLIMBING TECHNIQUES

b. Lie-back. This is a combination of both pull holds with your hands and friction holds with your feet.

LIE BACK TECHNIQUE

c. Push-Pull. As the name implies, this is when you use a push hold and a pull hold together.

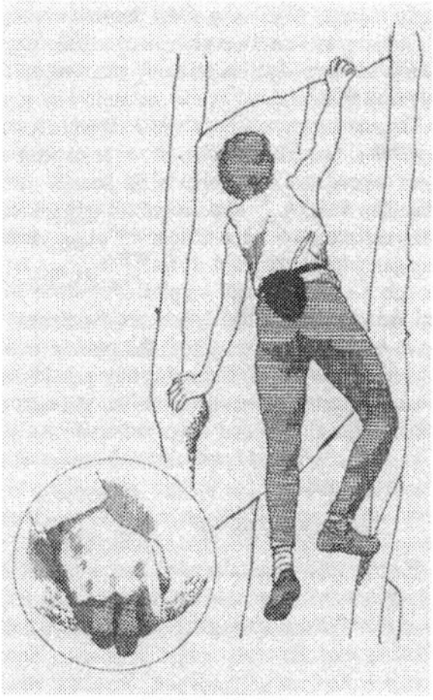

PUSH-PULL

d. Mantling. This is a technique where you continue to climb without moving your hands off a projection by pulling yourself up until your hands are at chest level and then invert your hands and push on the same projection.

MANTELING

MANTELING WITH FACE HOLD

e. Cross-Pressure in Cracks. This is a technique of putting both hands in the same crack and pulling your hands apart to hold/raise yourself.

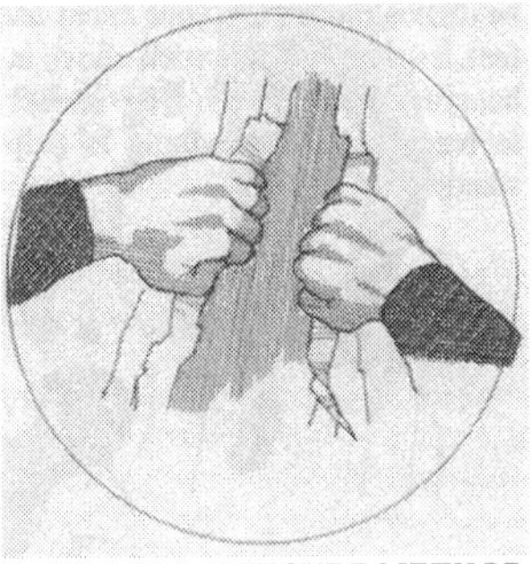

CROSS-PRESSURE METHOD

f. Inverted. Pull or push.

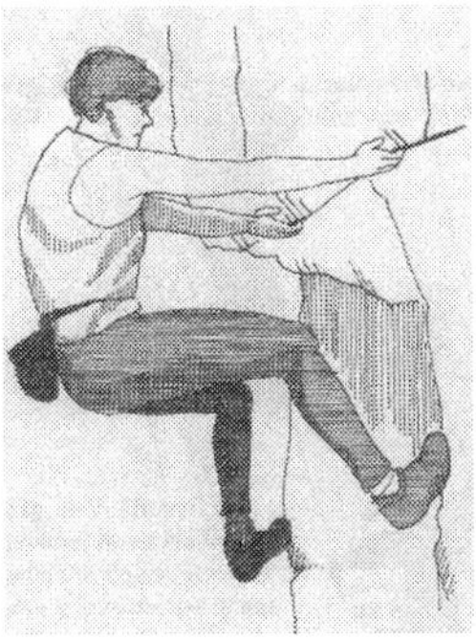

INVERTED TECHNIQUE USING COUNTERFORCE PRESSURE BETWEEN FEET AND HANDS

g. Pinch. As the name implies this is a grip used on tiny little nubbins.

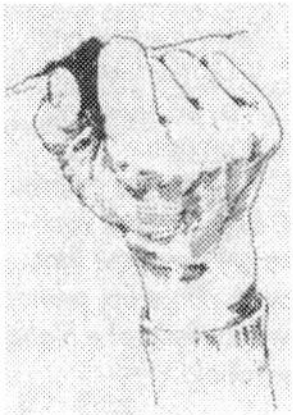

PINCH

h. Stemming. The spreading of arms or legs to maintain a proper body position. (i.e. usually used in a book or chimney.)

STEMMING

8. **GENERAL USE OF HOLDS**. How you use an individual hold is dependent on your experience level, or sometimes, your imagination. Here are some general guidelines.

a. Most handholds can be used as foot holds as you move up the rock.

b. Use all holds possible in order to conserve energy.

c. Even small projections may be used as holds.

d. Do not make use of your knees or elbows due to the reason that it is skin on bone and a slip could occur if pressure is exerted on them. Knees and elbows can be used with the extension of a limb jam.

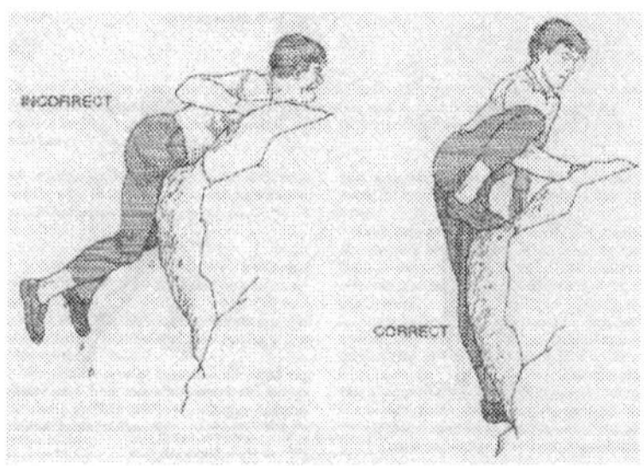

PROPER AND IMPROPER

9. **MOVEMENT ON SLAB**. Movement on slab is based on friction holds.

a. Use any and all irregularities in the slope to gain additional friction.

b. Traversing requires both hands and feet.

TRAVERSING

c. Descending steep slab may require turning inboard to face the slab and backing down.

d. The biggest mistake in slab climbing is leaning into the rock. Maintain maximum friction by keeping weight centered.

SLAB CLIMBING

10. BODY POSITION

a. The climber should climb with his body in balance by keeping his weight centered over and between his feet. Don't hug the rock. Don't over extend and become "spread-eagled". While climbing, keep in mind the acronym "CASHWORTH" for proper body position and movement. (WMC.C3F)

C - Conserve energy.

A - Always test holds.

S - Stand upright, on flexed joints.

H - Hands kept low; handholds should be waist to shoulder high.

W - Watch your feet.

O - On three points of contact.

R - Rhythmic movement.

T - Think ahead.

H - Heels kept low, lower than the toes.

UNITED STATES MARINE CORPS
Mountain Warfare Training Center
Bridgeport, California 93517-5001

WMC.C4
04/11/02

STUDENT OUTLINE

NATURAL AND ARTIFICIAL ANCHORS

TERMINAL LEARNING OBJECTIVE In a summer mountainous environment, establish an anchor system in accordance with the references. (WMC.C4)

ENABLING LEARNING OBJECTIVES.

(1) Without the aid of references and from a given list, choose the two types of anchors, in accordance with the references. (WMC.C4a)

(2) In a summer mountainous environment, establish a natural anchor point, in accordance with the references. (WMC.C4b)

OUTLINE

1. **TYPES OF ANCHORS.** (WMC.C4a)There are two types of anchors that we use. These two types are:

1) NATURAL
2) ARTIFICIAL

C4

2. **GENERAL CONSIDERATIONS FOR ANCHORS**. Anytime you employ natural or artificial anchors; there are special considerations that you must apply. These considerations apply both to the anchor itself and to the material being used to build the anchor. Some examples of these considerations are:

a. Whether using natural or artificial anchors, the installing unit must insure that the anchor is suitable for the load.

b. The anchor position must be relative to the direction of pull on the anchor.

c. The angle between the anchor points should not exceed 90 degrees. This is to ensure that no added stress is put upon the anchors, as well as the equipment being used to construct the anchor.

(1) To decrease the angle between anchor points, materials (i.e. sling ropes, web runners, Prusik cord, etc..) could be used to extend the anchor which will decrease the angle between anchor points.

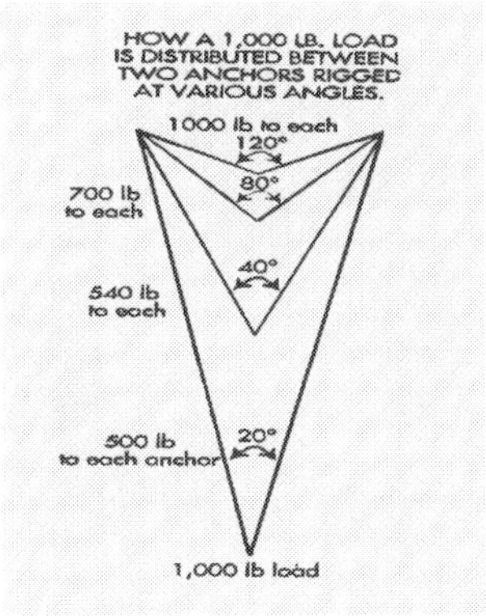

3. **NATURAL ANCHORS.** A natural feature is the preferred type of anchor point. Some examples and considerations are as followed:

a. Types of natural anchors.

(1) Trees.

(a) Select a tree that has not been chopped, burned or is rotten.

(b) The tree should be at least 6" in diameter and strong enough to support the intended load.

(c) Trees growing on rocky terrain should be treated with suspicion, since the roots normally are shallow and spread out along a relatively flat surface.

(2) Shrubs and Bushes.

(a) Select a shrub or bush that is alive and is not brittle, charred or loose.

(b) To avoid leverage, locate the central root and construct the anchor as near to the base as possible.

(3) Rocks and Boulders.

(a) Stability is of prime importance when considering a rock or boulder for an anchor. It must be strong enough and secure enough for the intended load.

(b) All surfaces of the rock or boulder should be inspected for any rough or sharp points. These areas must be padded to protect the rope from being abraded or cut.

(4) Spikes and Flakes.

(a) To check stability of a spike or flake, thump it with the heel of your hand. Anything that sounds hallow is suspicious.

(b) They should be checked for cracks or other signs of weathering that may impair their firmness.

(c) Sharp edges must be padded to protect the rope against cuts and abrasion.

(5) Threads and Chockstones.

(a) A thread is when the rock weathers or cracks to form a hole in the main wall. A chockstone is a rock wedged in a crack.

(b) Check a thread by thumping it with the heel of your hand. Make a visual inspection for cracks and weather. Common sense will prevail for choosing this type of anchor.

(c) When choosing a chockstone, ensure that it has substantial contact with the crack and that the stone's symmetry corresponds with the intended direction of pull.

b. Types of Natural High Tension Anchors. The following are types of Natural High Tension Anchor Systems used to construct installations. These can be tied on any suitable natural anchor point:

(1) Tree Wrap Anchor. The tree wrap is an anchor used to relieve tension on the actual knot itself. This system requires more rope when tied around a large anchor point.

(a) Tie a Figure 8 Loop on the standing end of the rope and wrap the 4-6 times around an anchor point.

(b) Attach a locking carabiner through the knot's loop and clip it onto the running end of the rope.

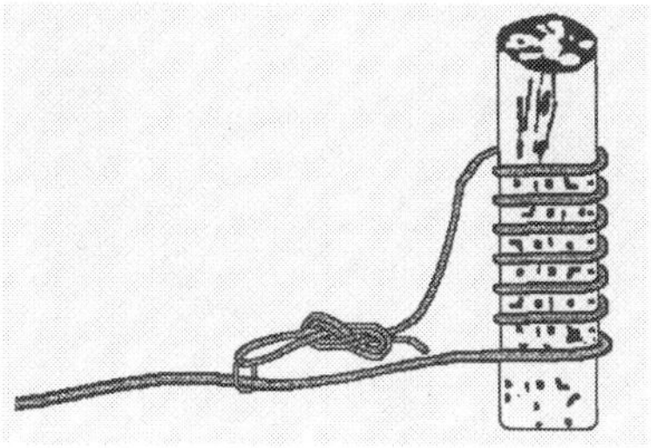

NOTE: If the anchor point to be wrapped is larger then 18' around, then three wraps will suffice.

(2) Swami Wrap Anchor. When tying the Swami Wrap, ensure that the joining knot is kept on the side of the anchor point. The Swami Wrap can be loosened under load if need be. This system can be substituted for a tree wrap when the installation's rope length is an issue.

(a) Select a suitable anchor point and tie a Swami Wrap around it.

(b) Clip a locking carabiner into as many wraps as possible. This will serve as the attachment point for the anchor system.

(3) Figure of 8 Anchor. The Figure of 8 Anchor is a quick and efficient system. This system cannot be loosened under load. It is tied in the following manner:

(a) Tie a Figure of 8 Loop on the standing end of the rope and wrap the knot around a suitable anchor point. Attach a locking carabiner through the knot's loop and clip it onto the running end of the rope.

(b) Before tensioning the anchor system, adjust the rope so the running end of the rope runs smoothly through the carabiner towards the direction of pull. This will prevent any lateral tension.

(b) If two ropes are used, the upper rope's carabiner has its gate upwards, and the lower rope's carabiner has it's gate downward.

c. Primary and Secondary Natural Anchor Point System. The two anchor points should be in line with the direction of pull. The primary anchor is the point nearest the running end while the secondary anchor point is directly behind the primary. It is constructed in the following manner:

(1) The primary anchor knot should be an around the object clove hitch. This is chosen for ease of untying the system after tension has been placed on the rope.

(2) The secondary knot is tied around a suitable anchor point ensuring that the rope is taut between the two anchor points.

4. **ARTIFICIAL ANCHOR SYSTEMS**. Any time we use anything other then a natural feature, we are using an artificial anchor point. Artificial anchors can be constructed in the ground, or on the rock itself. The following are artificial anchor systems:

a. Single Timber Deadman. This system is constructed in the ground and it requires considerable time and effort. The steps of it's construction are as follows:

(1) Dig a trench 6 feet long and 3 feet deep and wide enough to work in at a 90 degree angle to the direction of the pull.

(2) Dig another trench about 12 inches wide. This trench is dug so that it intersects the main trench at a right angle in the middle. The bottom of this trench should be parallel to the direction of pull and should join the bottom of the main trench.

(3) Take an anchoring device (i.e. log, engineer stakes, bundled up branches etc.) that is strong enough to support the intended load. The anchor is placed into the main trench and covered with dirt with the exception of that part of the anchor that joins the second trench. Stakes approximately 3 feet long should be driven approximately 1-1/2 feet into the ground between the dead-man and the slanted side of the trench to assist in holding the dead-man in place if the soil is soft.

b. Picket Hold Fast. The picket hold fast is an easier anchor to construct than the Deadman, and can be used almost anywhere. The strength of the picket system depends on the pickets and the soil or snow conditions. The picket system can be used for both high and non-tension systems. Construction is as follows:

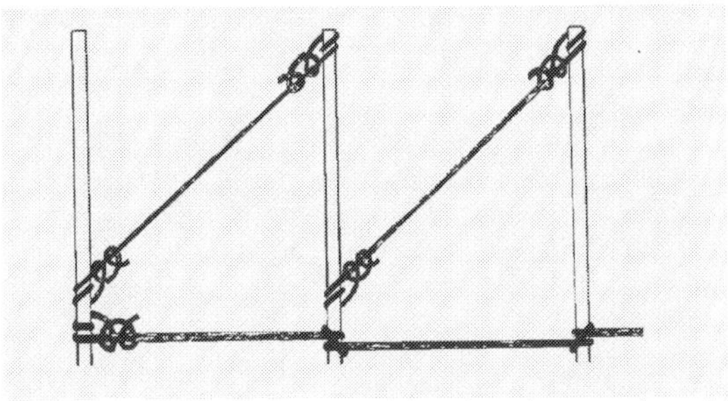

(1) Three stakes (i.e. logs, engineer stakes, etc.) are driven into the ground at a 30 degree angle against the direction of pull.

(2) The line of pickets should be driven in line with the direction of pull. The distance between the pickets can be anywhere from 3-12 feet apart depending on the terrain and soil conditions.

(3) Before tying any rope to the picket anchors, the base of the pickets must be buffed or padded.

(4) To tie a rope to the pickets, go to the furthest picket away from the cliff edge. Tie a round turn and two half hitches at the base of the picket. Take the running end of the rope and tie an over the object clove hitch to the base of the middle picket. Then with the running end of the rope tie an over the object clove hitch to the picket closest to the edge of the cliff face. Ensure that there is tension on the rope in between each picket.

(5) To tie off the pickets to themselves, go to the furthest picket from the cliff edge. And with a sling rope, tie a round turn and two half hitches at the base of that picket. With the running end of that sling rope tie a round turn and two half hitches at the top of the middle picket. Using a second sling rope tie a round turn and two half hitches to the base of the middle picket. Then with the running end of that sling rope tie a round turn and two half hitches to the top of the picket closest to the cliff edge.

c. <u>Equalized Anchor</u>. This system is built with a minimum of three pieces of protection. It can be tied with the standing end of a rope or by utilizing a practice coil.

(1) Construction with the Standing End of the Rope:

(a) Tie a Figure of 8 Loop in the standing end of the rope.

(b) Place a carabiner in each artificial anchor point and attach the knot in either of the outside carabiners.

(c) Clip the rope into the remaining carabiners.

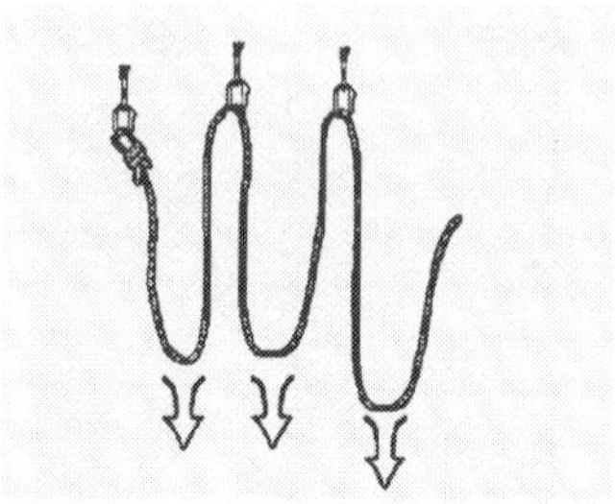

(d) A bight of rope is then pulled down after each carabiner into the anticipated direction of pull. With all three bights, tie an overhand knot around itself to include the running end of the rope.

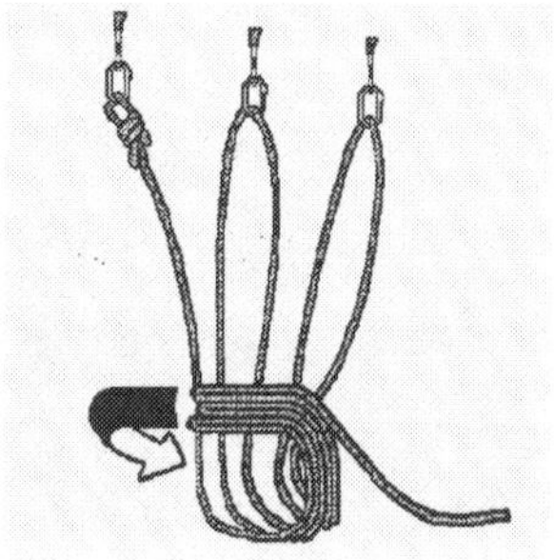

(e) With the running end of the rope coming from the overhand knot, tie a Figure of 8 Loop and attach it to the overhand knot's loop with a locking carabiner.

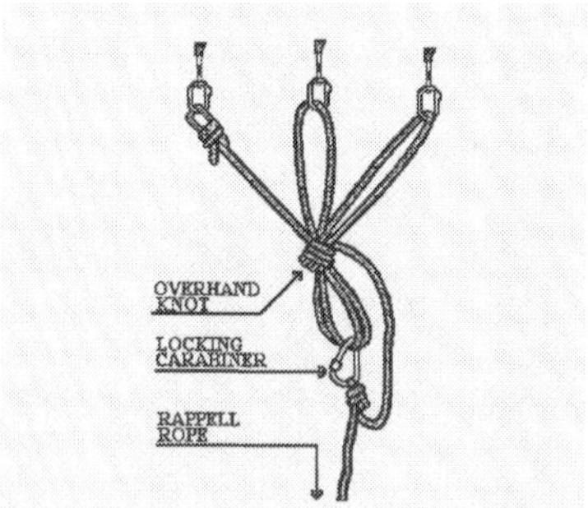

(f) Ensure that there is sufficient slack in the dead rope to prevent disfiguring the knot thus ruining the structural integrity of the knot.

(2) Construction with the Cordalette Method:

(a) Create an endless loop with a practice coil and clip it into each of the artificial anchor points.

(b) A bight of rope is then pulled down between each carabiner into the intended direction of pull.

(c) With all three bights, tie an overhand knot on itself.

(d) Place a locking carabiner into the overhand knot's loop. This will serve as the attachment point for the anchor system.

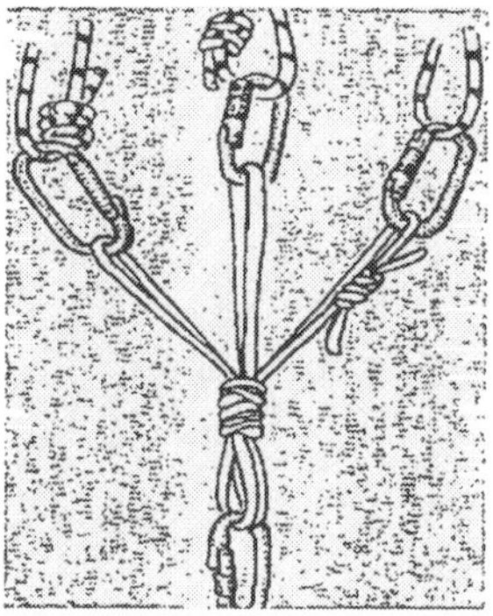

NOTE: If constructing a two-rope high-tension installation, a minimum of three artificial anchors per rope will be used.

UNITED STATES MARINE CORPS
Mountain Warfare Training Center
Bridgeport, California 93517-5001

SML
ACC
02/11/02

LESSON PLAN

SIT HARNESS

LEARNING OBJECTIVES

TERMINAL LEARNING OBJECTIVE: Without the aid of references, properly wear and maintain the MWTC sit harness, in accordance with the references.

ENABLING LEARNING OBJECTIVES

(1) Without the aid of references and given a diagram, name in writing the parts of a sit harness, in accordance with the references.

(2) Without the aid of references, wear a sit harness, in accordance with the references.

BODY

1. (5 Min) **NOMENCLATURE**

 a. Waist belt.

 b. Leg loops (adjustable).

 c. Buttocks straps (adjustable).

 d. Fastex buckle - 2.

 e. Doughnut.

 f. Equipment loops.

 g. D-ring (older models).

 h. Waist belt tie-in point.

 i. Crotch strap.

WEARING OF THE SIT

HARNESS

a. First, disconnect the fastex buckle at the rear of the harness.

b. Hold the harness in front of you, put your feet through the leg loops ensuring that the buckles on the leg loops are outboard, on your thighs.

c. Fasten the waist belt into the buckle, ensuring that it is a tight but comfortable fit. **You must ensure that the waist belt is threaded back through the buckle, this action locks the waist belt to the buckle. Failure to do this will cause the waist belt to slip through the buckle when it is under load.**

d. Adjust the leg loops so that they are high on your thigh, once you have them adjusted, get your buddy to clip the fastex buckle male to the most comfortable female fastex. Then adjust the buttock straps so as the leg loops are held up.

e. If all the above are done correctly, the harness should now be a comfortable but snug fit, after you have fitted it for the first time there is no need to go through the same procedure each time you put it on. Simply hold the harness in front of you, step into it and attach the waist belt to the buckle in the approved manner.

2. TYING IN THE ROPE (END OF ROPE)

a. Pass the end of the climbing rope up through the crotch strap, then through the doughnut and through the waist belt tie-in point.

b. Tie the rope using a retraceable figure eight, adjusting the knot to get it as close as possible to the body.

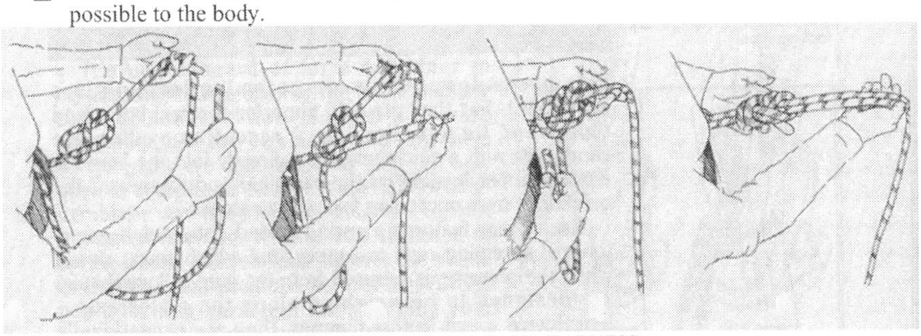

RETRACE FIGURE 8 IN SIT HARNESS

3. TYING IN THE ROPE (MIDDLE OF THE ROPE)

a. Take up a bight of rope and tie a figure eight loop, then take a steel locking carabiner and attach the carabiner to the harness by securing the crotch strap, doughnut, and the waist belt tie-in loop. Now attach the figure eight loop to the carabiner.

4. (5 Min) **CARE AND MAINTENANCE**

a. Avoid contact with chemicals, as this will damage the nylon.

b. Regularly inspect for signs of abrasions and normal wear. Pay particular attention to wear points such as the tie-in loops, buckles, and sewn joints.

c. Keep away from heat such as open flames, cigarettes, etc.

d. If soiled by grit and seawater, wash in lukewarm water with pure soap and allow to dry in a warm room away from direct heat.

e. Two to three years of life can be expected during normal climbing use.

f. It is recommended that a harness that has experienced a serious fall should be discarded.

g. Under no circumstances will you ever tie into the equipment rack of any sit harness as a belay, anchor point, etc.

UNITED STATES MARINE CORPS
Mountain Warfare Training Center
Bridgeport, California 93517-5001

FMST.07.33
05/02/01

STUDENT OUTLINE

ESTABLISHMENT OF RAPPEL POINTS AND RAPPELLING

TERMINAL LEARNING OBJECTIVE Given a mountainous environment, mountaineering equipment, helmet, sit harness, rappel gloves, 165 ft static rope, rescue 8, prusic cord, and one or more assistants, conduct rappelling operations to support mission requirements per the reference. (FMST.07.33)

ENABLING LEARNING OBJECTIVES.

(1) Without the aid of references, select from a given list the three criteria involved in site selection for a rappel site, in accordance with the references. (FMST.07.33a)

(2) In a summer mountainous environment, execute the commands used between a rappeller and his belay man when conducting a rappel, in accordance with the references. (FMST.07.33b)

(3) Without the aid of references, select from a given list three of the nine duties of the rappel point NCOIC, in accordance with the references. (FMST.07.33c)

(4) Without the aid of references, select from a given list three of the six duties of the first man down a rappel rope, in accordance with the references. (FMST.07.33d)

(5) Without the aid of references, select from a given list the three types of rappels, in accordance with the references. (FMST.07.33e)

(6) In a summer mountainous environment, rappel down a cliff face using a seat-hip rappel, in accordance with the references. (FMST.07.33g)

C6

OUTLINE

1. **INHERENT DANGER OF RAPPELLING**. Rappelling is inherently dangerous because rappellers rely totally on the equipment.

   a. To ensure a safe training evolution, two ropes will be utilized at MWTC.

   b. All the rappels taught at MWTC can be completed with a single rope if the situation arises.

   c. If utilizing a one-rope system with the carabiner wrap, the rope will be attached to the carabiner with two wraps vice one.

2. **SITE SELECTION**. (FMST.07.33a) when selecting a rappel site consider these three factors.

   a. There must be a good anchor. As previously taught in *NATURAL* and *ARTIFICIAL ANCHORS*, natural anchors are preferred.

   b. The rappel route down should be as free of obstacles (i.e., vegetation, debris) as possible.

   c. There must be suitable loading and unloading platforms.

NOTE: The evaluation of a site for the above factors should be made by the rappel point NCOIC, who should be the most experienced rappeller in the unit.

3. **DUTIES OF THE RAPPEL POINT NCOIC**. (FMST.07.33c)Once a rappelling site has been selected, one person will be appointed to each rappel lane as a rappel point NCOIC. These individuals should have experience as rappellers. The rappel point NCOIC has ten duties and responsibilities, which are:

   a. Ensures that the anchor points are sound and that the knots are properly tied.

   b. Ensures that loose rock and debris is cleared from the loading platform.

   c. Allows only one man on the loading platform at a time and ensures that the rappel point is run in an orderly manner.

   d. Ensures that each man is properly prepared for the particular rappel; i.e. gloves on, sleeves down, helmet secured, rappel seat tied correctly and secured properly.

   e. Attaches the rappeller to the rope and ensures the rappeller knows the proper braking position for that particular rappel.

   f. Ensures that the proper commands or signals are used.

   g. Dispatches each man down the rope.

h. The rappel point NCOIC will be the last man down the rope.

i. The rappel point NCOIC will ensure that the ropes are inspected after every 50 rappels.

j. The rappel point NCOIC will maintain a rope log.

4. **<u>TYPES/USE</u>**. (FMST.07.33e) The three types of rappels and when they are preferred to be used are:

a. <u>Hasty Rappel</u>: It is used when carrying loads down moderate slopes.

b. <u>Seat-Shoulder Rappel</u>: It is used for heavily laden troops over vertical to near vertical cliff faces.

c. <u>Seat-Hip Rappel</u>: It is used when carrying loads over vertical to near vertical faces.

5. **<u>HASTY RAPPEL</u>**. The hasty rappel is the easiest type of rappel to prepare for. It requires no equipment other than a rope and gloves.

a. <u>Conduct</u>. A hasty rappel is conducted in the following manner:

(1) Sleeves will be rolled down and gloves will be put on.

(2) Face slightly sideways.

(3) Place the rappel rope across your back, grasping it with both hands, palms forward, and arms extended.

(4) The hand nearest the anchor is the guide hand. The hand farthest from the anchor is the brake hand.

(5) Lean out at a moderate angle to the slope.

(6) Descend down the hill facing half-sideways, taking small steps and continually looking downhill while leading with the brake hand.

(7) Feet should not cross and the downhill foot should lead at all times.

b. <u>Braking</u>. The steps for braking during a hasty rappel are as follows:

(1) Bring the lower (brake) hand across the front of the chest to brake.

(2) At the same time, turn to face up toward the anchor point.

6. **SEAT-SHOULDER RAPPEL**. The seat-shoulder rappel relies on friction as the main effort of controlling the descent. It is very efficient for men with heavy packs because it provides support for heavy loads on the back.

a. Conduct. A seat-shoulder rappel is conducted in the following manner:

(1) Put on your rappel-seat, roll down your sleeves and put on your gloves.

NOTE: To avoid causing a possible injury, it is advised that the rappel seat be constructed at the rear of the rappel site loading platform. If the rappel seat is worn for too long, the rappel seat could loosen up enough to cause you to slip out of your seat while rappelling.

(2) The steel locking carabiner is placed on the rappel-seat so that the gate opens down and away, to prevent the gate from opening once the wraps are placed into the carabiner.

(3) Step-up to the rope with you're left shoulder facing the anchor.

(4) The rappel rope is attached to the rappeller's hard point carabiner as follows:

(a) Snap the rope into the locking carabiner.

(b) Taking slack from the standing (anchor) end of the rope, make one wrap with the rope around the body of the carabiner and back through the gate.

(c) Ensure that the locking nut of the carabiner is fastened to lock the carabiner closed.

NOTE: If you are using only one rope to rappel with due to the tactical situation or equipment availability, the procedures are the same, EXCEPT the individual will make two wraps around the body of the carabiner instead of one.

(5) Take the rope across your chest, over your left shoulder, diagonally across your pack and down to the right (brake) hand.

(6) Descend by walking down the cliff using the braking procedure to control the rate of descent. Look under your brake arm for possible obstacles to avoid.

b. Braking. The steps for braking during a seat-shoulder rappel are as follows:

(1) Lean back.

(2) Face directly uphill while bringing the brake hand across the chest.

7. **SEAT-HIP RAPPEL**. The seat-hip rappel is the most commonly used rappel.

a. Conduct. A seat-hip rappel is conducted in the following manner:

(1) Construct the rappel seat; roll down sleeves and put gloves on.

(2) The steel locking carabiner is placed on the rappel-seat so that the gate opens up and away.

(3) Step up to the rope with your left shoulder facing the anchor.

(4) The rappel rope is snapped into the carabiner as follows:

(a) Snap the rope into the locking carabiner.

(b) Taking slack from the standing (anchor) end of the rope, make one wrap with the rope around the body of the carabiner and through the gate again.

(c) Ensure that the locking nut of the carabiner is fastened to lock the carabiner closed.

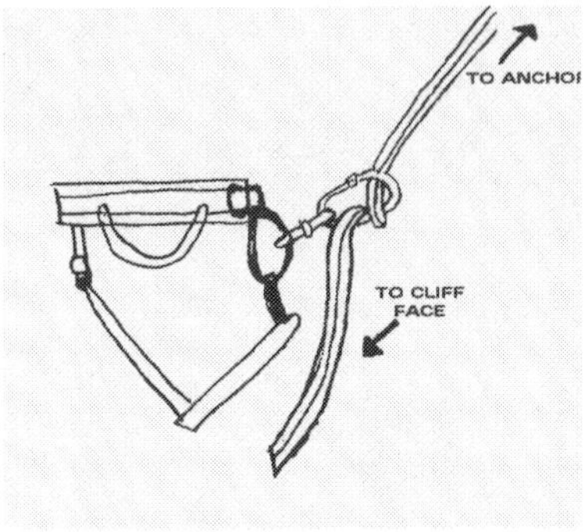

SEAT-HIP RAPPEL

NOTE: If you are using only one rope to rappel with due to the tactical situation or equipment availability, the procedures are the same, EXCEPT the individual will make two wraps around the body of the carabiner instead of one.

(5) The rappeller will grasp the running end of the rope with the brake (right) hand, palm down and turned slightly inboard, near the hip.

b. Braking. The steps in braking for a seat-hip rappel are as follows:

(1) Grasp the rope tightly with the brake hand.

(2) Take the brake hand and place it in the small of the back. This will create enough friction to stop all momentum.

NOTE: At no time will you bound or jump while you are descending. You "walk down" the cliff face using the proper braking procedure to control your rate of descent.

8. **SAFETY OF THE FIRST MAN DOWN**. To ensure the safety of the first man down, the rappeller will:

a. Before deploying the ropes, tie the two ends of the ropes together with an overhand knot.

b. Tie a friction knot on the standing end of the rappel rope with a Prusik cord.

c. Attach the rappel ropes to the harness's hard point with a locking carabiner.

d. Attach the Prusik cord to the harness's hard point with a locking carabiner. This cord is also referred to as a safety Prusik.

NOTE: If the Safety Prusik is too short, it can be extended with a web runner.

9. **DUTIES OF THE FIRST MAN DOWN**. (FMST.07.33d) The first man down the rope has specific duties. They are as follows:

a. Selects a smooth route down for the ropes.

b. Clears the route of loose rocks and debris.

c. The first man down will untie the overhand knot and straighten the ropes out once he reaches the bottom.

d. The first man down belays the next man down the rope.

(1) To belay correctly, the belay man will stand facing the cliff face, arms up with palms facing upward and over lapping. The rope will pass through the opening of the over lapping hand's thumb and index fingers.

(2) To stop a fallen rappeller, the belay man will grab the rope with closed fists and pull straight down bringing the forearms parallel to the deck.

e. Take charge of personnel as they arrive at the bottom to include appointing a belay man.

10. **RAPPELLING COMMANDS.** (FMST.07.33b) In order to conduct rappelling operations safely, it is essential that everyone understands the sequence of events. The following voice commands or rope tugs will be utilized:

VOICE COMMANDS	GIVEN BY	MEANING
"Lane # on Rappel"	Rappeller	I am ready to begin rappelling
"Lane # on Belay"	Belayer	I am on belay and you may begin to rappel
"Lane # off Rappel"	Rappeller	I have completed the rappel and am off the rope
"Lane # off Belay"	Belayer	I have completed the belaying of the rappeller

ROPE TUGS	GIVEN BY	MEANING
Three Tugs	Rappel Point NCOIC	The rappeller is ready to begin rappelling
Three Tugs	Belayer	I am on belay and the rappeller may begin to rappel
Three Tugs	Belayer	The rappeller is off the rappel rope

11. **TYING-OFF**. Occasionally, it may be necessary to stop during a rappel before reaching the bottom of a cliff. The following sequence is used:

a. The commands used in tying-off are as listed below:

COMMAND	GIVEN BY	MEANING
"Lane # Tying-off"	Rappeller	I am ready to tie-off, give me some slack.
"Lane # Tying-off"	Belayer	I have given you enough slack and you may tie-off.
"Lane # On rappel"	Rappeller	I have completed tying-off and I am ready to resume rappelling.
"Lane # On rappel"	Belayer	I am on belay, you may rappel.

b. The procedure used when tying-off is as follows:

(1) The rappeller gives the command, "Lane #, tying-off."

(2) The Belayer gives the rappeller slack and gives the command "Lane # Tying-off". He continues to hold the rope in the belay position, remaining alert and watching the rappeller.

(3) The rappeller quickly brings his brake hand to the twelve o'clock position, so that the running end and standing end are parallel.

(4) The rappeller grasps all ropes with his guide hand as close to the carabiner as possible.

(5) The rappeller steps over the rope so that the running end is going between his legs.

(6) The rappeller releases the rope with his brake hand then reaches under the running end, over the standing end. He then takes up a bight from the running end about two feet long and pulls it over the standing end and under the running end, forming a half hitch.

(7) He pulls the half hitch tightly against the guide hand.

(8) He works the half hitch down snugly against the carabiner while maintaining contact with the guide hand as long as possible.

NOTE*:* A loose half hitch could bind into the carabiner causing difficulties in clearing the knot out. Make sure that the first half hitch is dressed down tightly before moving it against the carabiner.

(9) Place another half hitch above the one already tied.

(10) To untie, reverse the steps. Remove the safety half hitch, then shrink the first half hitch down to a small loop by grabbing the running end of both ropes and pulling them straight out to the left. Place your right hand in the middle of the two bights until they are snug on all four fingers. Then remove one finger and make the bights snug on three fingers. Repeat this process until it's down to one finger and both the bights are equal.

(11) Now, grasp the running end with both hands and smartly jerk the running end of the ropes upward to pop the small loop out from the first half hitch. From this position, keep the guide hand around both of the ropes next to the carabiner; step back over the rope so that the running end is to your right side. Grab just the running end with your brake hand and quickly set the brake behind the small of your back then, readjust you guide hand onto the standing end. You now should be in the seat-hip rappel position.

NOTE: The wraps may bind up some after untying, where no further movement down is possible. To alleviate this, keep the brake on and force your body weight down to pop the wraps in the carabiner back to their intended position and then continue on with the rappel.

(12) The rappeller gives the command "Lane #, on rappel."

(13) The Belayer takes up the slack and gives the command "Lane #, on belay."

(14) The rappeller resumes to rappel down.

12. **RESCUE TECHNIQUES**. When conducting rappelling operations, the possibility of a rappeller getting caught on the rope due to either his clothing or equipment can occur. There are two types of rescues that can be performed to free the rappeller from this situation.

a. The Self-Rescue Technique.

(1) After realizing that you are caught up on the rope, check with your belay man to ensure that he has a solid brake set.

(2) Using a safety Prusik, tie a friction knot onto the rappel rope approximately an arms length above the area that is fouled. Anchor the other end of your safety Prusik to your rappel seats hard point. This is the same technique used as discussed earlier with the first man down the rope.

(3) Work the friction knot up until there is no tension on the rappel device. This will give you the necessary slack to free the malfunction.

(4) Once you have corrected the problem, continue to rappel down using the first man down method.

b. The Buddy Rescue Technique. Himself uses this method when the rappeller is unable to correct the problem.

(1) A rescuer will rappel down on another rope to the disabled rappeller and tie-off.

(2) The rescuer will then establish a safety Prusik onto the victim as taught in the self-rescue technique.

(3) After clearing the malfunction, continue with rappelling operations.

c. Rappelling Casualty Rescue. There are three different methods in which to get an injured rappeller to the bottom.

(1) The first method is to allow the belay man to lower the casualty by slowly releasing the tension on the rappel rope.

(2) The second method is used for critical injuries or when the belay man cannot properly control the casualty's descent.

(a) The rescuer will rappel down using another rope. Once the rescuer gets to the casualty, the rescuer will have his belay man brake him off, enabling the rescuer to use both hands.

(b) If necessary, the rescuer will perform the basic first aid needed. Once the casualty is ready to be lowered, the rescuer will call down to both of the belay men to simultaneously lower both men. The rescuer will hold onto the casualty the entire way down so that he doesn't bounce off the rock face.

(3) The third method is called a tandem rappel. Its used to rescue a casualty who has sustained an injury serious enough that he cannot operate the rappel device himself and requires the assistance of a second rappeller. This type of rescue can be conducted from the top of the cliff face or while on rappel, depending on where the injury occurred.

(a) If the injury occurs while on top of the cliff face, the following steps should be taken:

1. Take either a long sling rope or Prusik cord and tie a figure-eight knot offset so that one length of the cord is longer then the other.

2. Tie a figure-eight knot at each end.

3. Take a rappel device and attach it to the offset figure-eight loop with a carabiner and attach it to the rappel rope.

4. Next, attach the casualty's hard point to the short end of the sling rope/Prusik cord and the rescuer to the other end.

5. The rescuer will then utilize a safety Prusik in the same method as the first man down technique, except that the friction knot will be tied below the rappelling device and controlled with the brake hand.

6. The rescuer will then maneuver himself under the casualty to provide assistance and support.

7. At this point you are ready to rappel both rescuer and casualty at the same time.

NOTE: Your rappel device should be far enough in front of you so that the casualty will not be able to reach it.

(b) If the injury occurs during the rappel, the following steps are taken:

1. The rescuer will preset the same system on a separate rappel rope and rappel down to the casualty.

2. The rescuer will then take the short end of the sling rope and attach it to the casualty's hard point.

3. With a knife, the rescuer will cut the casualty's rappel rope away.

4. Rappel down to the bottom of the cliff face.

NOTE: THIS IS THE ONLY TIME THAT A KNIFE WILL BE USED DURING A RAPPEL. EXTREME CAUTION SHOULD BE USED. IT IS NOT TO BE DONE IN TRAINING, BUT ONLY DURING AN ACTUAL RESCUE AND AS A LAST RESORT.

13. **RETRIEVABLE RAPPELS**. Once a unit has rappelled down a vertical obstacle, it may be necessary to retrieve the rope(s). Depending on the height of the obstacle either one or two ropes will used to construct the rappel lane.

a. One-Rope Retrievable Rappel:

(1) Find the middle of the rope and place it directly behind a suitable anchor point.

(2) Join the pigtails of the rope with an overhand knot and deploy the rope down the obstacle.

(3) On one side of the rope in front of the anchor point, tie an over-the-object clove hitch onto a locking carabiner.

(4) On the other side of the rope, tie another over-the-object clove hitch and attach it to the same carabiner.

(5) The first man down will utilize a safety Prusik and untie the overhand knot after he reaches the bottom.

(6) All others conduct proper rappel procedures.

(7) The last man down will disconnect the carabiner from the rope, and ensuring that the middle of the rope is directly behind the anchor point, rappel down the rope.

(8) Pulling on either end of the rope then retrieves the rope.

b. Two-Rope Retrievable Rappel. This type of system is identical to the one rope retrievable rappel except for a few considerations.

(1) The reason for using two ropes vice one rope is that the height of the rappel is greater.

(2) When two ropes are used they should be joined together using a square knot. This knot will be placed directly behind the anchor point.

(3) All else remains the same as the one rope retrievable rappel except:

(a) The last man down will disconnect the carabiner from the rope and will move the joining knot as close to the vertical obstacle's edge as possible. This will prevent the knot from possibly getting caught up while retrieving.

(b) The last man down will then place a carabiner on the rope below the knot. This will enable him to know which line to pull for retrieval.

14. **EQUIPMENT**. Equipment should be worn in accordance with unit SOP. The unit has the responsibility to determine which methods it feels are most beneficial to the mission. Weapons should be worn across shoulder; muzzle down and away from the brake hand with a tight sling securely attached to the weapon. Weapons should also be dummy corded to the individuals.

UNITED STATES MARINE CORPS
Mountain Warfare Training Center
Bridgeport California, 93517-5001

WMC
04/11/02

STUDENT OUTLINE

TOP ROPING

LEARNING OBJECTIVES.

TERMINAL LEARNING OBJECTIVE. In a summer mountainous environment, and given a rock face and a belay man, conduct top roping with cartridge belt and rifle, in accordance with the references. (WMC)

ENABLING LEARNING OBJECTIVE

(1) In a summer mountainous environment, establish a belay point from the top of a cliff for use in a top rope climb, in accordance with the references. (WMC.a)

(2) In a summer mountainous environment, demonstrate the use of climbing commands while conducting a top rope, in accordance with the references. (WMC.b)

OUTLINE

1. **ESTABLISHING A BELAY STANCE FROM THE TOP**.

a. The Belayer will establish a sitting belay stance on the cliff head by:

(1) Constructing a suitable anchor with the standing end of the rope.

(2) With the direction of pull away from the anchor, tie a directional figure 8 loop near enough to the cliff's edge so that the climber can be observed, if possible.

(3) Tying a swami Wrap around oneself, clip a locking carabiner through all the rear wraps and clip a large locking carabiner through all the front wraps.

(4) Secure the rear locking carabiner into the directional figure of 8 loop.

(5) Secure the running end of the rope to the large locking carabiner utilizing a suitable friction belay (i.e., Munter hitch, stitch plate, etc.).

NOTE: Gloves **will not** be worn while belaying a climber during a top rope.

2. **SECURING THE CLIMBER TO THE ROPE**.

a. The climber will tie into the end of the top rope by:

(1) Constructing a Swami wrap around oneself.

(2) Tying a retrace figure 8 loop through all of the Swami's wraps or;

(3) A figure of 8 loop clipped into a locking carabiner secured through all the Swami's wraps.

b. Alternative methods.

(1) A bowline on a coil as discussed in ROPE MANAGEMENT or;

(2) Tie into the hard point of the harness as discussed in SIT HARNESS.

3. **ESTABLISHING A BELAY STANCE FROM THE BOTTOM**.

a. Belaying from the bottom is commonly referred to as a "yo-yo" or "Sling Shot" belay. This system can be constructed for either a direct or indirect belay.

(1) Construct a suitable top anchor utilizing either a pulley or one steel locking or two non-locking carabiners as the attachment point for the anchor.

(2) The rope will travel up from the bottom and pass through the attachment point then return to the bottom.

(3) The climber will be secured to one end of the rope and the Belayer from the other end.

4. **CLIMBER'S RESPONSIBILITIES**.

a. The climber ensures that the Belayer is anchored and on belay, by use of commands or prearranged signals prior to beginning the climb.

b. The climber will not out climb the belayer; this will cause slack in the rope between the belayer and the climber.

c. Avoid placing excess pressure/weight on belay man.

d. Weapons will be worn across shoulder, muzzle down and to the left, with a tight sling securely attached to the weapon.

5. **CLIMBING COMMANDS AND SIGNALS.** In order to conduct top rope operations safely, it is essential that everyone understands the sequence of events. The following voice commands or rope tugs will be utilized:

VOICE COMMANDS	GIVEN BY	MEANING
"Lane #, Up Rope"	Climber	Belayer needs to take in the slack.
"That's Me"	Climber	Excess slack has been taken up between us.
"Lane #, On Belay"	Belayer	I am on belay.
"Lane #, On Climb"	Climber	I am ready to climb.
"Climb Climber"	Belayer	You may begin to climb.

ROPE TUGS	GIVEN BY	MEANING
Three Tugs	Climber	Belayer needs to take in the slack.
Three Tugs	Belayer	I have taken up the slack and you may begin to climb.
Three Tugs	Climber	Climbing.

UNITED STATES MARINE CORPS
Mountain Warfare Training Center
Bridgeport, California 93517-5001

FMST.07.40
03/02/02

STUDENT OUTLINE

MOUNTAIN CASUALTY EVACUATIONS

TERMINAL LEARNING OBJECTIVES. In a summer mountainous environment and given a simulated/actual casualty, evacuate casualties, in accordance with the references. (FMST.07.40)

ENABLING LEARNING OBJECTIVES

(1) With the aid of references, state orally the general considerations for medevac procedures, in accordance with the references. (FMST.07.40a)

(2) In a summer mountainous environment, and with the aid of references, construct an expedient litter, in accordance with the references. (FMST.07.40b)

(3) In a summer mountainous environment, and with the aid of references, secure a simulated/actual casualty in a SKEDCO litter, in accordance with the references. (FMST.07.40c)

(4) In a summer mountainous environment, and with the aid of references, attach a belay line to a SKEDCO litter for movement up/down moderate slopes, in accordance with the references. (FMST.07.40d)

(5) In a summer mountainous environment, with the aid of references, and given a simulated/actual casualty, conduct a medevac using the barrow-boy method, in accordance with the references. (FMST.07.40e)

OUTLINE

1. **GENERAL CONSIDERATIONS** (FMST.07.40a) the general considerations are a set of guidelines that can be used no matter how serious the casualty is. They are remembered by a simple acronym, **APASSNGG.**

   a. Apply Essential First Aid. (i.e. splints, pressure bandage, etc.)

   b. Protect the Patient from the Elements. Provide the casualty with proper insulation from the ground. Ensure that he is warm and dry. If there are any natural hazards (i.e. rock fall, lighting, etc.) either move the casualty as quickly as possible or ensure that he is well protected.

   c. Avoid Unnecessary Handling of Patient.

   d. Select Easiest Route. Send scouts ahead if possible, to break trails.

   e. Set Up Relay Points and Warming Stations. If the route is long and arduous, set up relay points and warming stations with minimum amount of medical personnel at warming stations to:

      (1) Permit emergency treatment. Treat for shock, hemorrhage, or other conditions that may arise.

      (2) Reevaluate the patient constantly. If patient develops increased signs of shock or other symptoms during the evacuation, he may be retained at an emergency station until stable.

   f. Normal Litter Teams Must Be Augmented in Arduous Terrain.

   g. Give Litter Teams Specific Goals. This litter teams job is extremely tiring, both physically and mentally. The litter teams must be given realistic goals to work towards.

   h. Gear. Ensure all of the patient's gear is kept with him throughout the evacuation.

2. **ONE -MAN CARRIES**

   a. Sling Rope Carry. The sling rope carry requires two men and a 15-foot sling rope. One as the bearer, and an assistant to help in securing the casualty to the bearer. Conscious or unconscious casualties my be transported in this manner:

(1) Bearer kneels on all fours and the assistant places casualty face down on bearer's back ensuring the casualty's armpits are even with the bearer's shoulders.

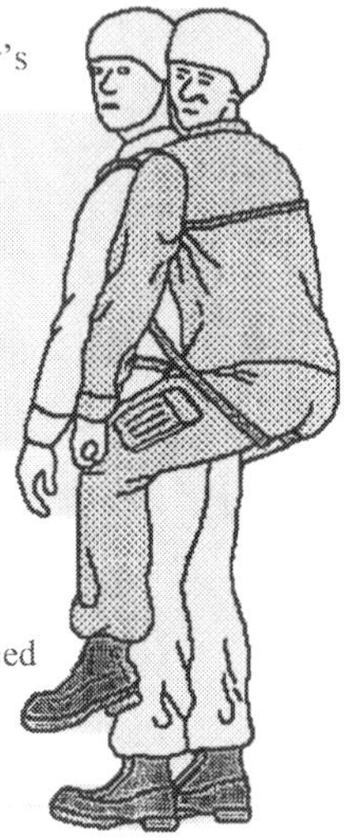

(2) He then finds the middle of the sling rope and places it between the casualty's shoulders and the ends of the sling rope are run under the casualties armpits, crossed and over the bearer's shoulders and under his arms.

(3) Then the ropes are run between the casualty's legs, around his thighs, and tied with a square knot with two overhands just above the bearer's belt buckle.

(4) Ensure the rope is tight. Padding, when available, should be placed where the rope passes over the bearer's shoulders and under the casualties thighs.

b. <u>The rope coil carry.</u> This requires a bearer and a rope coil. It can be used to carry a conscious or unconscious casualty.

(1) Position the casualty on his back.

(2) Separate the loops of the mountain coil into two approximately equal groups.

(3) Slip ½ of the coil over the casualty's left leg and ½ over his right leg so that the wraps holding the coil are in the casualty's crotch, the loops extending upward the armpits.

(4) The bearer lies on his back between the casualty's leg and slips his arms the loops. He then moves forward until the coil is extended. When using the rope coil the bearer ties the coil to himself vice slipping his arms through the loops.

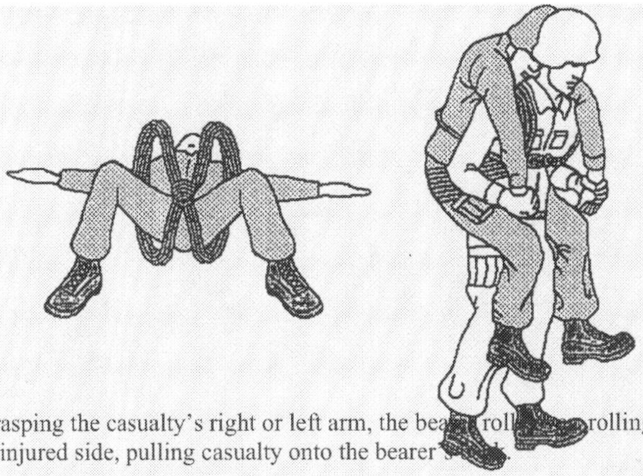

(5) Grasping the casualty's right or left arm, the bea[illegible] roll[illegible] rolling to the casualty's uninjured side, pulling casualty onto the bearer[illegible].

(6) Holding the casualty's wrists, the bearer carefully stands, using his legs to lift up and keeping his back as straight as possible.

NOTE: The length of the coils on the rope coil and the height of the bearer are to be considered. If the coils are too long and the bearer happens to be a shorter person, it will require the coils to be uncoiled and shortened. If this is not done, then the casualty will hang too low on the bearer's back and make it a very cumbersome evacuation. A sling rope harness can be used around the victim's back and the bearer's chest, which will free the bearer's hands.

3. **TWO-MAN CARRY**

a. Pole Carry. The pole carry method is a field expedient method and should be considered as a last resort only, when narrow ledges must be traversed; vegetation limits the bearers to a narrow trail. This method is difficult for the bearers and uncomfortable for the casualty. Two bearers, four sling ropes and a 12 foot pole - 3 inches in diameter, are required for this carry.

(1) The casualty is placed on his back in a sleeping bag or wrapped in a poncho or blanket, then placed on an insulated pad.

(2) One sling rope is placed under the casualty below the armpits and tied with a square knot across the casualty's chest.

(3) The second sling rope is tied in the same manner at the casualty's waist.

(4) The third sling rope is placed at the casualty's legs below the knee.

(5) The fourth sling rope is tied around the ankles.

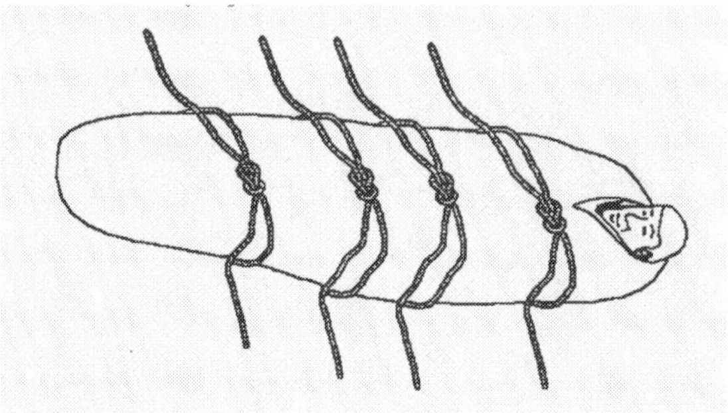

(6) The pole is placed along the casualty's length and secured using square knots with two overhands with the ends of the sling ropes. The square knots should be so tight that the overhands are tied onto themselves

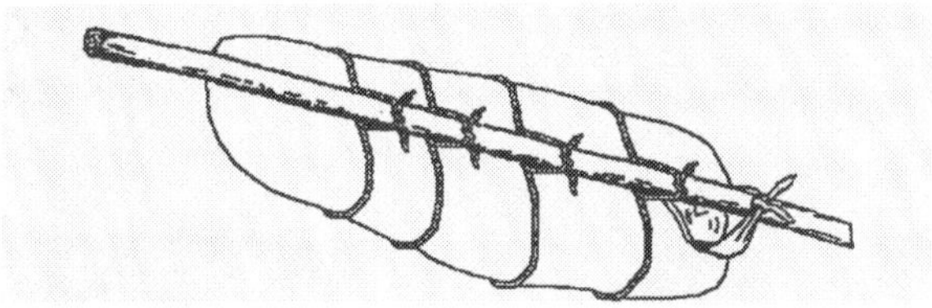

(7) The casualty should hang below the pole, as close to the pole as possible, to prevent swinging during movement.

(8) The casualty's head may be supported using a triangular bandage or a cartridge belt passed around the pole.

(9) For additional support and of movement, two additional bearers may be required, as well as a mountain coil.

(a) Mountain coil is split into two equal coils.

(b) Place knot of mountain coil under casualty's lower back.

(c) Additional bearers slip into each half of the hasty coil one on each side of casualty, aiding in support and movement of the casualty.

4. **EXPEDIENT LITTERS.** (FMST.07.40b) The two types of expedient litters that we will talk about are the alpine basket and the poncho litter.

a. Alpine Basket. To belay the alpine basket, the pre-rigs are attached to the bights formed coming through the loops.

(1) If barrow boy is to be used, the procedure previously discussed will be adhered to.

(2) If barrow boy is not used, then a tag line from the bottom must be implemented to keep the casualty away from the cliff face on the decent.

(3) Construction of the Alpine Basket:

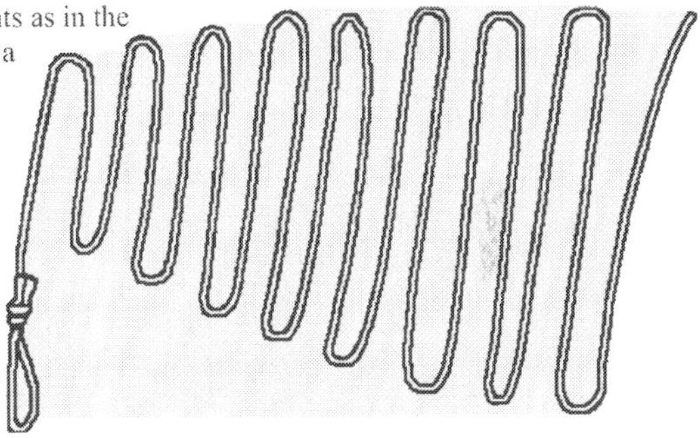

(a) Start by making the same amount of bights as in the rope litter, but start from one end and tie a figure-of-eight loop to run the first bight through.

(b) Place padding i.e., isopor mat, on top of the bights and then lay the casualty on the padding and bights.

(c) Start at the casualty's feet and pull the first bight up around the casualty's ankles and through the figure-of-eight loop tied into the starting end of the rope.

(d) Go to the opposite side of the casualty and pull up the second bight and pull it through the loop formed by the bight that was pulled through the figure-of-eight.

(e) Continue until you get to the casualty's armpits, bring the second to the last bight up over the casualty's shoulder and into the bight and then bring the last bight up over the casualty's other shoulder and into the last bight formed.

(f) Secure the end of the bight with an appropriate anchor knot.

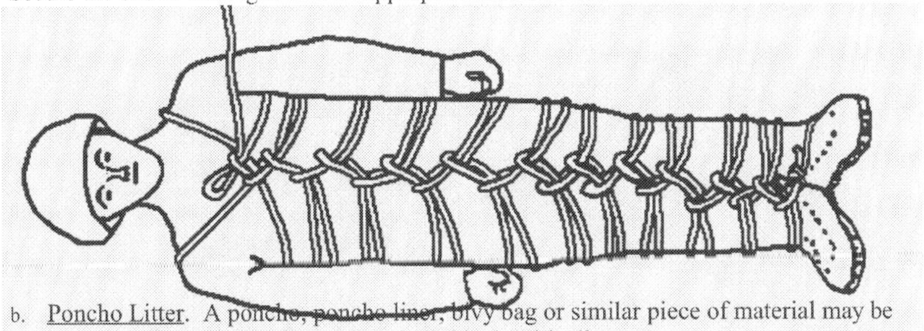

b. Poncho Litter. A poncho, poncho liner, bivy bag or similar piece of material may be used. In addition you will need six individuals with sling ropes.

(1) Lay poncho litter flat on the ground.

(2) Select six rocks about the size of a golf ball. Place one rock in each of the corners and one in the middle on each side in the middle of the litter. The rocks are placed on the under side of the poncho or like material. If a bivy bag is used the casualty should be zipped inside the bivy bag. The rocks should then be arranged in the same manner only on the inside below the zipper.

(3) Tie the sling rope together with an overhand knot. Take the middle of the rope and secure it around the rock with a clove hitch.

(4) An isomat may be laid on the poncho to help make the litter firmer.

(5) The casualty is then placed in the litter. The sling ropes are adjusted by feeding the pigtails of the over hand knot through itself to adjust for length. The loop is then put over the inboard shoulder of the carriers. Insure that the casualty is carried level.

5. **LITTERS.** There are two kinds of litters that are used for casualty evacuations in moderate to vertical terrain. The SKEDCO and the Stokes litter. Each has a variation of procedures for securing a casualty and rigging the litters for either raising or lowering. Let's first look at the SKEDCO litter.

(1) Securing a Casualty to the SKEDCO (FMST.07.40c)

(a) First unroll the litter. The litter must be re-rolled the opposite way to allow the litter to lay flat. Then lay the litter next to the casualty.

(b) If the casualty has any possibility of a spinal injury the Oregon Spine Splint must be used. Secure the splint to the casualty by use of the color-coded buckles. Experienced medical personal are recommended if spinal immobilization is necessary.

(c) Once the casualty is on the SKEDCO use the four body straps to secure the casualty to the litter. Unless injuries prevent, the casualty's arms should be at his sides to prevent further injuries to himself or the rescuers.

(d) Once the casualty is secured with the body straps the feet straps must be secured. The feet straps are secured last to ensure the casualty is in the proper position on the SKED. The feet can be positioned in three ways. The first position is feet together with the straps running on the outside of the feet. The second position is feet apart with the straps running on the inside of the feet. The last position is the feet stacked. This is the most uncomfortable position and not recommended for casualties with possible spinal cord injuries. Placing the heel of one foot on top of the toes of the other forms this position. This position will only be used for a casualty in confined spaces. Once the feet are positioned the feet straps must be secured. Start by bending the feet end of the SKED to form a plat form for the feet. Then loop the feet straps through the second grommets on each side.

(e) The last thing to do is to form the head end to protect the casualty's head. If possible the casualty should wear a helmet. Form the head end tying the pull strap up and secure it to the first body straps.

(2) Medevacing a Casualty. There are many ways to move a casualty once in the SKEDCO. However the medevac team must keep the general considerations in mind. The two methods that we will talk about next are the simplest and require the least amount of additional rigging

(a) The first way is to drag the casualty by the drag strap located at the head end of the SKED. You can also use the SKEDCO's carrying bag as a harness in conjunction with the pull strap and towing harness. If additional people are required, cordage can be added to the pull strap or the front carrying handles.

(b) The second way is to carry the casualty using the carrying handles. By using the set of four removable webbing the litter team can be augmented. To do this each pieces of webbing is tied to make an endless loop. Then pass a bight of the loop through one of the grommets to create additional handles.

(3) Rigging the SKED for Vertical Terrain and Helicopter Lift. In this type of terrain special requirements must be taken to raise or lower a casualty in the SKEDCO.

(a) Rigging the SKED for a Vertical Employment. A vertical raise or lower is used when moving a casualty on steep earth to avoid any further injury to the casualty. On vertical terrain the vertical raise or lower can be used if the terrain is not uniformed or there is a chance of rock fall. Ensure that the casualty's head is always above his feet.

(1) Identify the 30-foot piece of cordage that comes with the SKEDCO. Then tie a figure eight in the middle of the rope. If the rope is worn or missing, the same process can be done with two sling ropes.

(2) Next pass each end of the rope through the grommets at the head of the SKEDCO. Leaving approximately 1-2 feet of rope between the stretcher and the knot.

(3) Continue to feed each end through the grommets and the carrying handles towards the foot end of the SKEDCO. Pass the ends of the rope through the last grommets at the foot end and secure the two ends with a square knot without over hands.

(4) Bring the pigtails up and over the casualty's feet and pass the ends through the
a carrying handles towards the middle of the casualty. Then tie a square knot
b with two over hands.

c

(4) Rigging the SKEDCO for a Horizontal Employment. A horizontal raise or lower is preferred on uniformed vertical terrain. The horizontal employment allows the rescuer to assist the casualty on either a raise or lower. It also allows the rescuer the ability to monitor the casualty's condition and can easily treat the casualty if the need arises.

(a) Identify the two 4 inch nylon straps. They should be two lengths; one four inches shorter than the other. The shorter strap should be marked HEAD STRAPS.

(b) Insert one end of the head strap into a slot near the head of the litter. Then wrap the rest of the straps under the SKEDCO and pass the other end through the opposing slot. Do the same at the foot end of the SKEDCO with the other strap. Ensure that the strap runs smoothly under the SKEDCO.

(c) Connect the strap ends with the large locking caribiner that comes with the SKEDCO. If the caribiner is worn or missing, opposing Stubai 85 locking carabiners will suffice.

6. **STOKES LITTER**. The stokes litter is a litter that is constructed of metal tubing with a plastic covering. The litter is formed in a rectangular basket shape with mesh attached to the frame. Using the stokes for an evacuation (as with any evacuation) it should be padded for the casualty.

a. Securing the casualty to the stokes litter. In the event that the "seat belts" are missing from the stokes litter, sling ropes can be used to the lash the casualty. The steps involved are:

(1) Tie two sling ropes together using square knots and two overhands.

(2) Tie a stirrup hitch around ankles and feet, feed the two pigtails through the right angles of the stokes. Do not cross the ropes at the ankles.

(3) Lace the sling rope towards the casualty's head by passing the rope through the right angles (not over the top of the rails of the stokes).

(4) Secure the ends of the sling ropes by tying a clove hitch with two half hitches on the thick vertical bar located by the victim's shoulder.

7. **YOSEMITE PRE-RIG.** The title of this method, pre-rig implies the meaning ready to use, but in this case the rig must be constructed. Its purpose is for attaching a belay line to a litter and to make the litter easily adjusted. This is the one method that is used to secure a litter to a belay line.

a. Construction: This method normally requires four sling ropes. The steps are: (1) using one sling rope, tie a figure-of-eight loop with one tail.

(1) Take the remaining tail and run it through the window of the stokes litter or in the stirrups of the collapsible litter.

(2) Tie a Kragur knot onto the same sling rope.

(3) Repeat steps (1), (2), and (3) with the three remaining sling ropes.

(4) Suspend the litter to ensure that the comers are balanced.

8. **ASCENTS OR DESCENT OVER STEEP TO MODERATE SLOPES.** When the litter team is ascending or descending a slope they must consider the potential for further injury to the casualty or to themselves. If the risk of injury is high a belay line may be used to prevent injury to the casualty and the rescuers.

a. Preparing Casualty for Ascents or Decent over Steep to Moderate Terrain. This procedure will be depending on several things. Initially, site selection should contain the following features.

(1) Suitable anchor points.
(2) Clearance for casualty along the route
(3) Loading and unloading points.

b. Additional considerations.

(1) The casualty will always be rigged for vertical employment when on steep to moderate terrain.
(2) The smoothest possible route must be selected.
(3) Ensure that the casualty's head is above his feet.

c. Rescuers positions. There are two methods that can be used for the rescuers for moving a casualty in steep to moderate terrain.

(1) Two to four men will position themselves on each side of the litter. They can then carry the litter by the carrying handles. In steep terrain a second belay line may be used to assist the rescuers. We will discuses the belay line later in the chapter.

(2) The Caterpillar method will require as many personnel as possible. The personnel will split in half and position themselves on each side of the litter forming a tunnel. As the litter is raised or lowered each member will hand the litter to the next member in the tunnel. As the litter passes each person in the tunnel he will peel off and assume the lead either at the top or bottom of the tunnel. This will continue until the litter reaches its desired destination.

d. Belay Line. (FMST.07.40d) For belaying of a casualty, one rope will be used and from the top using one of two methods depending on the application.

(1) Body Belay. This method should only be used over moderate terrain. The belay man will establish a sitting position behind a suitable anchor (i.e., rock, tree, etc.) and pass the standing end of the rope behind his back. The running end of the rope will feed out from the belay man's right side. A figure of eight loop is tied to the end of the running end of the rope. It is then attached to the litter's figure eight loop with a locking carabiner. The belay man will then remove all of the slack between himself and the litter. The standing end of the rope should be stacked on the belay man's left side and run through his left side. As the casualty is lowered, the belay man will feed the rope from behind his back allowing it to run through his right hand. If the belay man needs to stop the casualty, he will clench the rope in his left hand, and

bring the rope to the center of his chest.

(2) Direct Belay. This method is the safest for either raising or lowering a casualty in either moderate to steep terrain.

(a) To Lower a Casualty. First a swami wrap will be tied around a suitable anchor point. Two locking carabiners will be clipped into all of the wraps of the swami wraps, gates up. A figure of eight loop is tied into the end of the static rope and attached to the litter with a locking carabiner. After all the slack has been taken up between the litter and the anchor, the rope must be tied through an appropriate belay device. The belay device is attached to the anchor through one of the two locking carabiners on the anchor. A safety (French) prussic will be tied to the running end of the rope and clipped into the second locking carabiner on the anchor. While the casualty is being lowered, one person will control the rope running through the belay device. The safety prussic will be controlled by a second person. Should the primary belay man lose control, the person operating the safety prussic simply lets go and the prussic will bind onto the rope, stopping the casualty.

(b) To Raise a Casualty. The anchors will be established in the same manner as discussed in lowering the casualty with one minor change. The one change is that instead of running the rope through a belay device, the rope will only run through a locking carabiner. The load will be raised by the use of a mule team. The mule must consist of as many people as possible. The mule team will raise the load in as straight a line from the anchor as possible. If the space does not permit a ninety-degree angle away from the anchor is also an option. The mule team will walk backward until the last man reaches his limit of advance. Once he reaches that limit he will peel off the end and return to the front of the mule team. This process is continued until the casualty reaches the top. If the load becomes unmanageable, the safety prussic will be allowed to bind on the rope while the mule team repositions themselves. If the person operating the safety prussic can not see the casualty a Point NCO will be in charge of communicating with the mule team.

(c) To Belay the Rescuers. If the route is too steep or the footing is poor the rescuers may need some assistance either on the raising or lowering of a casualty. If this is the case a separate belay line will be established for the rescuers. The anchor and the belay line are established in the same manner. The same anchor can be used if it is suitable for the load. The rescuers will then tie either around the chest bowlines or swami wraps. A figure eight loop will be tied into the end of the static rope and connected to the bottom rescuer with a locking carabiner. The other rescuers will connect themselves to the same rope with middle of the line prussic. They will be connected in this manner so that they can adjust their position to the casualty.

9. **BARROW BOY.** (FMST.07.40e) A barrow boy is no more than an assistant to the litter on vertical to near vertical cliff faces. The barrow boy can be used for either the Stokes or the

SKED litters. For this situation the Stokes litter should only be used in the horizontal position. However the SKED can be employed in either the horizontal or the vertical positions.

NOTE: Static load, lowered slowly, little friction, so don't need 2 ropes.

a. Rigging the Barrow Boy.

(1) First the rescuer must ensure that a suitable anchor has been established, a proper belay has been constructed, and that a safety prussic has been constructed.
(2) Then the rescuer must ensure that if an A-frame is used that it has been constructed and anchored properly.

(3) Next the rescuer will tie a rappel seat on. (a sit harness can also be used) Then he will ensure that an around the body bowline is tied onto the casualty. A figure eight will be tied on to the end to act as the casualty's safety.

(4) Then after running the running end of the rope though the carabiners or pulley of the A-frame he will tie a figure eight loop at the end of the static rope. He will then attach the figure eight loop to his hard point with a locking carabiner. Then he will tie a middle of the line prussic above the figure eight and attach it to the same locking carabiner in his hard point. This is called the adjustment prussic. It is used to adjust the position of the Barrow Boy in relation to the litter. Next he will take six to eight feet of slack from the end of the line figure eight and tie a directional figure of eight with direction of pull down. (Note: the prussic should be between the end of the line figure of eight and the directional figure of eight.) The directional figure of eight is the attaching point for the litter and the casualty's safety.

(5) Once the rescuer and the litter are secured, the belay man must take all the slack out of the system. The rescuers will the maneuver the litter through the apex of the A-frame with the help of the point NCO.

(6) Once onto the cliff face the rescuer will then position him self with his adjustment prussic so that he can be of most assistance to the litter on the raise or lower. The rescuer will pull the litter out away from the cliff face so that the casualty rides smoothly up or down the cliff face.

(7) The Point NCO will be in charge of the belay men or the mule team. He will also communicate with the rescuer about the rate of speed, if the rescuer needs to be stopped along the route, and when he reaches the top or bottom of the cliff.

10. **TANDEM LOWERING.** The tandem lowering system can be used for the walking wounded, POW's, or more serious casualties when situation would not permit using the barrow boy.

a. The assistant to the casualty should first tie a rappel seat on himself and then assist the casualty with his.

b. The assistant will take only one belay line and tie a end of the line figure 8 loop, and clip this into his rappel seat.

c. A directional figure 8 will be tied approximately 12 inches up the rope from the figure 8 loop (with the loop pointed down) and this will be clipped to the casualty's rappel seat.

d. If needed, safety prussic cords should be tied above the casualty's directional figure 8. These safety cords should be attached in the same manner as the first man down a rappel.

e. The casualty and the assistant will lower as one, with the assistant helping on the way down the cliff.

11. **OTHER CONSIDERATIONS.** All of the techniques we have discussed for the evacuation of a casualty from top to bottom can also be used on a suspension traverse or rope bridge, with a slight variation in the belay line. Two belay lines may be used for rope bridges and the suspension traverse, if they are available. No matter what type of litter is used, the individuals involved in the evacuation must ensure that the head is always uphill or not lower than the feet.

UNITED STATES MARINE CORPS
Mountain Warfare Training Center
Bridgeport, California 93517-5001

FMST.0
7.34
04/12/02

STUDENT HANDOUT

ONE-ROPE BRIDGE

LEARNING OBJECTIVES

TERMINAL LEARNING OBJECTIVES. In a summer mountainous environment, conduct bridging, in accordance with the references. (FMST.07.34)

ENABLING LEARNING OBJECTIVES

(1) Without the aid of references, list in writing the criteria for site selection for a one-rope bridge, in accordance with the reference. (FMST.07.34a)

(2) In a summer mountainous environment with an assault load , cross a one-rope bridge using the rappel seat method, in accordance with the reference. (FMST.07.34b)

(3) Without the aid of references, list in writing the rescue techniques, in accordance with the reference. (FMST.07.34c)

(4) In a summer mountainous environment transport a SKED stretcher across a one rope bridge, in accordance with the reference. (FMST.07.34d)

BODY

1. **SITE SELECTION**

a. The two criteria for site selection for a one-rope bridge are: (FMST.07.34a)

1) There must be suitable anchors on both sides of the stream.

2) The anchors must offer good loading and unloading platforms.

b. Other considerations involved are:

1) The site chosen for the initial crossing does not have to be at the location for the construction of the bridge, just as long as the rope can be taken to the selected site the

C9

crossing.

(2) The site chosen for the lead swimmer to cross should be as free as possible from obstacles in the water, such as large boulders, stumps or logs as discussed in *STREAM CROSSING.*

(3) The anchors must be close enough for the 150 foot coil to reach both near side and far side anchors. Keep in mind that it will take approximately 1/3 of the 150 foot rope for tightening and anchoring of the bridge.

2. **DISTANCE ESTIMATION**. The follow methods can be used to determine the distance between anchor points:

a. Azimuth Method. Shoot an azimuth to a point on the far side of the intended obstacle to cross. Then move LEFT or RIGHT (perpendicular to the azimuth) until you get a 15 degree offset of your previous azimuth. Next, measure the distance that you paced in feet from your first azimuth to your second azimuth, and multiply that distance by three. This total will give you the approximate distance across the obstacle in feet.

b. Unit Average Method. Get three Marines to judge (best guess) the distance across the intended obstacle. Add up the total accumulated distance, then divide by three. This will give an estimation of the distance across the obstacle.

Example:	Marine 1	240 ft
	Marine 2	230 ft
	Marine 3	250 ft
		720 ft

Divided by 3 = 240 ft

3. **ORGANIZATION**. The organization of construction is broken down into four groups. They are as follow:

a. The Bridging Team - consisting of the bridge NCO and another Marine.

b. The Safety Line Team - consisting of the lead swimmer and his belay man.

c. The Mule Team - consisting 3 Marines for a 9:1, and 6 Marines for a 3:1.

d. The Security Team – consisting of the rest of the party.

4. **CONSTRUCTION**. This bridge is ideal for squad and platoon sized units, for its quick construction and minimal amount of equipment required.

a. Once the site for the bridge has been designated by the bridge NCO, the Safety Line Team will move up stream from this site to enter the water. Their first step is to flake out the rope. Once this is accomplished, the lead swimmer's will take a bight of the rope and tie a figure 8 loop at the end that will be going to the far side, ensuring the knot has a 18"-24" loop. The lead swimmer's upstream arm will go in it, with the belay man tending his rope the lead swimmer will cross the never using the flow to assist him. (Ferry angle as taught in STREAM CROSSING)

b. At the same time as the lead swimmer is crossing, the bridging team will be preparing the near side anchor. The bridging team will tie a swammi wrap around the anchor, using a sling rope or practice coil, and ensuring the square knot is behind or on the side of the anchor. Once the anchor is secure, flake out the bridging line in order to send it across.

c. Once the lead swimmer is across he will move to the site where the far side anchor will be established and secure his rope. The belay man will then move down to the near side anchor, and as close to the waters edge as possible. He will attach the bridging line to the safety by tying a middle of the rope in a figure 8 knot to the safety line and an end of the rope figure 8 knot to the bridging line connecting the two ropes with an 85 carabiner. The lead swimmer will then pull the rope across. The belay man with the help of another Marine will belay the bridging line across keeping tension on the rope to keep the rope out of the water.

d. After the bridging line is across the lead swimmer will detach the bridging line from the safety line and secure it to the far side anchor using a tree wrap, wrapping from right to left, ensuring there are 3-5 wraps.

e. Once the bridging line is secure. The lead swimmer and the belay man will secure the safety line on both banks, ensuring the rope is creating a ferry angle (minimum 45 degrees). This will be the safety line for the Marines crossing the bridge.

NOTE: If a fall from a rope bridge would result in death or serious injury, then utilize a safety line that will be anchored parallel with the bridge and hand tightened. The Marine crossing will clip into both ropes prior crossing.

f. On the near side bank:

1. The bridging team will need the following gear to build the mechanical advantage:

    (a) Four steel locking carabiners & 3 -Three foot prusiks

2. Once the bridging line is secured on the far side, the bridging team will take one 85 and clip it into the near side anchor (swammi wrap). Then they will take the bridging line and clip into that 85, then using one three foot long prusik (16" loop) the bridging line will be secured to the 85 by tying a French Prusik, which will act as a braking knot. The bridging team will now pull the rope taut, and begin construction of the mechanical advantage.

3. The first step is to take one three foot prusik (16" loop) and tie a French prusik on the bridging line as far away from the anchor as possible. Then take one 85 and clip it to the loops of the French prusik.

4. Taking a bight from the running end of the rope clip it into the 85 hanging from the tails of the French prusik.

5. Then bring a bight back from the clipped in rope to the anchor, and taking one 85 clip it into the 85 created in step (2), take the bight of rope and clip into this 85.

6. Take one three foot prusik and tie a French prusik onto the bottom end of the bight clipped into the 85 created in step (3), take one 85 and clip it into the tail loops of the French prusik.

7. From the anchor, take a bight of rope from the 85 created in step (6) and clip it into the 85 created in step (6).

8. You have made approximately a 9:1 mechanical advantage.

NOTE: If the manpower is available the squad/platoon has the option of building a 3:1 mechanical advantage vise a 9: 1, the mule team will just have to add another body having no more then 3 bodies on a 9:1 and 6 bodies on a 3:1

g. Tensioning of the Bridge

1. Now that the bridge is built, the bridge NCO can call up the mule team to tension the bridge.

2. This procedure is facilitated by the mechanical advantage system that was just put into place. The braking knot (French prusik) is used to hold tension on the bridging line while the bridge is being tensioned.

3. The mule team will begin to pull on the running end of the rope coming out of the mechanical advantage. They will pull the rope straight back, trying to keep it in line with the bridge as best as possible.

4. The mule team will tension the rope as much as possible, the bridge NCO will monitor the system. Once the mule team cannot pull anymore tension, the bridge NCO will then have them hold the rope in place, and will reset the braking knot. Once the braking knot is set the bridge NCO will ask for slack from the mule team and cycle the system out. This process will continue until the bridge is tight.

5. On the last cycle the bridge NCO will set the brake knot and with the help of another Marine they will make a bight out of the running end and bring it around the tree while keeping tension.

6. Take a bight and make a complete round turn on the body of the 85 created in step (2). Last, the bridge NCO will tie two half hitches encompassing all the ropes just behind the anchor.

7. At this stage the bridge is tight and secured, the bridge NCO will now call up the remainder of the squad/platoon to cross. The bridge NCO will monitor the crossing of all Marines, and will be the last one to cross.

TRANSITION: Now that we have discussed the construction of the bridge, are there any questions over the material? Next we will discuss the crossing of the bridge.

5. **<u>CROSSING</u>.** (FMST.07.34b) The method used to cross is known as a horizontal traverse. This traverse can be accomplished in the following ways.

a. <u>Rappel Seat Method</u>

1) The Marine ties himself into a rappel seat and inserts a large steel locking carabiner with the gate facing down and away.

2) The Marine faces the bridge with his right hand towards the near anchor to snap into the bridge, once the 85 is locked onto the bridge rope, a helper flips the 85 over so that the locking nut screws down. A carrier rope with a girth hitch may be added to assist the Marine in mounting the rope.

3) The Marine hangs below the bridge from his rappel seat with his head pointing in the direction of the far anchor and allows his legs to hang free.

4) Pulling with his arms makes progress.

5) This method is the safest and therefore the preferred method.

6) If the Marine must take a pack across, he may wear it with the waistband secured: However, the preferred method is to have another carabiner (aluminum non-locking is sufficient) attached to the pack frame at the top and attach this to the bridge behind the Marine putting his legs through the shoulder straps and pulling the pack across with him.

7) One man at a time will cross, although one can load and another can unload concurrently.

8) Weapons will be worn across the shoulder; muzzle down with a tight sling securely attached to the weapon.

b. <u>Pulley Method.</u> This method is used when the one rope bridge is long, uphill, or speed is vital and the Marines crossing it have a lot of heavy personal equipment; M24OG's, radios, etc.

1) Equipment required. One pulley, four steel locking carabiners, and a hauling line twice the length of the obstacle.

2) Setting up the system. Construction of the first suspension point is done by attaching a pulley to the one rope bridge, one 85 carabiner is attached to the pulley, gate down, and a second 85 is attached to the first 85 with the gate facing the near side bank. Into this second 85, attach the hauling rope with a figure-8 loop. The figure-8 is placed halfway along the hauling rope and attached to the 85, which is then locked. Four feet down the line another figure-8 on a bight is placed into the second suspension point.

3) Connection. The Marine clips his 85 into the lower 85, his equipment is clipped into the second 85. The mule team starts to pull.

c. Helmets will be worn for all methods of crossing the bridge. Gloves are optional.

NOTE: There are additional methods of crossing a one rope bridge such as the Commando crawl, Monkey crawl and hand over hand techniques, however, they are not taught at MWTC due to safety reasons.

6. **<u>RESCUE TECHNIQUES</u>** If an individual is unable to complete the crossing on the one rope bridge using the rappel seat method of crossing (i.e. from a fall or exhaustion), a rescue will have to be made in the following manner: (FMST.07.34c)

a. Reach. First try to reach the victim by using an object such as a pole, your hand, if the victim is close enough, etc ...

b. Throw. If reaching the victim is impossible, try throwing a rope to the victim and have him attach the rope to himself, preferably his seat, and pull him back to the desired side of the installation.

c. Tow. If the victim is unable to catch or reach a rope being thrown to him or the victim is unconscious, tie a figure-of-eight loop into the middle of the safety line and then connect the rope to the rope bridge with a steel locking carabiner. The mule team will pull the carabiner up against the victim's seat and begin towing the victim to the desired side of the installation. The rope used to tow the victim should be twice the length of the span of the rope bridge (If necessary, ropes can be tied together to accommodate the span). This will allow towing from either side without having to throw or have a Marine carry the safety line back across the bridge each time a towing rescue is performed.

d. Go. If all else fails, the last option will be to go after the victim. The rescuer will move out onto the rope bridge with a safety line attached to himself at approximately eight feet from the end of the safety line. In the end of the safety line tie a figure-of-eight loop and insert a steel locking carabiner in the end of it. Once the rescuer has made contact with the victim. He will attach the steel locking carabiner to either the victim's seat (preferably the victim's carabiner that is attached to the rope bridge), or to the bridge itself. To ensuring that the carabiner is placed so that it pulls against the victim's carabiner. The mule team now starts to pull both the victim and the rescuer to the desired side of the installation.

e. Cut. If while crossing the rope over water the individual goes underwater and no other rescue technique can be employed, the rope will be cut.

7. RETRIEVING THE ONE ROPE BRIDGE

a. Before the bridge NCO sends the last Marine to cross, they must make the bridge retrievable.

b. The first step is to break down the mechanical advantage, ensuring the braking knot is set before doing so. The system will be broken down until the bridging line is attached only to the brake knot.

c. They will now take the rope around to the back of the anchor, and tie a slip figure 8, with the loop of the slip 8, take it and attach it to the bridging line just in back of the braking knot with a 85.

d. Now pull the running end coming out of the slip 8 until it tightens, and bites on the rope. Once the rope bites, tie a thumb knot behind the slip 8 as close to the slip figure 8 as possible.

e. The bridge NCO, with the help of another Marine, will now untie the swammi wrap, until all of the tension is on the bridging line.

f. The safety line will now be brought up, and attached to the end of the bridging line with an overhand knot.

g. The bridge NCO and the other. Marine will now cross the bridge.

NOTE: It is possible that the bridge may lose tension after breaking the bridge down. The packs of the last two Marines across can be sent over with other Marines, prior to the slip figure being tied.

h. After both Marines have crossed, the safely line will be brought up from down stream.

i. The tree wrap will now be taken off. Once the tree wrap is off, all knots and carabiners will be taken out.

j. The safety line will now be pulled, the bridge is retrieved.

UNITED STATES MARINE CORPS
Mountain Warfare Training Center
Bridgeport, California 93517-5001

WMC
04/12/02

STUDENT OUTLINE

SUMMER MOUNTAIN WARFIGHTING LOAD REQUIREMENTS

LESSON PURPOSE. The purpose of this period of instruction is to introduce the student to the different levels of gear requirements in accordance with MWTC's uniform codes. This lesson relates to mountain walking and tactical evolutions

OUTLINE

1. **BASIC UNIFORM REQUIREMENTS.**

a. Marines operating in mountainous terrain will generally wear the standard issued utility uniform. However, due to the wide range of temperatures, and sudden changes in weather in the mountains, Marines will find the need to be continually adding and removing layers. The amount of clothing worn may vary depending upon the severity of the weather, the activity level of the Marine, and the individual metabolism of the Marine. Every person in that unit should still maintain the same outer camouflage layer.

b. As part of the basic uniform, each man will be required to have in his possession, at all times, seven required pocket items. These seven items should be carried in the pockets of your utility uniform:

(1) Pocketknife

(2) Whistle

(3) Pressure Bandage

(4) Chapstick and sunscreen

(5) Sunglasses

C10

(6) Survival Kit and rations (fire starting material, signaling material, food gathering material, water procuring material, 1st aid, shelter material, and some high energy, lightweight snacks.)

(7) Notebook with pen/pencil

c. Some additional items that should be carried in your pockets at all times:

(1) Contact gloves

(2) Paracord (10 meters)

(3) Flashlight with tactical lens and spare batteries

(4) Chemlights

2. **ASSAULT LOAD.**

a. The Assault Load is equipment in addition to the basic uniform requirements. It is carried in the load-bearing vest (LBV), butt pack and the pack system. This is the equipment carried for short duration missions such as security patrols or during the assault. It is also carried at all times when away from the bivouac site.

(1) An extra insulating layer (Field jacket liner, wooly pully, etc.)

(2) Protective layer (ECWCS parka and trousers)

(3) LBV with 2 quarts of water and first aid kit

(4) Helmet

(5) Rations for the time away from your bivouac site

(6) Cold weather hat or balaclava

(7) Individual mountaineering gear

a. Sling rope

b. Carabiners

c. Rappelling Gloves

(8) Specialized mountaineering equipment

(9) Mission essential gear

3. **COMBAT LOAD**. The Combat Load is the equipment carried for longer duration missions.

The following items are in addition to the items already being carried in the Assault Load:

a. Sleeping bag with bivy bag

b. Isopor mat

c. Stove

d. Fuel bottle

e. Poncho (for expedient shelters or medevac purposes)

4. **EXISTENCE LOAD**. The Existence Load is to be packed for longer duration combat missions, where it is necessary to replace worn out items. These items could be uniforms, boots, personal hygiene gear, etc. The Existence Load should be brought to the forward elements once the situation allows.

5. **PACKING CONSIDERATIONS.** Because most Marines are familiar with how to pack a pack, these are general guidelines only.

a. When not wearing your protective layer or insulating layers, keep it handy at the top of the pack. When taking a break during a movement, you will be able to quickly don a layer to prevent getting chilled.

b. Keep your stove and fuel bottle in the outside pockets of your pack. They may leak and soak your equipment with fuel if stored inside your pack.

c. Keep the climbing rope in the rope bag. This will reduce the rope's exposure to petroleum products, ultra-violet light, cuts, and abrasions.

d. Ice ax is placed through the ice ax loops with spike up, adze outboard, and secured in place.

e. Crampons are attached to the outside of the pack with the tips covered and facing outboard.

Table of Contents

Chapter 1. The M9 Service Pistol

Chapter 2. Weapons Handling

Chapter 3. Fundamentals of Pistol Marksmanship

Chapter 6. Presentation of the M9 Service Pistol

Chapter 7. Pistol Engagement Techniques

Chapter 8. One-Handed Techniques

Chapter 9. Advanced Techniques

Chapter 10. Low Light and Darkness Techniques

Chapter 1
The M9 Service Pistol

The Marine Corps' uses the M9 service pistol as a defensive weapon. It is a semiautomatic, magazine-fed, recoil-operated, double-action pistol. The M9 service pistol's magazine holds 15 rounds and is chambered with a 9-mm ball (NATO M882 round). The M9 service pistol has a maximum effective range of 50 meters (54.7 yards).

Note: The procedures in this chapter are written for right-handed Marines; left-handed Marines reverse instructions as necessary.

Functional Capabilities

When the M9 service pistol's safety lever is moved to the firing position, it can be fired in either a single-action or double-action mode. It is designed to fire one round each time the trigger is pulled. When the last round is fired, the slide automatically locks to the rear.

Single-Action Mode

The single-action mode allows the pistol to be fired when the hammer is cocked; single action requires the hammer to be cocked to the rear before the trigger is pulled. The hammer can be manually cocked or mechanically cocked. The hammer is mechanically cocked after the first shot is fired. See figure 1-1.

Figure 1-1. Single-Action Mode.

Double-Action Mode

The double-action mode causes the hammer to move to the rear as the trigger is being pulled. See figure 1-2.

Figure 1-2. Double-Action Mode.

Nomenclature

See figure 1-3 on page 1-2.

Major Components

The M9's major components consist of the slide assembly, barrel assembly, and receiver. See figure 1-4 on page 1-3.

Slide Assembly

The slide assembly houses the firing pin, firing pin block, striker, extractor, and loaded chamber indicator, and it cocks the hammer during recoil.

Extractor

The extractor pulls the brass from the chamber after the round is fired.

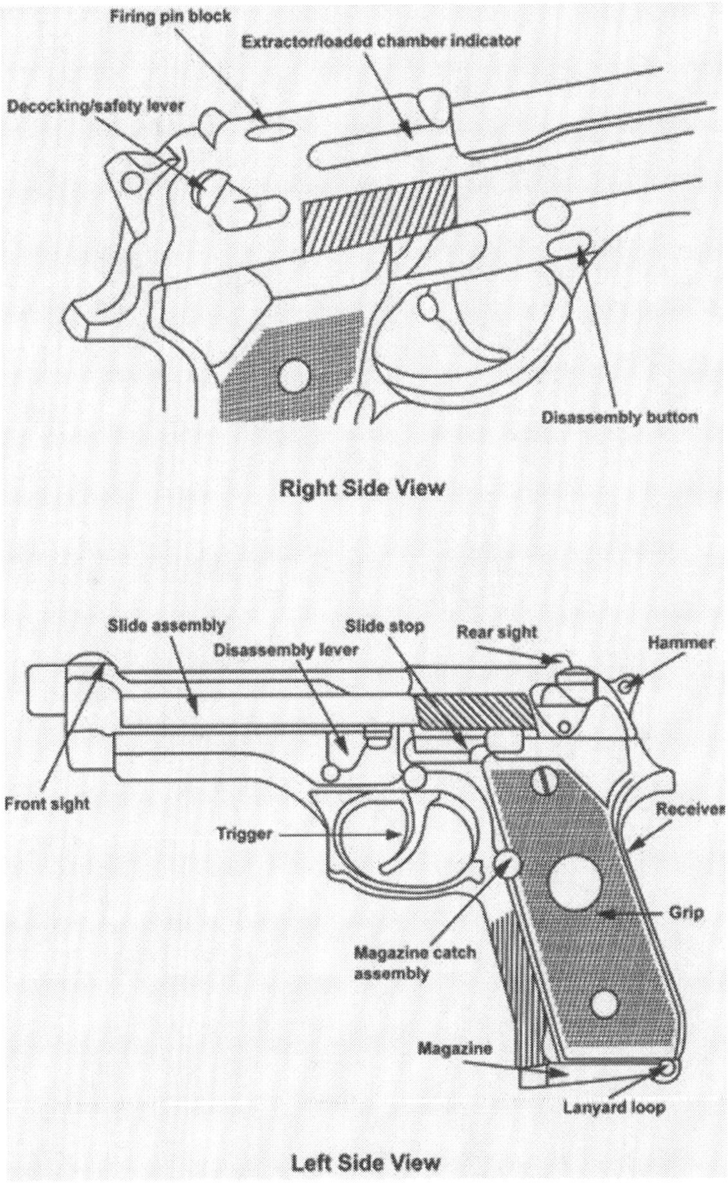

Figure 1-3. M9 Service Pistol.

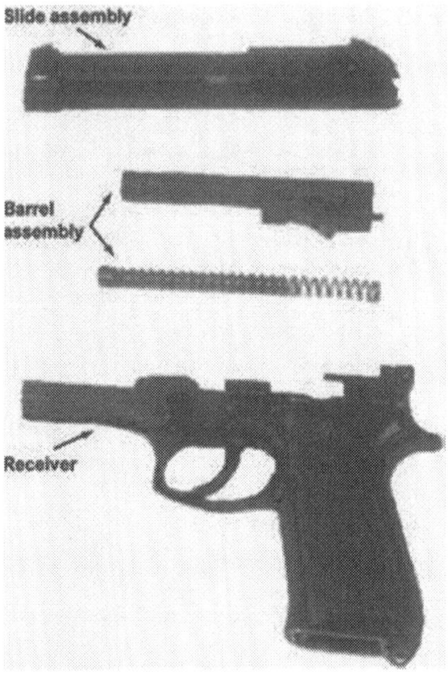

Figure 1-4. Major Components.

Loaded Chamber Indicator

When a round is in the chamber, the upper surface of the loaded chamber indicator protrudes from the right side of the slide. This protrusion can be felt with the finger, verifying that there is a round in the chamber.

Barrel Assembly

The barrel assembly houses the round for firing, directs the projectile, and locks the barrel in position during firing.

Receiver

The receiver supports the major components, controls the functioning of the pistol, and holds the magazine in place. The front and back straps of the grip are grooved vertically to ensure that the hand does not slip while firing. The receiver consists of the disassembly button, slide stop, and magazine catch assembly.

Disassembly Button

The disassembly button permits quick disassembly of the pistol.

Slide Stop

The slide stop holds the slide to the rear after the last round is fired. It can also be manually operated to lock the slide to the rear or release the slide.

Magazine Catch Assembly (Magazine Release Button)

The magazine catch assembly secures the magazine in place when loading, and it releases the magazine from the pistol when unloading. The magazine catch assembly is designed for both right- and left-handed Marines.

> *Note:* Reversal of the magazine catch assembly for left-handed Marines can be performed by a qualified armorer.

Safety Features

The safety features of the M9 service pistol include the decocking/safety lever, firing pin block, and half-cock notch.

Decocking/Safety Lever

The decocking/safety lever, commonly referred to as the safety, permits safe operation of the pistol by both right- and left-handed Marines. As the safety is moved to the safe (down) position, the firing pin striker moves out of alignment with the firing pin. This movement prevents the pistol from firing as the hammer moves forward.

> *Note:* In the fire (up) position, a red dot is visible, indicating that the pistol is ready to fire.

Firing Pin Block

The firing pin block rests in the firing pin notch and prevents movement of the firing pin until the trigger is pulled. As the trigger is pulled, the firing pin block moves up and out of the firing pin notch. This movement allows a round to be fired when the hammer strikes the firing pin.

Half-Cock Notch

The half-cock notch stops the forward movement of the hammer during a mechanical failure.

Cycle of Operation

There are eight steps in the cycle of operation for the M9 service pistol.

Firing

Once the safety is off and the trigger is pulled to the rear, the hammer falls on the firing pin, which strikes the primer and ignites the round. See figure 1-5.

Figure 1-5. Firing.

Unlocking

As the slide assembly moves to the rear, the locking block rotates out of the notches in the slide. See figure 1-6.

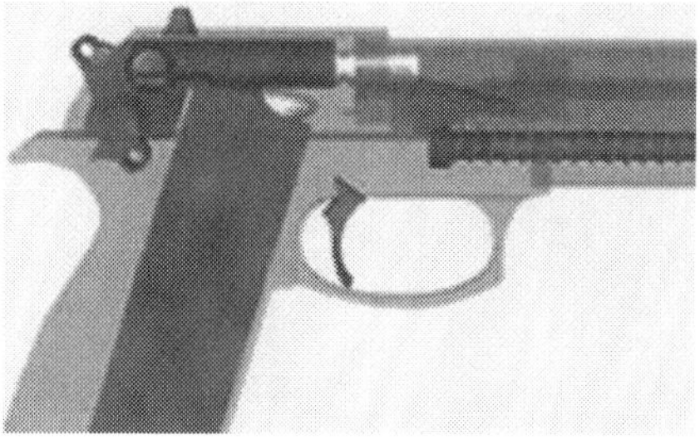

Figure 1-6. Unlocking.

Extracting

As the slide moves rearward, the extractor withdraws the cartridge case out of the chamber. See figure 1-7.

Figure 1-7. Extracting.

Ejecting

As the face of the slide passes over the ejector, the case strikes the ejector and it is knocked upward and outward through the ejection port. See figure 1-8.

Figure 1-8. Ejecting.

Cocking

As the slide moves rearward, the hammer is pushed back, allowing the sear to engage the hammer hooks, cock the hammer to the rear, and place the pistol in the single-action mode. See figure 1-9.

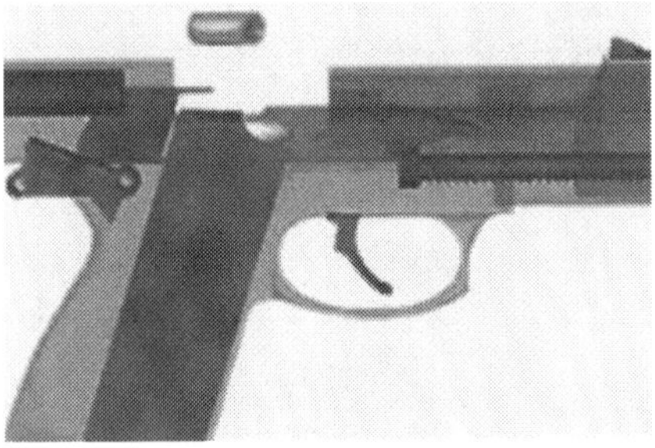

Figure 1-9. Cocking.

Feeding

The slide starts forward, pushed by the recoil spring. The face of the slide makes contact with the cartridge at the top of the magazine, stripping it from the magazine and pushing it toward the chamber. See figure 1-10.

Figure 1-10. Feeding.

Chambering

As the slide continues forward, it pushes the cartridge into the chamber. See figure 1-11.

Figure 1-11. Chambering.

Locking

As the slide assembly continues to move forward, the locking block lugs move into the locking block recesses on the right and left sides of the slide. See figure 1-12 on page 1-6.

Figure 1-12. Locking.

Ammunition

The only ammunition authorized for the M9 service pistol is the NATO M882 9-mm ball. Dummy ammunition can be used during training (a dummy round has a hole drilled in its side and contains no primer). See figure 1-13.

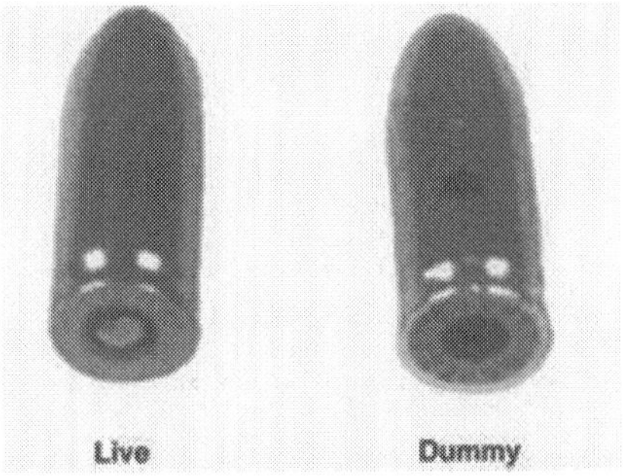

Figure 1-13. M9 Service Pistol Ammunition.

Do not open ammunition containers until the ammunition is to be used. Ammunition must be maintained in a high state of readiness. To care for ammunition—

- Keep ammunition dry and clean. If ammunition gets wet or dirty, wipe it off with a clean dry cloth.
- Wipe off light corrosion as soon as it is discovered. Never use ammunition that is heavily corroded, dented, or has the projectile pushed in.
- Do not expose ammunition to direct sunlight for long periods of time.
- Do not oil or grease ammunition. Dust or other abrasives can collect on greasy ammunition and may cause damage to the operating parts of the pistol. Oiled cartridges also produce excessive chamber pressure.

Wearing of the M9 Service Pistol's Gear

The proper placement of pistol gear helps ensure safety and aids the Marine in effectively handling and employing the pistol.

M12 Holster

The M12 holster consists of the holster, removable holster flap, and metal retaining clip. To check for proper placement of the holster, allow the right arm to hang freely. The holster should be slightly in front of the arm to permit easy access to the pistol upon presentation from the holster. See figure 1-14.

Note: In most cases, the holster is issued with the holster flap installed for a right-handed Marine. To convert the holster for a left-handed Marine, remove the metal retaining clip and install the clip on the opposite side of the holster.

Figure 1-14. The M12 Holster and Ammunition Pocket.

M1 Ammunition Pocket

The ammunition pocket attaches to the cartridge belt on the side opposite the holster in a position that best permits ready access for reloads. The magazine is stored in the ammunition pocket with the rounds down and pointed inboard.

M7 Shoulder Holster

The M7 shoulder holster consists of a holster with a thumb snap closure, shoulder strap, chest strap, and a belt retaining loop. The holster is positioned on the left side of the chest to provide easy and quick access with the right hand. See figure 1-15. The M7 holster comes fully assembled and has adjustable straps to accommodate each Marine's body size. The holster is available for right-handed Marines only, therefore, a left-handed Marine has to withdraw the pistol from the holster with the right hand and then transfer the pistol to the left hand before firing. (See chap. 8 for transferring the pistol from one hand to the other.) When the shoulder holster is worn properly—

Figure 1-15. M7 Shoulder Holster.

- The shoulder strap lays flat across the left shoulder with the shoulder pad directly on top of the shoulder.
- The chest strap attaches to the D-ring at the top of the holster and runs diagonally across the chest, underneath the right arm, and around the back where it attaches to the end of the shoulder strap.
- The belt retaining loop is at the bottom of the holster and attaches to the belt to stabilize the holster's position.
- The ammunition pocket attaches to the chest strap directly underneath the right arm.

Assault Holster

Some Marines (i.e., Marine security force, direct action platoon, and military police) are required to carry the assault holster. This holster has a retention strap that fastens over the top of the holster to retain the pistol. This type of holster generally has a thumb break on the retention strap that is disengaged to access the pistol. See figure 1-16.

Figure 1-16. Assault Holster.

Concealed Pistol Holster

Some Marines are required to carry a concealed pistol as part of their official duties. The primary consideration for placement of a concealed pistol holster is to ensure the pistol cannot be seen; therefore, the Marine must consider the type of clothing to be worn. A secondary consideration is to place the holster so the pistol can be presented easily. Typically, the best position for a concealed holster is just behind the strong side hip. This position best conceals the pistol while allowing it to be presented quickly. Another placement choice is in a shoulder holster, placing the pistol just under the weak side arm. See figures 1-17 through 1-20 on pages 1-8 and 1-9.

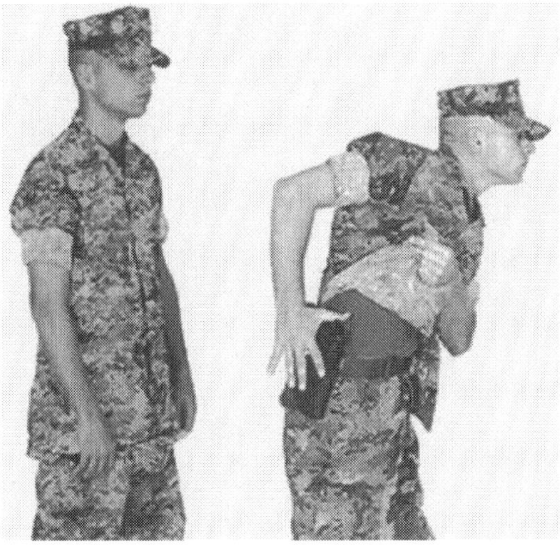

Figure 1-17. Wearing of the Concealed Pistol Holster—Utilities.

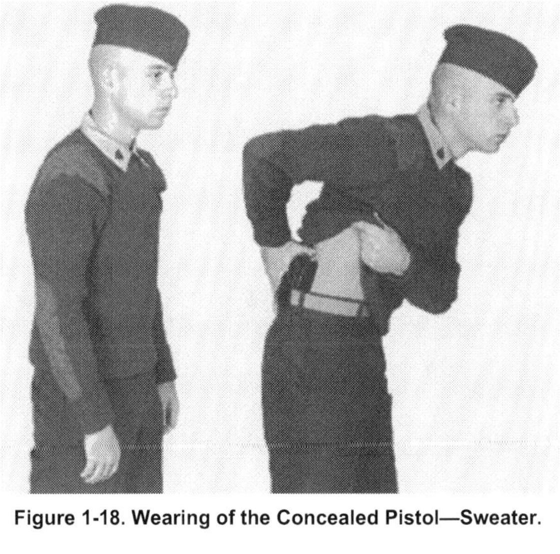

Figure 1-18. Wearing of the Concealed Pistol—Sweater.

Figure 1-19. Wearing of the Concealed Pistol Holster—Jacket.

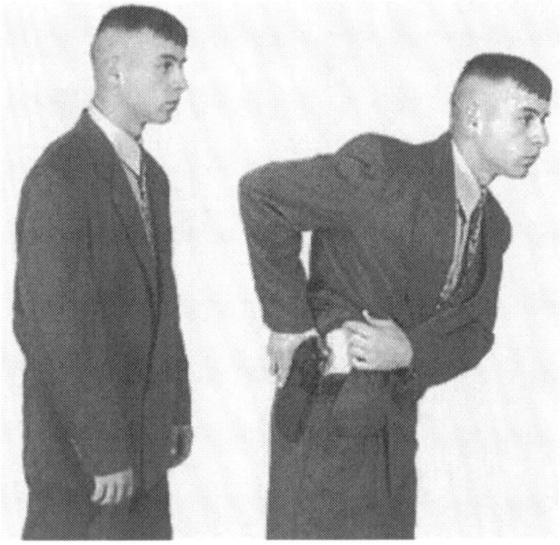

Figure 1-20. Wearing of the Concealed Pistol Holster—Civilian Attire.

Lanyard

The lanyard aids in pistol retention. It is issued in three sizes and is adjustable. The lanyard consists of a fabric cord, two cylindrical slip rings, and a metal clip that attaches the lanyard to the pistol's lanyard loop. See figure 1-21.

Figure 1-21. M9 Service Pistol (With Lanyard).

To don the lanyard—

- Adjust the slip rings so they are positioned flush with the lanyard's base (clip end).
- Place the right arm through the loop and place the loop over the head, resting on the left shoulder.
- Attach metal clip to the pistol's lanyard loop.
- Place the pistol in the holster.
- Use the left hand to hold the bottom slip ring against the base of the lanyard. Use the right hand to slide the top slip ring upward to position the loop of the lanyard under the arm. The lanyard should fit snugly against the body, but not restrict the Marine's movements.
- Tuck any excess cord behind the holster.
- Ensure that the lanyard is adjusted properly by removing the pistol from the holster and fully extending the right arm. The lanyard should be taut. Adjust as necessary.

Note: Re-adjust the lanyard if any equipment changes are made (e.g., flak jacket).

Firing the M9 Service Pistol While Wearing Gloves

Not all combat engagements occur during ideal weather conditions. During cold weather, the Marine may find it necessary to wear gloves to protect the fingers from frostbite and help prevent stiffening of the hands. Gloves may also be worn in mission-oriented protective posture conditions. Gloves provide protection to the hands, however, they also may interfere with the Marine's ability to engage targets effectively. The added bulk of the gloves may affect the Marine's ability to manipulate the safety, magazine release button, magazine, hammer, and slide stop/release. For example, the Marine may need to exert more pressure to engage the magazine release or slide release buttons to compensate for the thickness of the gloves.

The principles of target engagement (see chap. 7) do not change while wearing gloves, however, the specific ability to manipulate and control the trigger is greatly affected by the thickness of the gloves around the fingers. Wearing gloves reduce the Marine's ability to feel, which makes it difficult to apply trigger control when firing. The Marine may find that more pressure than normal must be applied with the trigger finger just to establish initial contact with the trigger. Once the Marine can "feel" the trigger through the gloves, then the pressure required to fire a shot can be applied. This action may increase the chances of firing the pistol prematurely due to excessive pressure on the trigger. Dry firing while wearing gloves allows the Marine to learn how to apply trigger control consistently and determines how much pressure is needed to effectively fire a shot.

While wearing gloves, the Marine may find it difficult to fire the pistol in the double-action mode due to the position of the trigger and the limited amount of space between the trigger and the trigger guard. Therefore, if the situation permits, the Marine may wish to thumbcock the pistol to fire in single-action mode. In single-action mode, there is more space between the trigger guard and the trigger, making it easier to position the finger on the trigger. However, the Marine's ability to thumbcock the pistol may also be hindered by the gloves' bulk. Therefore, to thumbcock the pistol while wearing gloves, the Marine may perform one of the following methods:

- For method one, loop a section of 550 cord (approximately 2 inches) through the loop located on the top rear portion of the hammer. The length of the cord should not interfere with the pistol's cycle of operation or with the Marine's ability to establish sight alignment. Once the cord is attached to the hammer, take the pistol off safe and pull downward on the cord to cock the hammer. See figure 1-22.

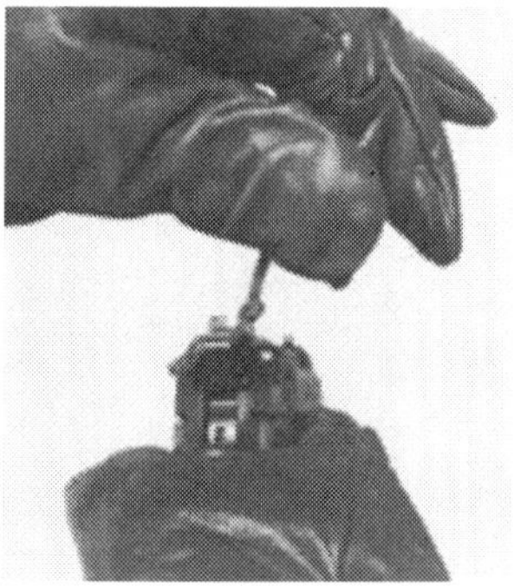

Figure 1-22. Cocking the Pistol with 550 Cord.

- For method two, take the pistol off safe, rotate the pistol inboard, and place the top of the hammer against a secure surface (e.g., cartridge belt, table top, heel of boot). Apply pressure on the pistol to keep the hammer in place and push downward on the pistol in one continuous motion to cock the hammer. See figure 1-23.

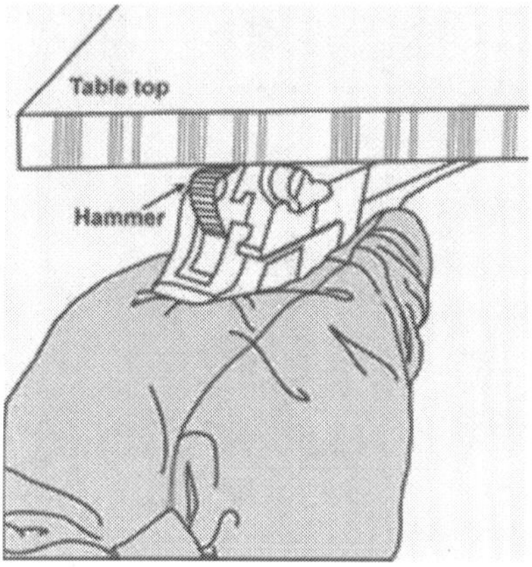

Figure 1-23. Cocking the Pistol on a Secure Surface.

Preventive Maintenance

If the M9 service pistol is to be effective, it must be maintained in a state of operational readiness at all times; therefore, maintenance of the M9 service pistol is a continuous effort. A clean, properly lubricated, well-maintained pistol will fire when needed.

Pistol Disassembly

Before disassembling the M9 service pistol, ensure that the pistol is in Condition 4. The pistol is in Condition 4 when the magazine is removed, the chamber is empty, the slide is forward, and the safety is on. To disassemble the pistol, perform the following steps in sequence:

- Hold the pistol in the right hand with the muzzle slightly elevated. Reach over the slide with the left hand and place the left index finger on the disassembly button and the left thumb on the disassembly lever. Press the disassembly button and hold it in place while rotating the disassembly lever downward until it stops.

Note: A left-handed Marine places the right thumb on the disassembly button and the right index finger on the disassembly lever.

- Pull the slide and barrel assembly forward and remove it while wrapping the fingers around the slide to hold the recoil spring and recoil spring guide in place.
- Turn the slide assembly over in the left hand until the recoil spring and recoil spring guide face up. Place the right thumb on the end of the recoil spring guide next to the locking block and compress the recoil spring and spring guide while lifting and removing them from the slide and barrel assembly. See figure 1-24. Allow the recoil spring to decompress slowly.

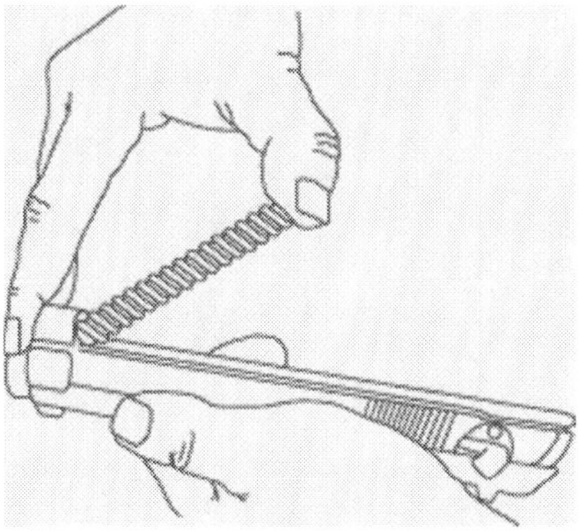

Figure 1-24. Removing the Recoil Spring and Recoil Spring Guide.

- Separate the recoil spring from the recoil spring guide.
- Push in on the locking block plunger with the right index finger while pushing the barrel forward slightly. Lift and remove the locking block and barrel assembly from the slide.

A Marine is not authorized to disassemble the pistol any further than the preceding steps. Any further disassembly must be performed by ordnance personnel. See figure 1-25.

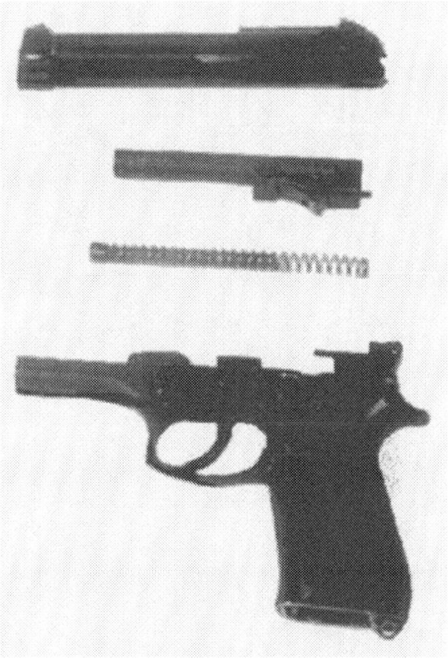

Figure 1-25. Disassembled M9 Service Pistol.

Disassembly of the Magazine

To disassembly the magazine, perform the following steps:

- Grip the magazine firmly in the left hand with the floorplate up and the thumb resting against the flat end of the floorplate.
- Release the floorplate by pushing down (with a blunt object; e.g., an ink pen) on the floorplate retainer stud in the center of the floorplate. At the same time, slide the floorplate a short distance forward with the thumb.
- Maintain the magazine spring pressure with the thumb and remove the floorplate from the magazine.
- Remove the floorplate retainer and magazine spring and follower from the magazine tube.

See figure 1-26.

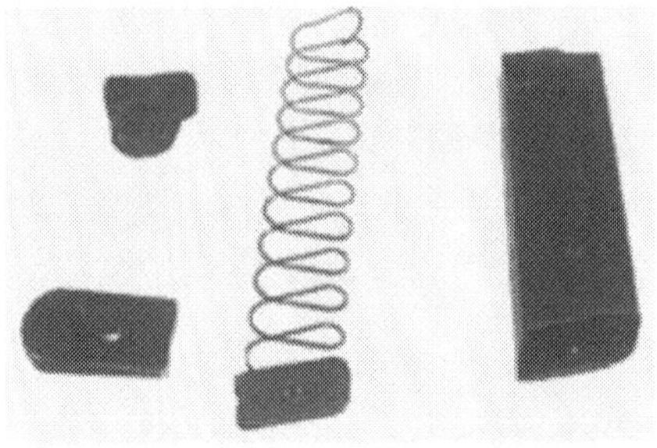

Figure 1-26. Disassembled Magazine.

Inspection of the Pistol

Once the pistol has been disassembled, it must be thoroughly inspected to ensure it is in a serviceable condition. Pistol inspection is continuous during the pistol's cleaning and reassembly:

Slide Assembly
- Check for free movement of the safety. Ensure the rear sight is secure.

Barrel Assembly
- Inspect the bore and chamber for pitting or obstructions.
- Check the locking block plunger for free movement of the locking block.
- Inspect the locking lugs for cracks and burrs.

Recoil Spring and Recoil Spring Guide
- Check the recoil spring for damage.
- Check that it is not bent.
- Check the recoil spring guide for straightness and smoothness.
- Check to be sure it is free of cracks and burrs.

Receiver Assembly
- Check for bends, chips, and cracks.
- Check for free movement of the slide stop and magazine catch assembly.
- Check the guide rails for excessive wear, burrs, cracks, or chips.

Magazine Assembly
- Check the spring and follower for damage.
- Ensure the lips of the magazine are not excessively bent and are free of cracks and burrs. The magazine tube should not be bent or dirty.

Cleaning and Lubricating the Pistol

Only authorized cleaning materials should be used to clean and lubricate the pistol. If these items are not issued with the pistol, they may be obtained from the armory. The following procedures are used to clean and lubricate the pistol—

Slide Assembly
- Clean the slide assembly with a cloth. A general purpose brush and cleaning lubricant protectant (CLP) can also assist in the removal of excess dirt and carbon buildup.
- Ensure the safety, bolt face, slide guides, and extractor are free of dirt and residue.
- Wipe dry with a cloth and apply a light coat of CLP.

Barrel Assembly
- Insert a bore brush with CLP into the chamber end of the barrel, ensuring that it completely clears the muzzle before it is pulled back through the bore.

Caution

Insert the bore brush through the chamber to prevent damage to the crown of the barrel.

- Repeat several times to loosen carbon deposits.
- Dry the barrel by pushing a swab through the bore.
- Repeat until a clean swab can be observed.
- Clean the locking block with a general purpose brush.
- Use the barrel brush to apply a light coat of CLP to the bore and chamber area and lubricate the exterior surfaces of the barrel and locking block.

Recoil Spring and Recoil Spring Guide
- Clean the recoil spring and recoil spring guide using CLP and a general purpose brush or cloth.
- Apply a light coat of CLP after wiping the recoil spring and recoil spring guide clean.

Receiver
- Wipe the receiver assembly clean with a cloth.
- Use a general purpose brush for areas that are hard to reach, paying special attention to the

disassembly lever, trigger, slide stop, hammer, and magazine release button.

- Apply a light coat of CLP.

Caution

Do not allow the hammer to fall with full force by pulling the trigger when the slide is removed. This can damage the receiver and hammer.

Magazine

- Clean the magazine tube and follower with CLP and a general purpose brush.
- Wipe the magazine spring, floorplate retainer, and floorplate clean with a cloth.
- Apply a light coat of CLP.

Pistol Re-assembly

After the M9 service pistol has been cleaned and lubricated, it must be properly re-assembled to ensure its serviceability. To re-assemble the M9 service pistol—

- Use the left hand to grasp the slide with the bottom facing up and the muzzle pointing toward the body. Use the right hand to grasp the barrel assembly with the locking block facing up. Use the index finger to push in the locking block plunger while placing the thumb on the base of the locking block.
- Insert the muzzle of the barrel assembly into the forward open end of the slide. At the same time, lower the rear of the barrel assembly by slightly moving the barrel downward. The locking block should fall into the notches of the slide assembly.
- Slip the recoil spring guide into the recoil spring.
- Insert the end of the recoil spring and recoil spring guide into the slide recoil spring housing. At the same time, compress the recoil spring and lower the spring guide until it is fully seated onto the locking block cutaway.
- Use the left hand to grasp the slide and barrel assembly, sights up, and wrap the fingers around the slide assembly to hold the recoil spring and guide in place. Align the slide assembly guide rails onto the receiver assembly guide rails.
- Push the slide rearward while pushing up on the slide stop with the thumb. Lock the slide to the rear while maintaining upward pressure on the slide stop. Rotate the disassembly lever upward. Listen for a click, an audible click indicates a positive lock.

Pistol Magazine Re-assembly

To re-assemble the magazine—

- Grip the magazine firmly in the left hand with the floorplate end up and the counting holes facing the Marine. Insert the follower into the magazine so the flat end of the follower is against the flat end of the magazine.
- Ensure the floorplate retainer is attached to the first curve of the bottom coil.
- Hold the spring upright with the right hand and insert the spring into the magazine tube so that the flat end of the floorplate retainer is against the flat end of the magazine.
- Push the magazine spring and floorplate retainer down with the right hand and hold it in place with the thumb of the left hand. Use the right hand to slide the floorplate over the side walls of the magazine until fully seated, which is indicated by an audible click.

Safety/Function Check

A safety/function check is performed after reassembling the M9 service pistol. Perform the following steps to ensure the pistol is operational:

- Ensure there is no ammunition in the chamber of the pistol.
- Ensure that the safety is in the safe position, then depress the slide stop, allowing the slide to return fully forward. At the same time, the hammer should fall to the full forward position.

- Pull and release the trigger. The firing pin block should move up and down but the hammer should not move.
- Place the safety in the fire position.
- Pull the trigger to check the double action. The hammer should cock and fall.
- Pull the trigger again and hold it to the rear. Use the fingers and thumb of the left hand to grasp the serrated sides of the slide just forward of the safety. Pull the slide to its rearmost position and release it while holding the trigger to the rear. Release the trigger, a click should be heard and the hammer should not fall.
- Pull the trigger to check the single action. The hammer should fall. Place the safety in the safe position.

If the safety/function check does not indicate an operational pistol, the Marine takes the pistol to organizational maintenance or the next authorized repair level.

User Serviceability Inspection

The Marine is responsible for performing a user serviceability inspection on the pistol prior to live fire. The user serviceability inspection ensures the pistol is in an acceptable operating condition. This inspection is not intended to replace the detailed pistol components inspection following disassembly or the limited technical inspection or pre-fire inspection conducted by a qualified armorer. To conduct a user serviceability inspection on the pistol, perform the following steps:

- Ensure the magazine release button is on the left side of the pistol for a right-handed Marine, the right side of the pistol for a left-handed Marine.
- Ensure the magazine seats into the magazine well when it is inserted and that it cannot be pulled out.
- Ensure the slide stays locked to the rear when the slide is pulled rearward with an empty magazine in the pistol.
- Ensure the magazine falls out freely when the magazine release button is depressed.
- Repeat the preceding four steps with the second magazine.
- Ensure the slide is locked to the rear before lubricating the spring guide, the top of the barrel just forward of the front sight, and the guide rails of the slide assembly behind the safety. With the muzzle pointed downward, work the slide several times and release.
- Inspect the pistol's external parts visually to ensure that there are no cracks or excessive wear.
- Perform a safety/function check of the pistol.

Pistol Maintenance in Adverse Conditions

Combat situations can place Marines in a variety of adverse conditions. Therefore, the M9 service pistol must be maintained properly to ensure its continued operation.

Extreme Cold

In extreme cold conditions, ensure that the following maintenance is performed:

- Clean and lubricate the pistol inside at room temperature, if possible.
- Apply a light coat of lubricant, arctic weather (LAW) to all functional parts.
- Always keep the pistol dry.
- Keep the pistol covered when moving from a warm to a cold area. This permit gradual cooling of the pistol and prevents freezing.
- Ensure that a hot pistol is not placed in snow or on ice.
- Keep snow out of the bore of the barrel. If snow should enter the bore, use a swab and cleaning rod to clean the bore before firing.

Hot, Wet Climates

Maintenance in hot, wet climates must be performed more frequently:

- Inspect hidden surfaces for corrosion. If corrosion is found, clean and lubricate.
- Remove handprints with a clean cloth in order to prevent corrosion.
- Dry the pistol with a cloth and lubricate it with CLP.
- Check ammunition and magazines frequently for corrosion. Disassemble and clean the magazines with CLP and wipe dry with a clean cloth. If necessary, clean ammunition with a dry cloth.

Hot, Dry Climates

In a hot, dry climate, dust and sand can cause stoppages and excessive wear on component contact surfaces during firing; therefore, keep the pistol covered whenever possible.

Corrosion is less likely to form on metal parts in a dry climate. Therefore, lightly lubricate internal working surfaces with CLP. Do not lubricate external parts of the pistol. Wipe excess lubricant from exposed surfaces. Do not lubricate internal components of the magazine.

Heavy Rain and Fording Operations

The following maintenance procedures are followed during periods of heavy rains or during fording operations:

- Always attempt to keep the pistol dry.
- Drain any water from the barrel prior to firing.
- Dry the bore with a swab and cleaning rod.
- Generously lubricate internal and external surfaces of the pistol with CLP.

Amphibious Conditions

If the pistol comes into contact with salt water, clean the pistol as soon as possible. If time does not permit cleaning in accordance with Technical Manual (TM) 1005A-10/1, *Operator's Manual, Pistol Semiautomatic, 9mm, M9*, then wash the pistol with fresh water

CHAPTER 2
WEAPONS HANDLING

Weapons handling is a method of providing consistent and standardized procedures for handling, operating, and employing the M9 service pistol. Understanding and applying the principles of weapons handling are critical to developing safe and consistent weapons skills. Strict adherence to training and diligent practice will make weapons handling instinctive. Mission accomplishment and survival during combat depend on a Marine's ability to react instinctively and with confidence.

Note: The procedures in this chapter are written for right-handed Marines; left-handed Marines should reverse directions as needed.

Safety Rules

The following safety rules are the foundation for responsible weapons handling. They must be observed at all times, both in training and combat.

Safety Rules

Rule 1: Treat every weapon as if it were loaded.
This rule is intended to prevent unintentional injury to personnel or damage to property from handling or transferring possession of a weapon.

Rule 2: Never point a weapon at anything you do not intend to shoot.
This rule enforces the importance of muzzle awareness and reinforces positive identification of the target.

Rule 3: Keep your finger straight and off the trigger until you are ready to fire.
This rule is intended to minimize the risk of firing the weapon negligently (when not firing, the trigger finger is straight along the receiver, outside of the trigger guard). This rule also reinforces positive identification of the target.

Rule 4: Keep weapon on safe until you intend to fire.
This rule enforces the use of the weapon's own safety feature and reinforces positive identification of the target.

Weapons Conditions

The M9 service pistol's level of readiness is defined by three specific conditions. The steps in the loading and unloading process take the pistol through the specific conditions that indicate the pistol's readiness for live fire. The Marine must understand and know the condition of his pistol at all times.

Condition 1. Magazine inserted, round in chamber, slide forward, and safety on.

Condition 2. Not applicable to the M9 service pistol.

Condition 3. Magazine inserted, chamber empty, slide forward, and safety on.

Condition 4. Magazine removed, chamber empty, slide forward, and safety on.

Determining a Weapon's Condition

There are two methods for determining the pistol's condition: checking the round indicator on the right side of the pistol and conducting a chamber check.

Checking the Round Indicator

When there is a round in the chamber, the upper surface of the extractor protrudes from the right side of the slide. The protrusion can be felt by sliding either the thumb or the index finger of the left hand over the top of the slide and across the extractor. See figure 2-1.

Conducting a Chamber Check

To conduct a chamber check—

- Point the pistol in a safe direction and grasp the pistol grip with the right hand.

Figure 2-1. Checking the Round Indicator.

- Place the trigger guard in the palm of the left hand. Use the thumb and index finger to grasp the forward end of the slide at the indentations under the front sight, behind the muzzle. See figure 2-2.

Figure 2-2. Placement of the Left Hand.

Caution

Ensure the muzzle does not cover the hand or fingers.

- Keep thumb in place around the pistol's backstrap and rotate fingers of the right hand over the top of the slide in front of the rear sight.
- Pull the slide to the rear by pushing forward with the right thumb and pulling back on the rear sight with the fingers. Use the left hand to steady the pistol and to assist in pulling the slide to the rear.
- Use the right hand to hold the slide to the rear (just enough to visually inspect the chamber for a round). Physically check for a round by inserting a finger of the right hand into the chamber area. See figure 2-3.

Figure 2-3. Chamber Check.

Note: At night or in low light conditions, the Marine's visibility is reduced; therefore, the Marine will have to rely on the physical check with the finger to determine if a round is in the chamber.

Caution

Pulling the slide too far to the rear while inspecting the chamber may cause a double feed or the ejection of a round.

- Remove the finger from the chamber and release tension on both hands to allow the slide to go forward. Ensure that the slide is all the way forward.

Weapons Commands

Weapons commands direct the Marine to safely load, unload, and employ the M9 service pistol. Six commands are used in weapons handling:

"*Load*"
This command is used to take a weapon from Condition 4 to Condition 3.

"*Make Ready*"
This command is used to take a weapon from Condition 3 to Condition 1.

"*Fire*"
This command is used to engage targets.

"*Cease Fire*"
This command is used to cease target engagement.

"*Unload*"
This command is used to take a weapon from any condition to Condition 4.

"*Unload, Show Clear*"
This command is used to require a second individual to check the weapon to verify that no ammunition is present before the weapon is put into Condition 4.

Loading the Pistol

Perform the following steps to load the pistol (take the pistol from Condition 4 to Condition 3):

- Ensure the pistol is on safe.
- Use the right hand to grip the pistol grip firmly. Ensure that the pistol is pointed in a safe direction, bring the trigger guard to the right of eye level and cant the pistol so the magazine well faces inboard at approximately a 45-degree angle to the deck. Draw the right elbow in to facilitate control of the pistol.
- Use the left hand to remove a filled magazine from the ammunition pocket. Index the magazine by sliding the index finger along the forward edge of the magazine. See figure 2-4.
- Insert the filled magazine into the magazine well by guiding it with the index finger and, with the fingers extended, pushing it in with the heel of the hand until it is fully seated. Do not relinquish contact with the magazine until it is fully seated. See figure 2-5.

Figure 2-4. Indexing the Magazine.

Figure 2-5. Seating the Magazine.

Making the Pistol Ready

Perform the following steps to take the pistol from Condition 3 to Condition 1:

- Firmly grip the pistol grip with the right hand. Ensure that the pistol is pointed in a safe direction and the slide is in its forward position.
- Rotate the magazine well outboard to facilitate pulling the slide to the rear. With the fingers and thumb of the left hand, grasp the serrated sides of the slide just forward of the safety. See figure 2-6.

Caution

Ensure the muzzle does not cover the hand or fingers.

- Pull the slide to its rearmost position by pushing forward with the right hand while pulling back on the slide with the left hand.
- Release the slide, this strips a round from the magazine and chambers it as the slide moves forward.
- Ensure the pistol remains on safe.
- Conduct a chamber check to ensure a round is in the chamber.

Note: A chamber check may be conducted at any time to check the pistol's condition.

Fire

Perform the following steps to fire the pistol:

- Keep trigger finger straight and use the right thumb to take the pistol off safe.
- Place the trigger finger on the trigger and apply pressure to the trigger until the shot is fired.

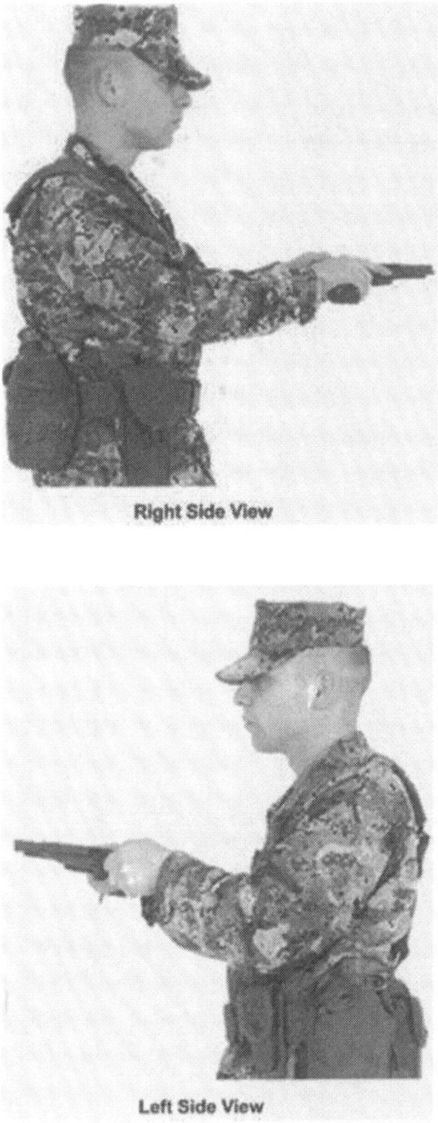

Figure 2-6. Grasping the Slide to Make Ready.

Cease Fire

Perform the following steps to execute a cease fire of the pistol:

- Remove the finger from the trigger and place it straight along the receiver.
- Place the pistol on safe without breaking the grip of the right hand.
- Assume a carry or transport position.

Unloading the Pistol

Perform the following steps to take the pistol from any condition to Condition 4:

- Use the right hand to grip the pistol firmly. Ensure that the pistol is on safe.
- Rotate the pistol so the magazine well is pointed inboard and angled down.

  Note: The angle of the magazine well must allow the magazine to fall freely from the well once the magazine release button is engaged.

- Depress the magazine release button to remove the magazine from the pistol. Catch the magazine with the left hand and retain it.
- Push upward on the slide stop with the right thumb and maintain pressure. Rotate the weapon so the chamber is outboard.

  Note: A left-handed Marine pushes upward on the slide stop with the left index finger.

- Reach over the top of the pistol with the left hand and grasp the slide serrations with the thumb and index finger. The left hand should partially cover the ejection port so it is positioned to catch an ejected round.
- Point the pistol in a safe direction and fully retract the slide and lock it to the rear. At the same time, catch the ejected round with the left hand. See figure 2-7.

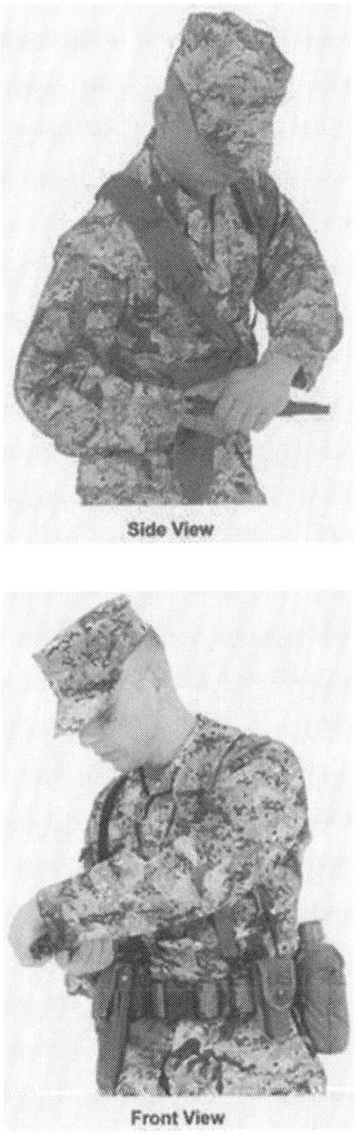

Figure 2-7. Catching the Ejected Round.

- Rotate the pistol so the inside of the chamber can be seen. Visually inspect the chamber to ensure it is empty.
- Press the slide stop to release the slide and observe it going forward on an empty chamber.

Unloading and Showing the Pistol Clear

Perform the following steps to take the pistol from any condition to Condition 4. See figure 2-8.

Figure 2-8. Unload, Show Clear.

- Use the right hand to grip the pistol firmly. Ensure that the pistol is on safe.
- Rotate the pistol so the magazine well is pointed inboard and angled down.

Note: The angle of the magazine well must allow the magazine to fall freely from the well once the magazine release button is engaged.

- Depress the magazine release button to remove the magazine from the pistol. Catch the magazine with the left hand and retain it.
- Push upward on the slide stop with the right thumb and maintain pressure. Rotate the weapon so the chamber is outboard.

Note: A left-handed Marine pushes upward on the slide stop with the left index finger.

- Reach over the top of the pistol with the left hand and grasp the slide serrations with the thumb and index finger. The left hand should partially cover the ejection port so it is positioned to catch an ejected round.
- Point the pistol in a safe direction and fully retract slide and lock it to the rear. At the same time, catch the ejected round with the left hand.
- Rotate the pistol so the inside of the chamber can be seen. Visually inspect the chamber to ensure it is empty.
- Bring the pistol to the administrative transport and have another Marine visually inspect the chamber to ensure that—
  - The chamber is empty, no ammunition is present, and the magazine is removed.
  - The pistol is on safe.
- Acknowledge that the pistol is clear.
- Press the slide stop to release the slide and observe it going forward on an empty chamber.

Emptying the Magazine

Once the pistol is unloaded, the pistol magazine can be emptied of ammunition. To empty the magazine, perform the following steps:

- Hold the magazine upright with the back of the magazine tube against the palm of the hand.
- Push the top round forward with the thumb and catch it with the other hand as it is removed.
- Repeat until the magazine is empty.

Filling the Magazine

Prior to loading the pistol, the pistol magazine must be filled with the prescribed number of rounds of ammunition. See figure 2-9. To fill the magazine, perform the following steps:

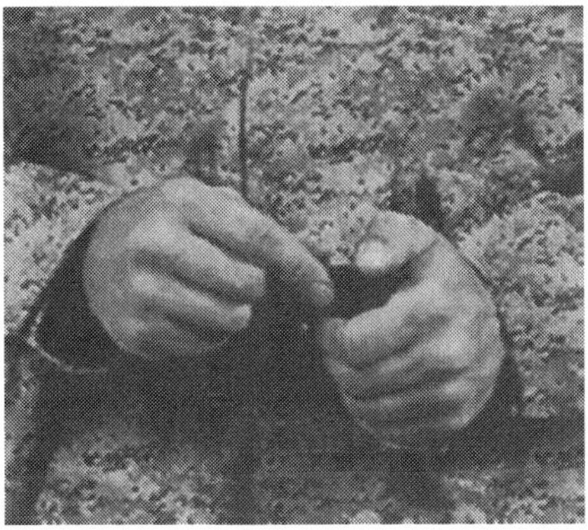

Figure 2-9. Filling the Magazine.

- Hold the magazine with the back of the magazine against the palm of the hand and the follower up.
- Use the other hand to place a round (primer end first) on the follower in front of the magazine lips.
- Press down on the round and slide the round completely back under the lips. The thumb or finger may push down on the back of the round to assist movement. The base of the round should be flush with the back of the magazine.

Repeat this procedure until the magazine is filled with the appropriate number of rounds. Holes on the back of the magazine allow the visual counting of rounds in five-round increments.

Reloading the Pistol

The Marine's ability to reload the pistol quickly improves his chance for success on the battlefield.

Dry Reload

A dry reload is conducted when the pistol runs out of ammunition during engagement and the slide locks to the rear. See figure 2-10. Perform the following steps to conduct a dry reload with the slide locked to the rear:

> *Note:* The pistol is not placed on safe during a dry reload; the trigger finger is taken out of the trigger guard and placed straight along the side of the receiver.

- Seek cover, if the situation permits.

Figure 2-10. Pistol Ran Dry of Ammunition.

- Retain the firing grip with the right hand and pull the pistol in close to the body to facilitate control. Bring the trigger guard to the right of eye level and cant the pistol so the magazine well is facing inboard at approximately a 45-degree angle to the deck. See figure 2-11.

Figure 2-11. Bring the Pistol Toward the Body.

- Press the magazine release button and let the empty magazine fall to the deck. At the same time, unfasten the ammunition pocket to withdraw a filled magazine. See figure 2-12.

Note: The primary objective during a dry reload is to get the pistol back in action as quickly as possible. Following engagement, retrieve the magazine before moving.

- Grasp the magazine by curling the middle finger and thumb of the left hand around the base of the magazine, with the index finger straight along the ammunition pocket. See figure 2-13.

Figure 2-12. Release Magazine and Unfasten Ammunition Pocket.

Figure 2-13. Grasping the Magazine.

- Index the magazine: as the magazine is being withdrawn from the pocket, the index finger should be along the front of the magazine. See figure 2-14.

Figure 2-14. Indexing the Magazine.

- Rotate the hand up so the magazine is aligned with the magazine well.
- Glance quickly at the magazine well, insert the magazine into the magazine well (see fig. 2-15).

Figure 2-15. Glance at the Magazine Well.

- Bring the eyes back on target at the same time as the heel of the left hand seats the magazine, do not relinquish contact with the magazine. See figure 2-16.

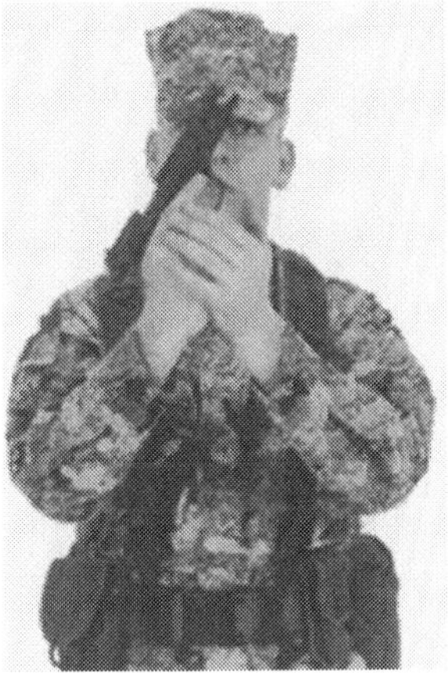

Figure 2-16. Seating the Magazine.

- Roll both hands inward to establish a two-handed grip and press the slide release with the left thumb to allow the slide to move forward, chambering the first round, and present the pistol to the target. See figure 2-17.

Note: A left-handed Marine presses the slide release with his trigger finger.

Figure 2-17. Present Pistol Back to Target.

Figure 2-18. Facilitating Control of Pistol.

Condition 1 Reload

In a Condition 1 reload, a partially-filled magazine is removed from the pistol and replaced with a fully filled magazine. A Condition 1 reload is performed when there is a lull in the action or whenever deemed necessary by the Marine. To perform a Condition 1 reload—

- Retain the firing grip with the right hand and pull the pistol in close to the body to facilitate control. Keep the pistol pointed in the direction of the likely threat. See figure 2-18.
- Withdraw a filled magazine from the ammunition pocket with the left hand. Index the magazine and bring it up to the left of eye level. See figure 2-19.

Figure 2-19. Withdrawing and Indexing a Filled Magazine.

- Slide the index finger to the side of the magazine to grasp the magazine between the index and middle fingers. See figure 2-20.

Figure 2-20. Grasping a Filled Magazine.

- Raise the pistol and bring the trigger guard to the right of eye level and cant the pistol so that the magazine well is facing inboard at approximately a 45-degree angle to the deck.

Note: The angle of the magazine well must allow the magazine to fall freely from the well once the magazine release button is engaged.

- Press the magazine release button with the right thumb to eject the partially-filled magazine from the magazine well. Grasp the magazine between the index finger and thumb. See figure 2-21.

Figure 2-21. Removing a Partially-Filled Magazine.

- Insert the filled magazine into the magazine well (see fig. 2-22). Use the heel of the hand to ensure it is fully seated (see fig. 2-23).

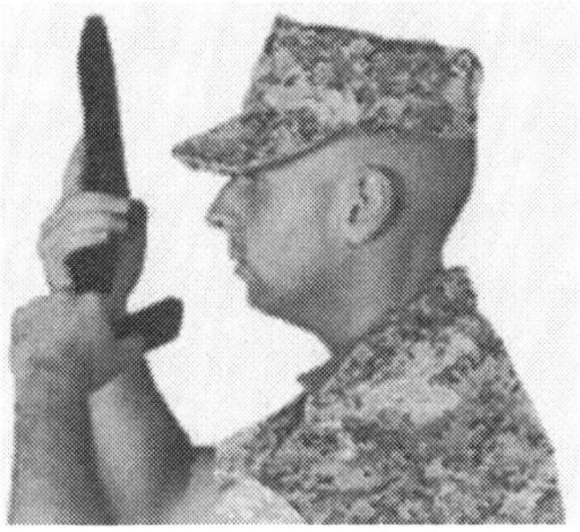

Figure 2-22. Inserting a Filled Magazine.

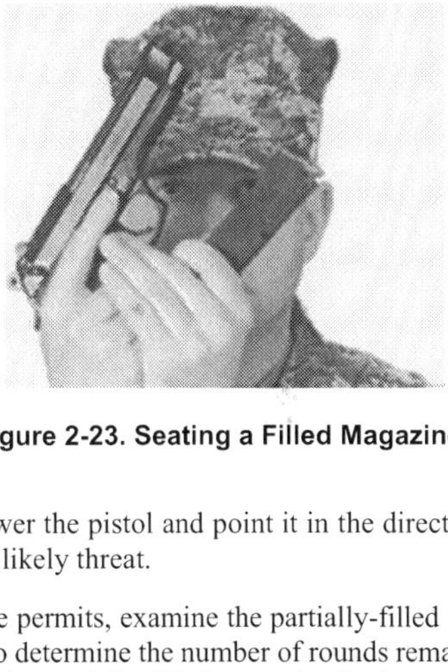

Figure 2-23. Seating a Filled Magazine.

- Lower the pistol and point it in the direction of the likely threat.

If time permits, examine the partially-filled magazine to determine the number of rounds remaining. Stow the partially-filled magazine in the ammunition pocket for later use.

Reloading Considerations

If possible, take cover before reloading. Always reload before leaving cover to take advantage of the protection.

When reloading, the first priority is to reload the pistol quickly so that it is ready to fire. During a reload, the Marine focuses on reloading only—not on the enemy.

The next priority is for the Marine to retain the magazine during the reload. However, the combat situation may dictate dropping the magazine to the deck when performing a reload (i.e., dry reload). If time permits (i.e., Condition 1 reload), the Marine picks the magazine up or secures the magazine (e.g., ammunition pocket, flak jacket) before moving to another location.

Remedial Action

The M9 service pistol is an effective and extremely reliable weapon. Proper care and preventive maintenance usually ensures the pistol's serviceability. However, stoppages, while infrequent, do occur. To keep the pistol in action, stoppages must be cleared as quickly as possible through remedial action. A malfunction cannot be corrected through remedial action by the Marine.

Stoppage

A stoppage is an unintentional interruption in the cycle of operation; e.g., the slide not moving forward completely. A stoppage is normally discovered when the pistol will not fire. Most stoppages can be prevented by proper care, cleaning, and lubrication of the pistol.

Many stoppages of the M9 service pistol are caused by shooter error. The Marine must be aware of shooter-induced stoppages in order to avoid them or to quickly identify and correct the stoppage and return the pistol to action. In a shooter-induced stoppage, the Marine—

- Fails to make ready.
- Fails to take the pistol off safe prior to firing.
- Engages safety while firing.
- Engages magazine release button while firing.
- Engages slide stop while firing (particularly prevalent when firing with an Isosceles grip).
- Fails to reset the trigger.
- Fails to recognize the pistol has run dry and the slide has locked to the rear.

Malfunction

A malfunction is a failure of the pistol to fire satisfactorily or to perform as designed (e.g., a broken front sight that does not affect the functioning of the pistol). A malfunction does not necessarily cause an interruption in the cycle of operation. When a malfunction occurs, the pistol must be repaired by an armorer.

Remedial Action

There is no one set of procedures (i.e., immediate action) that can be performed to clear all or even most of the stoppages that can occur with the M9 service pistol. Therefore, remedial action requires investigating the cause of the stoppage, clearing the stoppage, and returning the pistol to operation. When performing remedial action, the Marine should seek cover if the tactical situation permits. Once a pistol ceases to fire, the Marine must visually or physically observe the pistol to identify the problem before it can be cleared:

> *Note:* The steps taken to clear the pistol are based on what is observed.

- Remove the finger from the trigger and place it straight along the receiver.
- Bring the pistol in close to the body and in a position to observe the chamber.
- Pull the slide to the rear while observing the chamber area to identify the stoppage. See figure 2-24 on page 2-14.

> *Note:* Ensure the pistol does not move to safe when pulling the slide to the rear.

Figure 2-24. Observing Chamber.

- Correct the stoppage:
  - If there is a round in the magazine but not in the chamber (see fig. 2-25), the slide is released and a round is observed being chambered.

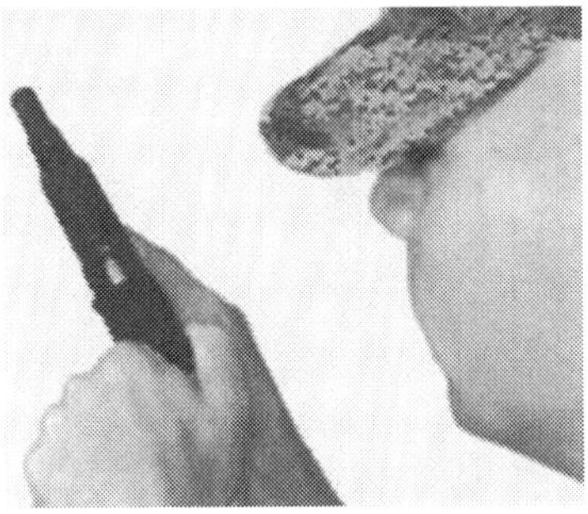

Figure 2-25. Round in Magazine but Not in Chamber.

  - If a round being chambered is not observed, the bottom of the magazine is tapped to seat it properly, and the slide is racked to the rear. See figure 2-26.
  - If there is no round in the magazine or chamber, a reload is conducted. See figure 2-27.

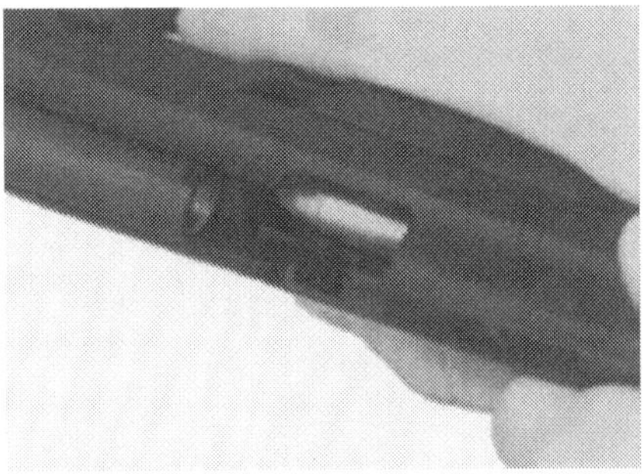

Figure 2-26. Round Not Being Chambered.

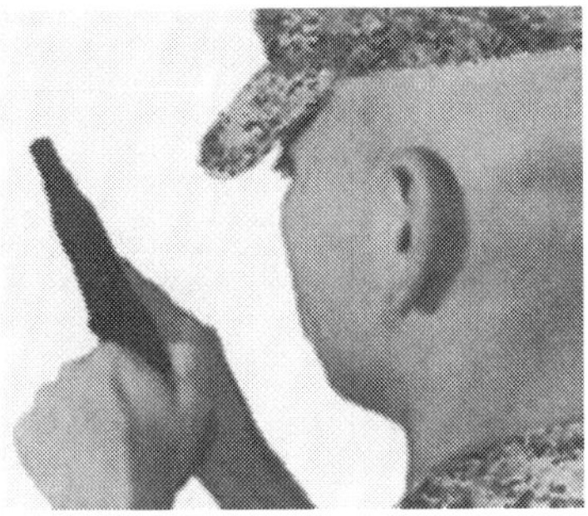

Figure 2-27. No Round in Magazine or Chamber.

- Fire the pistol.

Audible Pop or Reduced Recoil

WARNING

When an audible pop or reduced recoil is experienced, the Marine DOES NOT perform remedial action unless he is in a combat environment. An audible pop occurs when only a portion of the propellant is ignited. It is normally identifiable by reduced recoil and the pistol will not cycle. Sometimes, it is accompanied by excessive smoke escaping from the chamber area.

Training Environment

If an audible pop or reduced recoil is experienced during firing, cease fire immediately. Do not apply remedial action; instead, perform the following steps:

- Remove the finger from the trigger and place it straight along the receiver.
- Point the pistol down range.
- Place the pistol on safe.
- Raise a hand to receive assistance from available range personnel.

Combat Environment

The tactical situation may dictate correction of an audible pop or reduced recoil. To clear the pistol, perform the following steps:

- Remove the finger from the trigger and place it straight along the receiver.
- Seek cover if the tactical situation permits.
- Unload the pistol, but leave the slide locked to the rear.
- Insert something into the bore and clear the obstruction.
- Observe the barrel for cracks or bulges.
- Reload the pistol.

Weapons Carries

As the threat level increases, so should the Marine's readiness for engagement. Weapons carries are designed to place the Marine in a state of increased readiness as the threat level increases. There are two carries with the pistol: the Alert and the Ready. The carries permit quick engagement when necessary.

Alert

The Alert is used when enemy contact is likely (probable). See figure 2-28. The Marine performs the following steps to assume the Alert:

Figure 2-28. Alert.

- Ensure the pistol is on safe.
- Grasp the pistol grip firmly with two hands. The trigger finger is straight and the right thumb is on the safety and in a position to

- Extend the arms down at approximately a 45-degree angle to the body or bend the elbows. See figure 2-29.
- The muzzle of the pistol is pointed in the likely direction of the threat.

Ready

The Ready is used when there is no target, but contact with the enemy is imminent. The Marine performs the following steps to assume the Ready (see fig. 2-30):

- Ensure the pistol is on safe.
- Grasp the pistol firmly with two hands. The trigger finger is straight and the right thumb is on the safety and in a position to operate it.
- Extend arms and raise the pistol to just below eye level so a clear field of view is maintained.
- Point the muzzle of the pistol in the direction of enemy contact.

Weapons Transports

The M9 service pistol is transported in either the holster transport or the administrative transport.

Holster Transport

The holster transport is the most common method of carrying the pistol because it can be transported safely in the holster. This transport is used when there is no immediate threat (enemy contact is remote). See figure 2-31. To transport the pistol in the holster:

- Point the pistol in a safe direction.

WARNING

Ensure the pistol is pointed in a safe direction at all times and does not cover any part of the body while holstering.

Figure 2-29. Alert—Close Quarters.

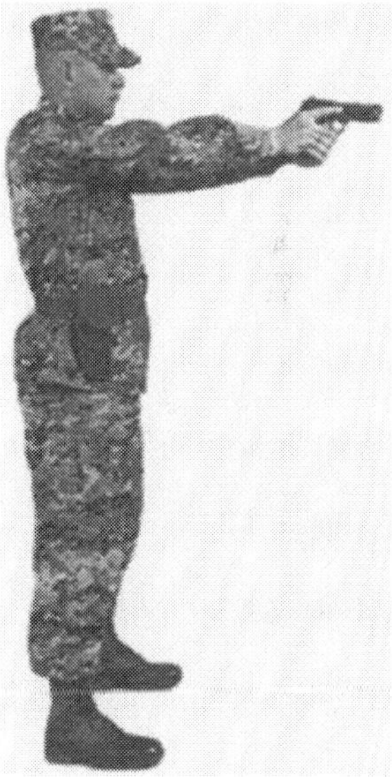

Figure 2-30. Ready.

Figure 2-31. Holster Transport.

- Ensure that the safety is on, the slide is forward, and the trigger finger is straight.
- Use the right hand to firmly grip the pistol grip and place the pistol in the holster:
  - Lift the flap of the holster with the left hand.
  - Look down at the holster, bring the pistol back to a position above the holster, and rotate the muzzle down into the holster.
- Push the pistol snugly into the holster and fasten the flap with the right hand.

Administrative Transport

The administrative transport is used to transport the pistol when the Marine does not have a holster. See figure 2-32. The Marine performs the following steps to assume the administrative transport:

- Establish a firm grip around the pistol grip with the right hand.

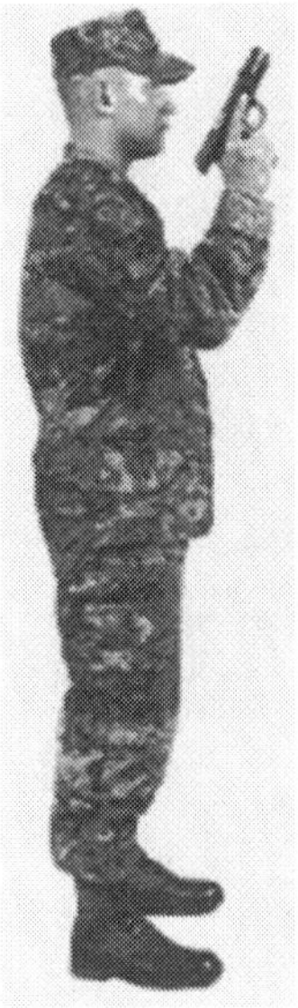

Figure 2-32. Administrative Transport.

- Ensure the pistol is on safe, the magazine is removed, the slide is locked to the rear, and the trigger finger is straight along the receiver.
- Bend the elbow to approximately a 45-degree angle so the pistol is positioned near shoulder level. The wrist should be straight so the pistol's muzzle points up.

Combat Mindset

In a combat environment, the Marine must be constantly prepared to engage targets. When a target presents itself, there may be little time to react. The target must be engaged quickly and accurately. It is not enough to simply know marksmanship techniques, the Marine must be able to react instinctively.

The development of a combat mindset can be associated with the carries and holster transport for the pistol. The use of each carry/transport is dictated by the perceived level of threat. Each carry and transport should signify a stage of mental and physical preparedness for combat. The intensity of the Marine's mental and physical preparation depends on the likelihood of enemy contact.

Physical and Mental Preparation

Physical Preparation

In combat, targets can present themselves without warning. Therefore, it is essential for the Marine to maintain proper balance and control of the pistol at all times so that the pistol can be presented quickly and the target engaged accurately.

Speed alone does not equate to effective target engagement. The Marine should fire only as fast as he can fire accurately, never exceeding his physical ability to apply the fundamentals of marksmanship. To be effective in combat, the Marine must train to perfect the physical skills of shooting so they become second nature. The more physical skills that can be performed automatically, the more concentration that can be given to the mental side of target engagement.

Mental Preparation

While combat is unpredictable and constantly changing, the Marine can prepare mentally for the contingencies of the operational setting and confrontation with a threat. The stress of combat, coupled with the limited time available to engage targets, requires concentration on the mental aspects of target engagement; e.g., identification of targets, shoot/no-shoot decisionmaking, and the selection and use of cover. Minimizing stress and maximizing the limited time available to engage targets can be accomplished by—

- Knowing the combat environment and being constantly aware of the surroundings (e.g., terrain, available cover, possible threats) enables the Marine to quickly present the pistol and accurately engage targets.
- Identifying and evaluating possible courses of action and developing potential plans for target engagement that will be appropriate to the combat situation.
- Instilling confidence in a Marine's ability to fire well-aimed shots in the stress of a combat situation. A key factor in a Marine's level of confidence is the degree to which he has mastered the tactics, techniques, and procedures of pistol marksmanship.

Threat Levels

No Immediate Threat

When there is no immediate threat, the Marine assumes the holster transport. The pistol should be in Condition 1. This is the lowest level of awareness for the Marine in a combat environment, but the Marine must stay alert and aware of any nearby activity. To prepare for target engagement at this level, the Marine must—

- Be aware of likely areas of enemy contact.
- Be aware of the condition of his pistol.
- Establish a plan or course of action to present the pistol to a target should a target appear.
- Mentally review appropriate actions such as reloading and remedial action.

Contact Likely (Probable)

If enemy contact is likely (probable), the Marine assumes the Alert. When enemy contact is likely, the Marine should—

- Expect enemy contact and be constantly prepared to present the pistol.
- Search the entire area for indications of enemy targets and for suitable terrain features that offer cover and concealment. The Marine should avoid restricting the search to a single terrain feature because this hinders awareness to a sector of the battlefield and to approach by the enemy.
- Be mentally prepared for contact. Plan a course of action for immediate response to a target. Modify the plan of action as needed.

- Be physically prepared to fire. Maintain proper balance at all times. Avoid self-induced physical fatigue. For example, do not grip the pistol so tightly that fingers, hands, and arms tire from carrying the pistol.

Contact Imminent

When contact with an enemy target is imminent, assume the Ready. In this carry, the Marine is at the highest level of awareness and is constantly searching for and expecting a target. To fire well-aimed shots upon target detection, the Marine must be at the peak of his mental preparation, all distractions must be eliminated, and his focus must be on firing an accurate shot. In the Ready, the Marine must—

- Keep the pistol oriented in the general direction of observation (eyes, muzzle, target).
- Maintain a clear field of view above the pistol sights until the target is detected.
- Be mentally and physically prepared to engage the target. The Marine must be ready to:
  - Identify the target.
  - Sweep the safety.
  - Apply the fundamentals of marksmanship.
- Move only as fast as he is capable of delivering well-aimed shots, ensuring that speed of engagement does not exceed his physical abilities.
- Search the entire area for indications of enemy targets, lowering the pistol enough to observe a clear field of view of the area.

Transferring the Pistol

The Marine's ability to transfer a pistol to another Marine is critical to safe weapons handling. There are two methods for transferring the pistol from one Marine to another: show clear transfer and condition unknown transfer. Each transfer is performed based on the operational environment/ combat situation.

Show Clear Transfer

To transfer the pistol—

- Grasp the pistol firmly in the right hand while ensuring that the pistol is on safe.
- Remove and retain the magazine.
- Lock the slide to the rear and catch the round if there is a round in the chamber.
- Inspect the chamber visually to ensure it is empty and leave the slide locked to the rear. To transfer the pistol—
  - If the receiving Marine is to the right: Cradle the trigger guard in the palm of the left hand and wrap the fingers around the top of the pistol. Release the firing grip.
  - If the receiving Marine is to the left: With the left hand, grasp the slide of the pistol with the thumb over the slide and the fingers underneath. Release the firing grip.
- Ensure that the muzzle points up at a 45-degree angle in a safe direction and the chamber exposed. Hand the pistol to the other Marine, grip first. See figure 2-33.

Figure 2-33. Show Clear Transfer.

- The receiving Marine—
  - Grasps the pistol grip with the trigger finger straight along the receiver.
  - Inspects the chamber visually to ensure it is empty.
  - Ensures the pistol is on safe.

Condition Unknown Transfer

To transfer the pistol—

- Grasp the pistol firmly in the right hand while ensuring that the pistol is on safe. To transfer the pistol—
  - If the receiving Marine is to the right: Cradle the trigger guard in the palm of the left hand and wrap the fingers around the top of the pistol. Release the firing grip. See figure 2-34.

Figure 2-34. Condition Unknown Transfer to the Right.

  - If the receiving Marine is to the left: With the left hand, grasp the slide of the pistol with the thumb over the slide and the fingers underneath. Release the firing grip. See figure 2-35.

Figure 2-35. Condition Unknown Transfer to the Left.

- Ensure that the muzzle points up at a 45-degree angle in a safe direction. Hand the pistol to the receiving Marine, grip first.
- The receiving Marine—
  - Grasps the pistol grip with the trigger finger straight along the receiver.
  - Ensures the pistol is on safe.
  - Conducts a chamber check to determine the condition of the pistol.
  - Remove the magazine and count the number of rounds in the magazine by using the counting holes, if time permits. Re-insert the magazine into the magazine well ensuring it is fully seated.

CHAPTER 3
FUNDAMENTALS OF PISTOL MARKSMANSHIP

The fundamentals of pistol marksmanship are aiming, trigger control, and breath control. Understanding and applying the basic pistol marksmanship fundamentals ensures the Marine's effectiveness in target engagement. The fundamentals must be continually studied and practiced because they are the means by which accurate shots are placed on target. A Marine with a solid foundation in the fundamentals of marksmanship will be successful in the application of these fundamentals during combat.

Aiming

Maintaining the correct relationship between the pistol sights is essential for accurate target engagement. Because of the short distance between the pistol sights, a small error in their alignment causes a considerable error at the target.

Sight Alignment

Sight alignment is the relationship between the front sight and rear sight with respect to the aiming eye. Correct sight alignment is the front sight centered in the rear sight notch with the top edge of the front sight level **aligned** with the top edge of the rear sight. There should be equal space on either side of the front sight. See figure 3-1.

Figure 3-1. Sight Alignment.

Establishing Sight Alignment

The pistol is fired without benefit of bone support; therefore, the pistol is in constant motion. The Marine must understand this, yet continually strive to align the sights. To fire accurately, the sights must be aligned when the shot breaks.

Grip

The grip is key to acquiring sight alignment. If the grip is correct, the front and rear sights should align naturally. Dry fire during presentation of the M9 service pistol aids in obtaining a grip that allows sight alignment to be acquired consistently.

Controlled Muscular Tension

There must be enough controlled muscular tension in the grip, wrists, and forearms to hold the pistol steady and level the barrel to maintain sight alignment. Consistent tension stabilizes the sights and maintains sight alignment.

Sight Picture

Sight picture is the placement of the front sight in relation to the target while maintaining sight alignment. See figure 3-2 on page 3-2.

Because the pistol is constantly moving, sight picture is acquired within an aiming area that is located center mass on the target. The aiming area allows for movement of the sights on the target while maintaining sight alignment. Each Marine defines an acceptable aiming area within his own ability to stabilize the sights. Time, distance to the target, and personal ability affect dictate the aiming area. As the Marine becomes more proficient with the pistol, the aiming area becomes more precise.

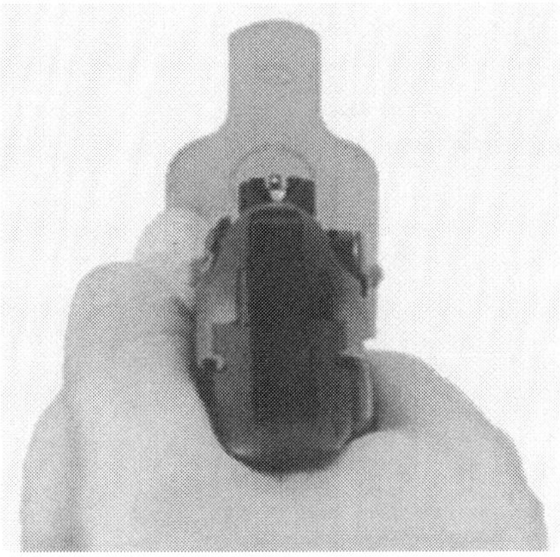

Figure 3-2. Sight Picture.

The aiming area is determined by the Marine's stability of hold. The proper grip stabilizes the sights so sight alignment can be maintained, but the sights move continuously within the aiming area of the target. The Marine understands the pistol's movement and learns to apply trigger control as he is obtaining sight alignment/sight picture within the aiming area so the shot breaks the moment sight picture is established.

Relationship Between the Eye and the Sights

The human eye can focus clearly on only one object at a time. The Marine must focus on the top edge of the front sight and fire the shot while maintaining the relationship between the front and rear sights within the aiming area. Focusing on the top edge of the front sight rather than the target keeps the front sight clear and distinct, which allows the Marine to detect minor variations in sight alignment. Secondary vision allows the Marine to see the target (although slightly blurred) and maintain sight picture within his aiming area.

Trigger Control

Trigger control is the Marine's skillful manipulation of the trigger that causes the pistol to fire while maintaining sight alignment and sight picture. Proper trigger control aids in maintaining sight alignment while the shot is fired.

Sight Alignment and Trigger Control

Aiming and trigger control are mutually supportive—one cannot be performed without the other. Sight alignment and trigger control must be performed simultaneously to fire an accurate shot. As pressure is applied to the trigger, the sights may move, causing them to be misaligned. To fire accurate shots, the sights must be aligned when the shot breaks. Trigger control can actually assist in aligning the sights. With proper trigger finger placement and consistent muscular tension applied to the grip, the sights can be controlled as the trigger is moved to the rear. If the sights move extensively while pressing the trigger, this can indicate an improper grip or inconsistency in the muscular tension being applied to the grip.

Grip

A firm grip is essential for good trigger control. The grip is established before applying trigger control and is maintained throughout the firing process. To establish the grip, the hand is placed around the pistol grip in a location that allows the trigger finger to move the trigger straight to the rear while maintaining sight alignment. Once the grip is established, it should be firm enough to allow manipulation of the trigger while maintaining sight alignment. The pressure applied to the grip must be equal to or more than the pressure required to move the trigger to the rear. If the pressure is not applied correctly, the sights move as the trigger is pressed to the rear and sight alignment is disturbed.

Trigger Finger Placement

Once the grip is established, the finger is placed on the trigger. Placement of the finger should be natural and allow free movement of the trigger finger. A natural trigger finger placement allows the trigger to be moved straight to the rear while maintaining sight alignment. If the finger presses the trigger to the side, it can cause an error in sight alignment and shot placement.

Each Marine must experiment with finger placement in order to determine effective placement on the trigger. Once established, effective trigger finger placement allows the trigger to be consistently moved straight to the rear while maintaining sight alignment.

Types of Trigger Control

Uninterrupted Trigger Control

During uninterrupted trigger control, the Marine applies a steady, unchanging pressure to the trigger until the shot is fired. Uninterrupted trigger control is particularly effective at close range, when the target area is large, and when stability of hold is not critical for accuracy. To apply uninterrupted trigger control, apply pressure on the trigger while maintaining focus on the top edge of the front sight. Continue pressure on the trigger to begin moving the trigger straight to the rear while obtaining sight alignment and sight picture. Move the trigger straight to the rear in a single, smooth motion with no hesitation.

Interrupted Trigger Control

Interrupted trigger control is particularly effective at longer ranges, when the target is small, and when stability of hold is critical to maintaining sight picture in the aiming area. This method is also used if the pistol sights move outside the aiming area when applying trigger control. If the sight picture is outside the aiming area, the Marine stops and holds the rearward movement on the trigger until sight picture is re-established. When sight picture is re-established, the rearward movement of the trigger is continued until the shot is fired.

Breath Control

Breathing causes movement of the chest, abdomen, and shoulders, which causes the pistol sights to move vertically while attempting to aim and fire. Therefore, it is necessary to stop breathing for a period of time while firing a shot or a series of shots.

The object of breath control is to stop breathing just long enough to fire the shot while maintaining sight alignment, stabilizing the sights, and establishing the sight picture. To be consistent, the breath should be held at the same point in the breathing cycle; i.e., the natural respiratory pause.

Breathing should not be stopped for too long because it has adverse visual and physical effects. Holding the breath longer than is comfortable results in a lack of oxygen that causes vision to deteriorate and then affects the ability to focus on the sights.

Application of Marksmanship Fundamentals in Field Firing

Compression of the Fundamentals

Pistol engagements typically occur over close distances and are short in duration. Because an immediate response to the threat is required, the application of the fundamentals must be a conditioned response that is executable in a compressed time. The goal of successful, quick target engagement is the application of the fundamentals of marksmanship while firing the shot the moment weapon presentation is complete.

The time required to move the trigger to the rear while acquiring and maintaining sight alignment and sight picture is unique to each Marine and is based on his capabilities. Each Marine should know his abilities and fire only as quickly as he is capable of firing accurately. The Marine must not exceed his shooting skills in an effort to get rounds off quickly.

Aiming

In field firing, the fundamentals are applied in a compressed time so sight alignment and sight picture are achieved as the shot is fired. Although the target must be quickly engaged in combat, sight alignment is still the first priority: strive for a clear front sight. Distance to the target and the size of the target affects sight alignment as follows:

- As the distance to the target increases and the size of the target decreases, sight alignment becomes more critical to target engagement. Accurate sight picture/sight alignment cannot be compromised for speed.
- Sight alignment is critical to the effective engagement of smaller targets such as partially exposed targets.
- As the distance to the target decreases, perfect sight alignment is not as critical, but there must be a relationship between the sights and their placement on the target within the aiming area to ensure accuracy.

Trigger Control

Proper trigger control aids in maintaining sight alignment while the shot is fired. As pressure is applied to the trigger, the sights may move, causing them to be misaligned. Therefore, the Marine is constantly re-aligning the sights as pressure is applied to the trigger. Sight alignment and trigger control must be performed simultaneously to fire an accurate shot.

Breath Control

During combat, the Marine's breathing and heart rate often increase due to physical exertion or the stress of battle. The key to breath control in field firing is to stop breathing just long enough to fire an accurate shot or a series of shots.

Chapter 4
Pistol Firing Positions and Grip

The M9 service pistol is fired from the standing, kneeling, and prone positions. Each firing position may be adapted to either a Weaver or Isosceles variation, each possessing a distinct advantage in combat. The Weaver variation stabilizes the pistol sights. The Isosceles variation manages recoil. The advantages apply whether the Marine is firing in the standing, kneeling, or prone position. The Marine must select and assume a stable firing position that provides a solid foundation for accurate shooting while meeting the demands of the combat situation

Note: The procedures in this chapter are written for right-handed Marines; left-handed Marines should reverse directions as needed.

The Pistol Firing Position

During combat, the Marine selects a firing position based on mobility, observation of the enemy, and stability.

Mobility

A firing position must provide mobility should the Marine need to move. The standing position permits maximum mobility because it is quickly assumed and easily maneuvered from and it permits lateral mobility to engage widely dispersed targets. The prone position provides limited mobility because it is the most time-consuming position to get into and out of and it lacks the lateral mobility required to engage dispersed targets.

Observation of the Enemy

A firing position should allow observation of the enemy while minimizing the Marine's exposure. During combat there can be many obstructions to a clear field of view. Terrain features such as vegetation, earth contours, and manmade structures often dictate the firing position. The prone position normally allows the least exposure, but it usually provides a limited field of view. Kneeling may provide a wider field of view, but generally provides less concealment.

Stability

A solid firing position establishes a stable foundation for target engagement and provides accurate and consistent shooting. The definition of a stable position is one in which the body is positioned so as to resist forces that cause motion (i.e., recoil and movement of the pistol sights). The standing position is the least stable firing position, while the prone is the most stable firing position. A consistent, stable position is assumed for two distinct purposes:

- Minimize the pistol's movement in order to control the sights and to deliver accurate fire on a target.
- Minimize the affects of recoil in order to recover the sights to the same area on the target.

During combat, it may be necessary to engage the same target more than once to eliminate it. If the firing position is stable, the pistol's sights should recover to the same area on the target, allowing rapid re-enactment. Distributing the body's weight to balance the position also stabilizes the pistol and allows better management of recoil. A pistol firing position is stabilized through controlled muscular tension. Because the pistol is fired without benefit of bone support, muscular tension is needed in the body to stabilize the position and the pistol sights. Muscular tension must follow these guidelines—

- A consistent amount of muscular tension is needed to hold the pistol steady so the sights are aligned with the aiming eye and the target.

- Controlled and consistent tension in the body provides resistance that the Marine uses to manage recoil and bring the sights back on target quicker. However, too much tension can cause strain or trembling.
- Muscular tension is correct when the Marine can control the pistol before, during, and after firing the shot.

Pistol Firing Grip

A proper grip is one that provides maximum control of the pistol before, during, and after firing. It stabilizes the pistol sights before firing, allows trigger control to be applied during firing, and manages recoil after firing.

Before Firing: Stabilize the Sights

To fire an accurate shot, the sights must be stabilized prior to and as the bullet exits the muzzle of the pistol. A proper grip controls the alignment of the pistol's sights and stabilizes the sights so an accurate shot may be fired. The following guidelines apply to the establishment of a proper grip—

- There must be muscular tension in the wrist and forearms. Consistent muscular tension in the wrist, forearms, and grip helps maintain sight alignment by reducing the movement in the grip that can cause movement in the sights. The grip is correct when it allows the sights to be naturally aligned to the aiming eye.
- When establishing a two-handed grip, equal pressure must be applied with both hands. Consistent, equal pressure from both hands stabilizes the sights and allows them to be aligned and level with respect to the aiming eye.
- The hand grips as high on the back strap as possible, bringing the centerline of the bore as close as possible to a parallel line with the bones in the forearm.

During Firing: Allow Trigger Control

The Marine's grip provides a foundation for the movement of the trigger finger. The trigger finger applies positive pressure on the trigger as an independent action, completely free of the other muscles of the gripping hand. The Marine should not apply excessive pressure to the web of the hand touching the pistol's back strap because excessive pressure interferes with the manipulation of the trigger by the trigger finger.

After Firing: Manage Recoil

Once a shot is fired, the pistol recoils, disturbing alignment of the sights. A proper grip facilitates a quick recovery from recoil so the sights quickly return to the same area on the target. The Marine's grip determines the following during recoil:

- The amount the muzzle climbs during recoil depends on the amount of controlled muscular tension in the grip and wrists applied to stabilize the pistol and create consistency in resistance to recoil. Controlled muscular tension allows the sights to recover consistently back on target within a minimum amount of time.
- Firm, equal pressure must be applied to the grip with both hands to ensure that the pistol does not slip during recoil. An improper grip or lack of controlled muscular tension causes the pistol to move after the shot is fired, disrupting sight alignment and requiring the Marine to re-establish his grip.
- Locking out the wrist of the firing hand, similar to executing a punch, helps provide resistance to recoil and speed recovery. The elbows should be at a position slightly less than locked out to help absorb the recoil and aid in recoil management.

Withdrawing the Pistol From the Holster

The firing grip is not established in the holster, however, a proper firing grip can be assumed quickly if the pistol is withdrawn from the holster correctly. The pistol should be withdrawn from the holster in one continuous, fluid motion:

- Place the heel of the left hand at the center of the torso with the fingers extended toward the target. At the same time, unfasten and release the D-ring with the right hand. See figure 4-1.

Figure 4-1. Withdrawing the Pistol From the Holster—Step 1.

Note: The left hand is placed on the torso in a position that allows a two-handed firing grip to be established in a minimum amount of movement.

- Place the right thumb on the forward edge of the lower portion of the holster and the fingers around the back edge of the holster, keeping the trigger finger straight. See figure 4-2.

Note: The left hand may assist in holding the holster flap up.

Figure 4-2. Withdrawing the Pistol From the Holster—Step 2.

- Slide the hand up the holster until the fingers come in contact with the pistol grip. At the same time, keep the thumb above the pistol to guide the holster flap up. See figure 4-3 on page 4-4.
- Grasp the pistol grip with the fingers and draw the pistol straight up. Continue withdrawing the pistol while moving the thumb to a position on the safety. See figure 4-4 on page 4-4.

Note: This hand placement allows a firing grip to be established once the thumb disengages the safety. Any adjustments made to the firing grip after the safety is disengaged should be minor.

Figure 4-3. Withdrawing the Pistol From the Holster—Step 3.

Figure 4-4. Withdrawing the Pistol From the Holster—Step 4.

- Once the muzzle clears the holster, rotate the muzzle forward to clear the body. Ensure the muzzle is pointed in a safe direction.
- Establish a two-handed grip on the pistol by joining the left hand with the right hand in the front of the torso. See figure 4-5.

Caution

Ensure the muzzle does not cover the left hand when establishing the two-handed grip.

Weaver and Isosceles Variations

The size of the target, distance to the target, time, and type of engagement needed (i.e., two shots, single precision shot) determine whether to fire in the Weaver or Isosceles variation. The Weaver and Isosceles variations each consist of three firing positions: standing, kneeling, and prone. The firing position chosen is based on the combat situation and the Marine's body configuration, and

Figure 4-5. Withdrawing the Pistol From the Holster—Step 5.

it must permit balance, control, and stability during firing. In addition, there is a specific firing grip that supports each position and the combination of the grip and the body configuration is key to establishing the variation. See table 4-1 for a list of advantages and disadvantages of both the Weaver and the Isosceles variations.

At longer ranges, the target is smaller, a more precise shot is required to eliminate the target, and any small movement moves the sights off the target; therefore, the pistol must be steadied. Although the Weaver variation is effective at any distance, some aspects of the position make it more effective for long range or precision shots on small or partially exposed targets; i.e., sight alignment and sight picture are easier for stability of hold.

The Isosceles variation is effective at any distance, however, some aspects of the position make it more effective for close range engagements. When confronted with a target, the natural physical reaction is to face the target and push out with the arms. This makes the Isosceles variation advantageous for quick engagements at close range. When a target is at close range, it must be engaged quickly before it engages the Marine. The management of recoil is a bigger factor in close-range engagements because it is more likely that multiple shots will be fired to eliminate the target and the sights have to recover quickly back on target. However, at close ranges, the target is larger so stability of hold is not as important because it is easier to hold the sights on target and sight picture is not as critical.

Standing Position

The standing position is the most often employed position during a pistol engagement due to the short distance of the engagement and the nature of combat. When properly assumed, the standing position provides a stable base for firing, a clear field of view, and excellent mobility. The standing position can be adapted to either the Weaver variation or the Isosceles variation.

Weaver Standing Position

The key to successful employment of the Weaver variation is the body's angle to the target and the push-pull pressure applied to the grip. See figure 4-6.

Table 4-1. Advantages and Disadvantages.

Variation	Advantages	Disadvantages
Weaver	Additional balance, control, and stability of hold during firing due to placement of the arms (left arm bent, pistol is in close to the body). Easier to maintain sight picture because the pistol's foundation is steady.	Recoil has a greater impact due to the hand placement on the pistol (some of the pistol grip is exposed and pressure is applied in two different directions around the pistol). Recovery of the sights back on target may take longer since recoil is affected.
Isosceles	Recoil has a lesser impact because muscular tension and grip pressure are evenly distributed around the pistol. Allows quicker recovery of the sights on target.	Stability of hold is degraded due to the pistol being further from the body without support. Since it is harder to steady the pistol, acquiring sight picture becomes more difficult.

To assume the Weaver standing position—

- Face the target and make a half turn to the right, keeping the pistol oriented toward the target. This orients the body at approximately a 40 to 60 degree angle oblique to the target. The shoulders are angled to the target, the left shoulder forward of the right. The feet are about shoulder-width apart, the left foot forward of the right.
- Grip the pistol grip firmly with the right hand. Place the right thumb on the safety.
- Keep the shoulders at a 40 to 60 degree angle oblique to the target and raise the right arm and extend it across the body toward the target. Ensure that the right shoulder does not roll forward or turn toward the target.
- Extend the left arm to the target, bending the left elbow to join the left and right hands.

Figure 4-6. Weaver Standing Position.

The left elbow is inverted and tucked in toward the body so the left arm supports the pistol.

Note: The angle of the body determines how much the elbow bends.

- Establish a two-handed firing grip in the Weaver variation. See figure 4-7.
- Place the palm of the left hand over the front of the right hand so the palm covers the curled fingers of the right hand. The trigger guard should rest in the "V" formed by the left thumb and forefinger. The knuckles of the left hand are just outboard of the trigger guard. A portion of the pistol grip is exposed.
- Rest the trigger finger naturally, straight and outside of the trigger guard, so the finger can be moved quickly and easily to the trigger.
- Rest the left thumb against the receiver so that both thumbs are on the left side of the pistol. Once the safety is disengaged with the right thumb, the left thumb is placed over the right thumb and positive pressure is applied to hold the right thumb in place.
- Apply rearward pressure with the left hand and forward pressure with the right hand to achieve a "push-pull" grip. Isometric tension (push-pull) stabilizes the pistol during firing.

Figure 4-7. Weaver Grip.

- Lean forward slightly and apply muscular tension throughout the body to stabilize the position and manage recoil. The muscular tension in the upper body is not symmetrical due to the "push-pull" tension applied on the grip.
- Keep the head erect so the aiming eye can look through the sights.

Isosceles Standing Position

The key to the Isosceles variation is that the body is squared to the target and equal pressure is applied on the pistol from the grip. To assume the Isosceles standing position—

- Face the target with feet approximately shoulder width apart. The shoulders are squared to the target.
- Establish a two-handed firing grip in the Isosceles variation. See figure 4-8.

Figure 4-8. Isosceles Grip.

- Grip the pistol grip firmly with the right hand. Place the right thumb on the safety.
- Place the heel of the left hand on the exposed portion of the pistol grip in the pocket formed by the fingertips and heel of the right hand. There should be maximum contact between the pistol grip and the hands. Wrap the fingers of the left hand over the fingers of the right hand. Ensure both thumbs rest on the left side of the pistol and point toward the target.
- Apply equal pressure on both sides of the pistol to allow for the best management of recoil.

Note: Ensure the left thumb does not apply excessive pressure to the slide stop or slide.

- Rest the trigger finger naturally, straight and outside of the trigger guard, so the finger can be moved quickly and easily to the trigger.

Notes: Index finger of the left hand may or may not rest on the front of the trigger guard.

Marines with large hands must ensure that their right thumb does not rest on the slide stop, preventing reliable pistol operation.

- Elevate and extend the arms toward the target.
- Roll the shoulders forward and shift the body weight slightly forward to stabilize the position and better manage recoil. The left foot may be slightly forward of the right foot to balance the position. There should be an equal amount of muscular tension on both sides of the body to best manage recoil.
- Tuck the head between the shoulders; the head is extended forward but kept erect so the aiming eye can see through the sights. See figure 4-9.

Figure 4-9. Isosceles Standing Position.

Kneeling Position

The kneeling position offers a smaller exposure than the standing position and greater stability. Increased stability makes the kneeling position effective for longer range shooting. It does not, however, offer as much mobility for quick reaction as the standing position. The kneeling position can be quickly assumed and allows firing from various types of cover. Depending on the cover and the need for observation, the kneeling position may be adapted to a high kneeling, a medium kneeling, a low kneeling, or a two-knee kneeling position. The kneeling positions can be adapted to either the Weaver variation or the Isosceles variation.

Weaver Kneeling Position

The advantage of the Weaver variation of the kneeling position is that it provides bone support due to the left elbow's placement on the knee. The Weaver variation further enables firing from the side of cover while exposing less of the body to a threat. To assume the Weaver kneeling position, the following steps are basic to all adaptations:

- Make a half turn to the right, drop the right foot back or step forward with the left foot, and place the right knee on the deck. The body is positioned at a 40 to 60 degree angle oblique to the target.
- Blade the shoulders at a 40 to 60 degree oblique angle to the target, the left shoulder forward of the right.
- Extend the arms toward the target.
- Bend forward at the waist to better manage recoil.
- Place the flat part of the upper left arm, just above the elbow, in firm contact with the flat surface formed on top of the bent knee. The point of the left elbow extends just slightly past the left knee. However, depending on the need for stability or observation of the enemy, the elbow does not have to rest on the knee.

The following steps provide specifics for each adaptation:

- High kneeling: the toes of the right foot are curled and in contact with the deck or the inside of the foot may be in contact with the deck. Depending on the need for observation, the buttocks may or may not rest on the right heel. The left leg is bent at the knee; the shin straight up and down. The left foot is flat on the deck. See figure 4-10.

Figure 4-10. Weaver High Kneeling.

- Medium kneeling: the right ankle is straight with the foot stretched out and the bootlaces in contact with the deck. The left leg is bent at the knee, the left foot flat on the deck. The right shin may be angled to the body to create a tripod of support for the position. See figure 4-11.

Figure 4-11. Weaver Medium Kneeling.

- Low kneeling: the right ankle is turned so the outside of the foot is in contact with the deck and the buttocks are in contact with the inside of the foot. The right shin may be angled to the body to create a tripod of support for the position. See figure 4-12.

Figure 4-12. Weaver Low Kneeling.

- Two-knee: drop both knees onto the deck. The toes may be curled to get into and out of the position quickly. Depending on the need for observation of the enemy, the buttocks may or may not rest on the heels. See figure 4-13.

Isosceles Kneeling Position

The advantage of the Isosceles variation is that it enables the Marine to fire over the top of cover while exposing less of the body to a threat. To assume the Isosceles kneeling position, the following steps are basic to all adaptations:

- Drop the right foot back or step forward with the left foot and place the right knee on the deck.
- Square the shoulders to the target.
- Extend the arms toward the target.
- Lean forward with the shoulders rolled forward and the head tucked between the shoulders to better manage recoil.

Figure 4-13. Weaver Two-Knee Kneeling.

The following steps provide specifics for each adaptation:

- High kneeling: the toes of the right foot are curled and in contact with the deck or the inside of the foot may be in contact with the deck. Depending on the need for observation, the buttocks may or may not rest on the right heel. The left leg is bent at the knee, the shin is straight up and down. The left foot is flat on the deck. See figure 4-14.

Figure 4-14. Isosceles High Kneeling.

- Medium kneeling: the right ankle is straight with the foot stretched out and the bootlaces in contact with the deck. The left leg is bent at the knee, the left foot is flat on the deck. The right shin may be angled to the body to create a tripod of support for the position. See figure 4-15.

Figure 4-15. Isosceles Medium Kneeling.

- Low kneeling: the right ankle is turned so the outside of the foot is in contact with the deck and the buttocks are in contact with the inside of the foot. The right shin may be angled to the body to create a tripod of support for the position. See figure 4-16.

Figure 4-16. Isosceles Low Kneeling.

- Two-knee: drop both knees onto the deck. The toes may be curled to get into and out of the position quickly. Depending on the need for observation of the enemy, the buttocks may or may not rest on the heels. See figure 4-17.

Figure 4-17. Isosceles Two-Knee.

Prone Position

The prone position is easily assumed, stable, and presents a small target to the enemy. Since the prone position places most of the body on the deck, it offers great stability for long range shooting. However, it is the least mobile of the firing positions and may restrict the field of view. The prone position can be adapted to either the Weaver variation or the Isosceles variation.

Weaver Prone Position

The Weaver variation of the prone position produces a cocked leg position by angling the body to the target and cocking the leg to support the position. The Weaver prone position is ideal for firing from behind cover. See figure 4-18.

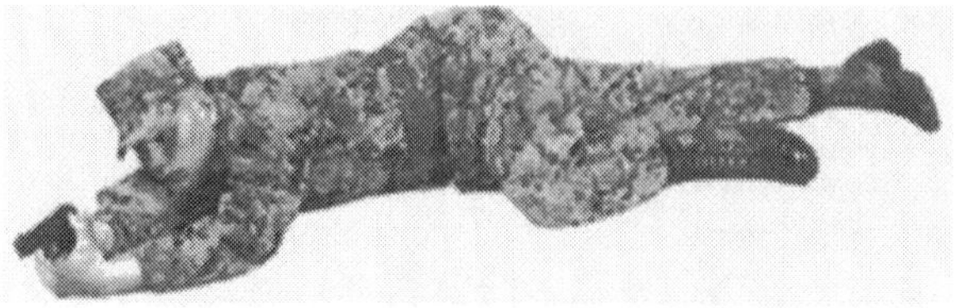

Figure 4-18. Weaver Prone Position.

To assume the Weaver prone position, perform the following steps:

- Face the target and make a half turn to the right (this places the body at a 40 to 60 degree oblique to the target). Grip the pistol in the right hand, placing the pistol in a position that facilitates control of the weapon. Ensure the pistol is pointed in a safe direction and does not cover any portion of the body.
- Move the body to the deck by using either the squat or drop method, keeping the body at a 40 to 60 degree oblique to the target.

  Squat Method
  - Squat down and place the left hand on the deck.
  - Kick both feet backward and come down on the right side of the body with the right arm extended toward the target.
  - Ensure the pistol does not cover the body or the left hand. See figure 4-19.

  Drop Method
  - Drop to a kneeling position.
  - Place the left hand on the deck in front of the body, push the pistol out toward the target.
  - Roll the right side of the body onto the deck.
  - Ensure the pistol does not cover the body or the left hand. See figure 4-20 on page 4-12.
- Bring the left knee up to support the firing position and to raise the diaphragm off the deck so as not to interfere with breathing. The inside of the knee rests on the deck. The knee is drawn up to provide maximum stability for the position.
- Establish a two handed-firing grip on the pistol.

Figure 4-19. Squat Method.

- Place the left elbow on the ground for stability. For maximum stability, strive to keep the grip firmly placed on the deck.
- The head may rest against the right arm so the pistol sights can be aligned. The head may be canted as long as the aiming eye can look directly through the sights. Strive to keep the pistol sights as level as possible while acquiring sight alignment.

To make minor increases in elevation, keep the left hand in place and firmly on the deck and raise the right hand to achieve the desired elevation. However, contact between the right and left hands must be maintained to stabilize the pistol. See figure 4-21 on page 4-12. (There is a tradeoff between obtaining the needed elevation and losing stability, so the Marine must strike a balance between the two.)

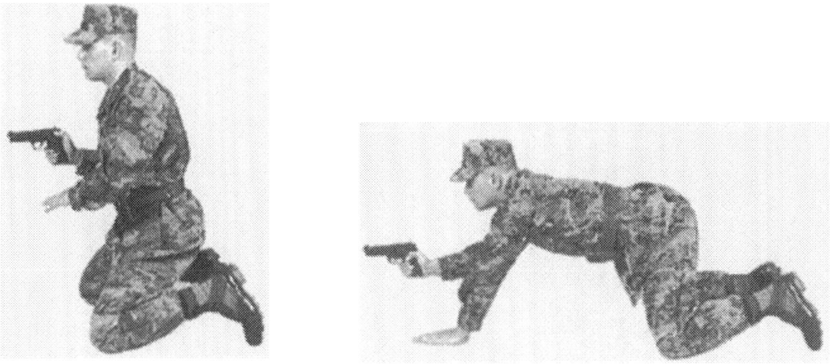

Figure 4-20. Drop Method.

Figure 4-21. Increasing Elevation (Minor Adjustments).

Isosceles Prone Position

The Isosceles variation of the prone position produces a straight leg position. See figure 4-22.

To assume the Isosceles prone position:

- Stand facing the target. Grip the pistol in the right hand, placing the pistol in a position that facilitates control of the weapon. Ensure the pistol is pointed in a safe direction and does not cover any portion of the body.
- Move the body to the deck by using either the squat or drop method.

  Squat Method
  - Squat down and place the left hand on the deck.
  - Kick both feet backward and come down on the right side of the body with the right arm extended toward the target.
  - Ensure the pistol does not cover the body or the left hand. See figure 4-23.

  Drop Method
  - Drop to a kneeling position.
  - Place the left hand on the deck in front of the body.

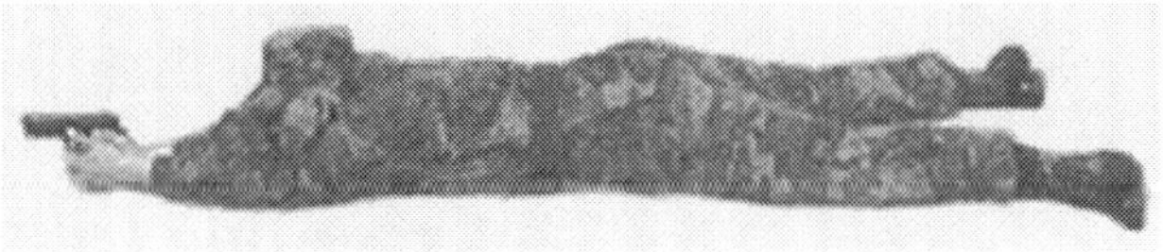

Figure 4-22. Isosceles Prone Position.

Figure 4-23. Squat Method.

  - Push the pistol out toward the target, and roll the right side of the body onto the deck.
  - Ensure the pistol does not cover the body or the left hand. See figures 4-24.
- Establish a two-handed firing grip on the pistol.
- Spread the legs to a position that provides maximum stability. The insteps of both feet may be flat on the deck or the toes may be curled and dug into the deck.
- Keep the pistol sights as level as possible while acquiring sight alignment. Keep the head in a position to allow the aiming eye to look directly through the sights.
- When wearing a helmet, the head may be canted slightly and rest against the right arm to push the helmet from the eyes so the sights can be aligned. Likewise, the pistol may be canted outboard to allow the aiming eye to look directly through the sights.

To make minor increases in elevation, keep the left hand in place and firmly on the deck and raise the right hand to achieve the desired elevation. However, contact between the right and left hands must be maintained to stabilize the pistol. See figure 4-25 on page 4-14. (There is a tradeoff between obtaining the needed elevation and losing stability, so the Marine must strike a balance between the two.)

Natural Body Alignment

The body must be properly aligned to the target so the sights fall naturally on the target when the pistol is presented. It takes a combination of body alignment and consistent muscular tension to

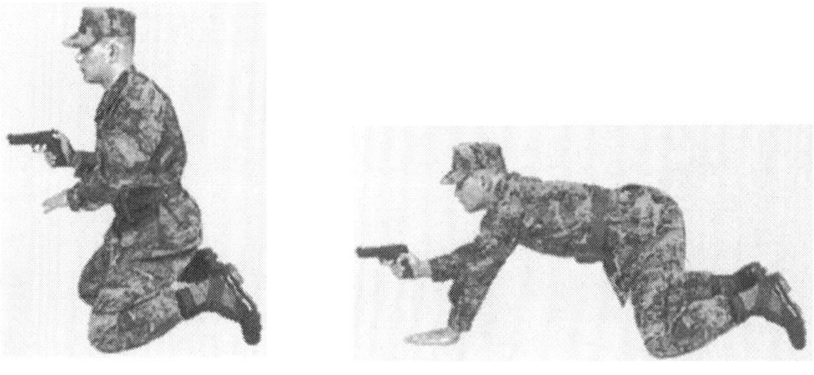

Figure 4-24. Drop Method.

Figure 4-25. Increasing Elevation (Minor Adjustments).

ensure the sights fall naturally to the same area of the target every time the pistol is presented. The Marine can execute the following to check natural body alignment and to ensure the sights center on the aiming area:

- Orient the body to a target and establish a variation of the standing position and a two-handed firing grip on the pistol. Aim in on the target.
- Close the eyes and take a deep breath.
- Open the eyes and see where the pistol sights are in relation to the target. If the pistol sights are right or left of the target, move the feet to adjust the position right or left.

Note: Do not force the sights onto the target area by moving the arms; this increases the muscular tension on one side of the body, disturbs balance, and makes recoil harder to manage.

If the sights are significantly out of alignment when the pistol is at eye level, it may be an indication of a poor grip. When the grip is correct (to include the muscular tension in the grip, wrist, and forearms), the sights should align to the point that only minor adjustments are needed to align the sights to the aiming eye.

Repeat the preceding steps as necessary. Body alignment and muscular tension are correct when the sights are naturally placed in the same area on the target every time the Marine aims on the target.

Chapter 5
Use of Cover and Concealment

On the battlefield, a firing position that allows maximum observation of the enemy as well as cover and concealment is a necessity. A good position provides a solid foundation for the pistol, maximizes the use of cover to provide protection from enemy fire, allows mobility, and provides observation of the enemy. Where possible, the cover should be used to provide additional support for the position.

When contact with the enemy is made, it is important to seek cover as quickly as possible. Cover is anything that provides protection from enemy fire. Cover should be, at a minimum, thick enough to stop small arms fire and high enough to protect the Marine when firing from behind it. The effective use of cover enables engagement of enemy targets while affording protection from enemy fire. Cover can also be effectively used to conceal the Marine from enemy view while searching for targets.

Concealment is anything that hides a Marine from enemy view; however, it might not afford protection. Concealment can be provided by brush, trees, etc.

> *Note:* The procedures in this chapter are written for right-handed Marines; left-handed Marines must reverse directions as needed.

Cover Materials

Natural cover (rocks, logs, rubble, etc.) is best because it is hard to detect. But any material (including buildings, structures, etc.) that may protect an individual from small arms fire can be used for cover. Some common types of cover material are as follows.

Dirt

The best type of cover is dirt packed to a minimum thickness of 18 inches.

Cinder Blocks

Cinder blocks used as foundations for houses or walls can be used for cover, but they can be penetrated. They are also brittle and can shatter upon impact from small arms fire, which can cause injury by secondary fragmentation.

Trees, Logs, and Telephone Poles

Wood is a relatively dense material; therefore, it offers good covering protection since bullets have a tendency to fragment as they penetrate. Live trees have a greater resistance to bullet penetration than dead wood. Wood that has been treated with creosote, such as telephone poles and railroad ties, offers better protection from projectiles than untreated wood, but it still does not ensure protection from small arms fire.

Sandbags

Sandbags can be used for cover. However, they should always be packed tightly and tamped down to increase their density. If loosely filled or moist, a bullet can more easily penetrate the sandbag. Doubling or overlapping sandbags also increases their protective qualities.

Considerations for Firing From Cover

Adjusting the Firing Position

Cover can provide additional support for the firing position. The firing position is adjusted to ensure stability, mobility, and observation of the enemy.

The Marine chooses a firing position based on his height in relation to the height of the cover. The firing position must minimize exposure to the enemy but allow observation of the area.

Although the cover provides additional support, the Marine continues to apply the same amount of muscular tension in the grip, wrist, and forearms. Muscular tension is still necessary to stabilize the pistol sights and manage recovery.

Because the sights are higher than the muzzle of the pistol, the Marine must ensure that the muzzle of the pistol clears the cover as he obtains sight alignment/sight picture on the target. The closer the Marine is to the cover, the easier it is to ensure the muzzle clears the cover.

Applying the Weaver and Isosceles Variations

The type of cover may dictate which variation of a firing position will be the most effective. The position should provide the Marine with the maximum amount of stability and control and allow the Marine to manage recoil effectively in order to recover on target.

Weaver Variation

The Weaver variation exposes less of the body from behind cover due to the angle of the body. The Weaver variation may be better suited for firing from behind the right or left side of cover. For example, the Weaver prone is ideal for firing from behind a log.

Isosceles Variation

The Isosceles variation is good for firing over the top of cover (e.g., a window). The Isosceles prone is ideal for firing from behind narrow cover (e.g., a telephone pole).

Keeping the Body Behind Cover

Avoid inadvertent exposure of any part of the body. Be especially aware of the top of the head, elbows, knees, or any other body part that may extend beyond the cover.

Log or Curb

When firing from behind a log or curb, the Marine must present the lowest possible silhouette and may use the log or curb for maximum support of the position. For maximum protection, muzzle clearance is kept as close as possible to the top of the log or curb. The Marine may fire from either the side or the top of the log, depending on cover and concealment (see fig. 5-1). The Marine fires over the top of a curb when it is used for cover.

Figure 5-1. Firing Around a Log.

Wall or Barricade

Firing is done from either the side or over the top of a wall or barricade. See figure 5-2.

Figure 5-2. Firing From a Barricade.

Window

If the Marine has not been detected by the enemy, he should use the side of the window or the window sill for support. Ideally, it is best to fire from the corner of the window sill when using the window for support. If there is little chance for detection or the shot can be made without support, the Marine should remain back and to the side of the window opening so the pistol does not protrude and his body is concealed by the shadows/darkness of the room. If the Marine is positioned too close to the window, his body provides a silhouette to the enemy.

Vehicle

In many combat situations, particularly in urban environments, a vehicle may be the best form of cover. When using a vehicle for cover, the engine block provides the most protection from small arms fire. The Marine establishes a position behind the front wheel or front door jamb so the engine block is between him and the target. See figure 5-3.

From this position, the Marine fires over the hood of the car or underneath the car from behind the wheel. At the back of the car the axle and the

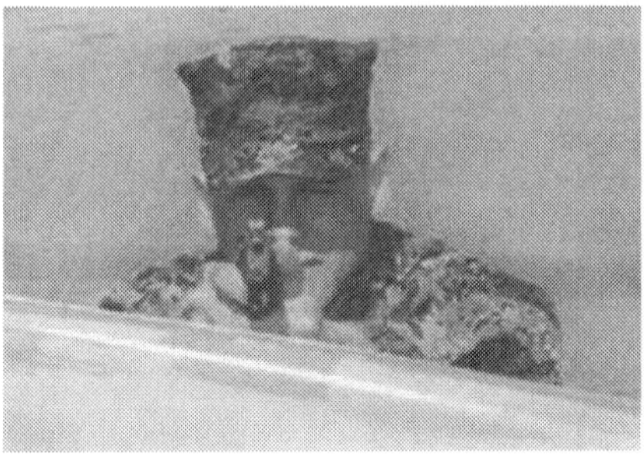

Figure 5-3. Firing Over the Hood of a Vehicle.

wheel provide the only cover. If the Marine must shoot from the back of the car, he must position himself directly behind the wheel as much as possible. See figure 5-4.

Figure 5-4. Firing From the Back of a Vehicle.

Providing Support for the Position and the Pistol

Support helps stabilize both the firing position and the pistol and enable the Marine to maintain sight alignment and sight picture.

The forearms or hands can contact the support to stabilize the pistol. The Marine may rest the pistol on or against the support as long as the support does not interfere or affect the pistol's cycle of operation. See figure 5-5.

Figure 5-5. Hand Resting on Support.

When firing over the top of cover, the Marine can establish a supported position and stabilize the position by resting the trigger guard or the magazine on the cover. The pistol may be pushed up against the support so the "V" formed by the receiver and the front of the trigger guard rests firmly against the support. See figure 5-6.

Figure 5-6. Trigger Guard Resting Against Support.

When firing around the sides of cover, the Marine can establish support and stabilize the position by placing the back of the hand or arm against the cover. Avoid placing the slide of the pistol against the cover because it can interfere with the pistol's cycle of operation. However, the pistol can be canted and placed against the cover so the trigger guard or the "V" formed between the receiver and trigger guard rests against the support. This position enables the Marine to expose less of himself to the enemy. See figure 5-7.

Figure 5-7. Hand Resting Against Side of Support.

When using a vehicle for cover, the Marine can establish additional support for the pistol by positioning himself in the car behind the door jamb (frame of door) and placing his hands or pistol against the "V" formed by the open door and door frame. See figure 5-8.

When shooting from the left side of cover, the Marine still uses his right hand and eye. See figure 5-9. He may have to cant his head and the pistol to the left to establish sight alignment. For right-handed Marines, shooting from the left side of cover may expose more of the Marine to the enemy than shooting from the right side.

Figure 5-8. Firing From Behind the Door Jamb of a Vehicle.

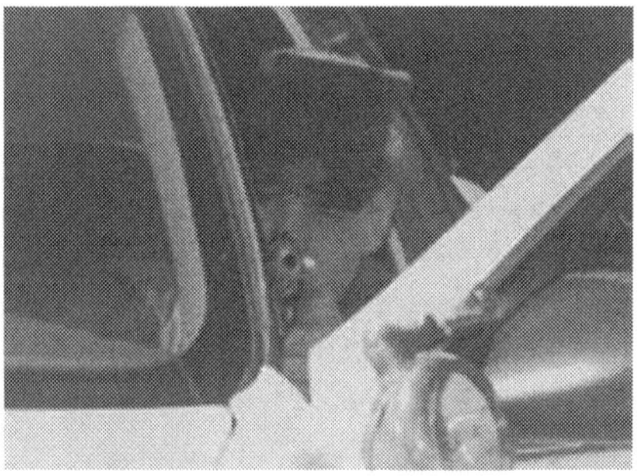

Figure 5-9. Firing From the Left Side of Cover.

Changing Positions

If the Marine has been firing from cover and has to reload or clear a stoppage behind cover he should attempt to resume firing from a different position. The enemy is aware of the Marine's current position and will be ready to engage him once he re-appears.

Moving

In combat, the Marine must be constantly aware of his surroundings and the available cover should enemy contact occur during movement from one position to another.

When moving from cover to cover, the Marine selects the next cover location and plans the route of movement before leaving his present position. This is done by quickly looking from behind cover to ensure the area is clear, ensuring the head and eyes are exposed for as short a time as possible. If necessary, the Marine should conduct a Condition 1 reload before moving from cover. Once the Marine is committed to moving, he must focus on the move until cover is reassumed.

Supported Firing Positions

During combat, the Marine may not have the time to assume a perfect firing position. He must know instinctively that his position is correct rather than follow a regimented sequence of movements to ensure its correctness. With training, the Marine can assume stable firing positions quickly and instinctively by incorporating the use of cover for support. Support provides foundations for the firing position; which, in turn, provides support for the pistol. To maximize the support provided by the position, the firing position should be adjusted to fit or conform to the shape of the cover.

A supported firing position minimizes exposure to the enemy, maximizes the stability of the pistol and protection from fire, and provides observation of the enemy. Any stable support may be used (e.g., logs, sandbags, walls). The surrounding environment dictates the support and position.

The size, distance to the target, and time affect the need for stability and recovery in the selection of a supported firing position. For example, if the target is a great distance from the Marine, he may sacrifice some of his ability to manage recoil in order to assume a supported firing position that provides him the additional stability needed to fire accurately at long range. But, recovery may be more important for the Marine if he must fire multiple shots on target quickly; therefore, he may sacrifice some stability in his supported firing position in order to engage a target with multiple shots.

Supported Prone

The supported prone position presents the lowest silhouette and provides maximum protection from enemy fire. The supported prone position can be assumed behind a tree, a wall, or almost any type and size of cover. It is flexible and allows firing of the pistol from all sides. To assume the supported prone position and maximize the use of cover, the position is kept as low as possible to ensure no part of the body is exposed to the enemy. If the cover is narrow, the Marine positions his body directly behind cover and keeps his legs together. The Marine's body is in line with the pistol and directly behind the pistol. The Isosceles prone position presents a smaller target to the enemy and more body mass to absorb recoil. The Weaver prone position is ideally suited to fire from around cover (e.g., log) because of the angle of the body, but this position presents a larger target to the enemy. See figure 5-10.

Figure 5-10. Weaver Prone.

Supported Kneeling

If the prone position cannot be used because of the height of the support, the Marine may use the supported kneeling position. The kneeling position allows firing of the pistol from all sides. This position may be altered to maximize the use of cover or support by assuming a variation of the kneeling position (high, medium, low, or two-knee). The kneeling position provides more mobility than the prone position. See figure 5-11.

Caution

The Marine must not indicate to the enemy his position with his exposed knee. If the Marine changes knees while exposed to the enemy, he has indicated the direction of his next shot to the enemy.

Figure 5-11. Supported Kneeling.

Supported Standing

When use of the support is maximized, the supported standing position can be as stable as the supported kneeling or prone position. The supported standing position provides greater mobility than the other positions and usually provides greater observation of the enemy. In the standing position, the Marine must not allow the placement of his foot to indicate to the enemy his position behind cover. See figure 5-12.

Figure 5-12. Supported Standing.

Locate and Engage Targets From Behind Cover

To locate targets when behind cover or to ensure the area is clear before moving, the Marine must expose himself to the enemy. There are two techniques used to locate and engage targets from behind cover: the pie and the rollout. These techniques minimize the Marine's exposure to enemy fire while placing him in a position to engage targets or to move to another location if necessary. Both techniques are used in the kneeling and standing positions.

Pie Technique

The Weaver position is the most effective position when executing the pie technique because the position of the Marine's body minimizes exposure to the enemy. See figure 5-13.

Figure 5-13. Pie Technique.

To perform the pie technique—

- Stay behind cover while moving back and away from the leading edge of the cover. The surroundings and situation dictate the distance the Marine moves back and away from the cover. Generally, the further back the Marine is from cover, the greater his area of observation; staying too close to cover decreases the area of observation.
- Assume a firing position and lower the pistol sights enough to have a clear field of view, aiming in on the leading edge of the cover.
- Take small side steps and slowly move out from behind cover, covering the field of view with the aiming eye and muzzle of the pistol. Wherever the eyes move, the muzzle should move (eyes, muzzle, target).
- Continue moving out from cover until a target is identified or the area is found to be clear. If a target is identified, sweep the safety off, place the finger on the trigger, and engage the target.

Note: If a target is identified before moving out from cover, the pistol should be thumb-cocked and off safe before moving out.

Rollout Technique

In this technique, the Isosceles position is the most effective position when executing the rollout technique because the position of the body allows the Marine to better maintain his balance. See figure 5-14. To perform the rollout technique—

- Stay behind cover, move back, and position the body so it is in line with the leading edge of the cover, ensuring that no part of the body extends beyond cover.
- Assume firing position and come to the Ready, ensuring the muzzle is just behind cover.
- Cant the head and pistol slightly and roll the upper body out to the side enough to obtain a clear field of view. Keep the feet in place and push up on the ball of one foot to facilitate rolling out.
- Continue to roll out from cover until a target is identified or the area is found to be clear. If a target is identified, sweep the safety off, place the finger on the trigger, and engage the target.

Figure 5-14. Rollout Technique.

Note: If a target is identified before moving out from cover, the pistol should be thumb-cocked and off safe before moving out.

Combining the Pie and Rollout Techniques

In some situations, it may be necessary to utilize both the pie and rollout technique in order to search an entire area for targets (i.e., corner of a building, doorway). Changing from one technique to another may allow the Marine to minimize his exposure to the enemy and reduce the time he is exposed to enemy fire.

CHAPTER 6
PRESENTATION OF THE M9 SERVICE PISTOL

In combat, targets present themselves with little or no warning. The Marine must have the ability to react quickly and to effectively and efficiently present the pistol, whether the pistol is in the holster or at a carry. To successfully engage a combat target with the M9 service pistol, the Marine must master pistol presentation from the carries and transport, pistol presentation while assuming a firing position, and search and assess techniques.

> *Note:* The procedures in this chapter are written for right-handed Marines; left-handed Marines should reverse directions as needed.

Sight Alignment/Sight Picture

Pistol presentation drills and dry fire help the Marine achieve a consistent grip and rapid presentation and aid in quickly acquiring sight alignment and sight picture. The Marine always executes the following steps:

- Identify the target and quickly present the pistol to the target while simultaneously sweeping the safety with the thumb of the right hand. (Disengaging the safety with the right thumb ensures the trigger is not pulled before taking the pistol off safe.)
- Shift the focus from the target to the front sight to obtain sight alignment while presenting the pistol. As the front sight breaks the plane of vision, acquire the front sight and begin to apply trigger control as sight picture is acquired.
- Apply trigger pressure until the shot is fired.

Presentation From the Carries and Holsters

Presentation From the Ready

To present the pistol from the Ready, the Marine performs the following steps in sequence. When a target appears—

- Sweep the safety with the thumb of the right hand, place the trigger finger on the trigger, and raise the arms to bring the pistol to the target.

  Note: If the Marine wishes to thumbcock the pistol for a single action shot, the pistol is thumbcocked with the left thumb after the safety is swept with the right thumb. The grip of the left hand may have to be broken to thumbcock the pistol; re-establish the grip after thumbcocking.

- Acquire sight alignment and sight picture within the aiming area and apply trigger pressure until the shot is fired.

Presentation From the Alert

To present the pistol from the Alert, the Marine performs the following steps in sequence. When a target appears—

- Sweep the safety with the thumb of the right hand, place the trigger finger on the trigger, and bring the pistol to the target:
  - If the arms are straight, raise the arms to a 45-degree angle with the deck.
  - If the arms are bent, straighten the arms out toward the target.

Note: If the Marine wishes to thumbcock the pistol for a single action shot, the pistol is thumbcocked with the left thumb after the safety is swept with the right thumb. The grip of the left hand may have to be broken to thumbcock the pistol; re-establish the grip after thumbcocking.

- Acquire sight alignment and sight picture within the aiming area and apply trigger pressure until the shot is fired.

Presentation From a Holster Transport

The pistol is presented from the holster in one continuous, fluid motion.

M12 Holster

To present the pistol from the M12 holster transport, perform the following steps in sequence once the target appears:

- Place the heel of the left hand at the center of the torso with the fingers extended toward the target. (The placement of the left hand allows a two-handed grip to be established in a minimum amount of movement.) At the same time, unfasten and release the D-ring with the right hand.
- Use the right hand to place the thumb on the forward edge of the holster and the fingers around the back edge of the holster, keeping the trigger finger straight.
- Slide the hand up the holster until the fingers come in contact with the pistol grip. At the same time, keep the thumb above the pistol to guide the holster flap up.
- Grasp the pistol grip with the fingers and draw the pistol straight up. Continue withdrawing the pistol while moving the thumb to a position on the safety.

Note: Hand placement should allow the firing grip to be established once the thumb disengages the safety. Any adjustments made to the firing grip after the safety has been disengaged should be minor.

- Once the muzzle clears the holster, rotate the muzzle forward while sweeping the safety.
- Establish a two-handed grip on the pistol by joining the left hand with the right hand in front of the torso. At the same time, start to straighten the pistol out toward the target. See figure 6-1.

Figure 6-1. Presentation From the M12 Holster—Step 1.

- Continue moving the pistol toward the target and, at the same time, place the trigger finger on the trigger, acquire sight alignment and sight picture within the aiming area, and apply trigger pressure until the shot is fired. See figure 6-2.

Figure 6-2. Presentation From the M12 Holster—Step 2.

When time permits to thumbcock the pistol for a single action shot (e.g., a long-range engagement), the Marine presents the pistol from the holster by performing the following steps once the target appears:

- Withdraw the pistol from the holster.
- Sweep the safety with the thumb of the right hand.
- Join the left and right hands and thumbcock the pistol with the left thumb.
- Establish a two-handed grip on the pistol.
- Straighten the pistol out toward the target and, at the same time, place the trigger finger on the trigger, acquire sight alignment and sight picture within the aiming area, and apply trigger pressure until the shot is fired.

M7 Shoulder Holster

The M9 service pistol fits very snugly in the M7 shoulder holster because the holster was originally designed for the M1911A1 .45-caliber pistol, which has a more slim, round design, particularly around the trigger guard. A firm grip is required when holstering and withdrawing the M9 service pistol from the M7 shoulder holster. To present the pistol from the M7 shoulder holster, perform the following steps:

- Unsnap the thumb snap closure with the left hand.
- Wrap the fingers of the right hand around the pistol grip and rest the thumb on top of the inside of the holster. See figure 6-3.

Figure 6-3. Presentation From the M7 Shoulder Holster—Step 1.

- Grasp the pistol grip firmly and draw the pistol up and away from the holster while rotating the thumb in a position to operate the safety.
- Continue withdrawing the pistol until the muzzle clears the holster and rotate the muzzle toward the target. See figure 6-4.

Figure 6-4. Presentation From the M7 Shoulder Holster—Step 2.

- Sweep the safety with the thumb of the right hand while starting to punch the pistol out toward the target.
- Establish a two-handed grip on the pistol by joining the right hand with the left hand in the center of the torso. See figure 6-5.
- Continue punching the pistol out and, at the same time, place the trigger finger on the trigger, establish sight alignment and sight picture within the aiming area, and continue trigger pressure until the shot is fired.

Assault Holster

To present the pistol from the assault holster, perform the following in one, continuous motion:

Figure 6-5. Presentation From the M7 Shoulder Holster—Step 3.

- Bring the heel of the hand down on the hammer of the pistol so that it disengages the thumb break. See figure 6-6.
- Sweep the hand forward in a small circular motion, slide the hand up the holster, and bring the fingers up under the pistol grip.

Note: Depending on the type of retention strap, this circular motion releases the retention strap or pushes the retention strap forward and out of the way.

Figure 6-6. Presentation From the Assault Holster—Step 1.

- Grasp the pistol grip and draw the pistol straight up and out of the holster while establishing a firing grip and positioning the thumb on the safety to operate it. See figure 6-7.

Figure 6-7. Presentation From the Assault Holster—Step 2.

Concealed Pistol Holster

Depending on the clothing worn by the Marine, there are two methods for accessing the concealed pistol holster so that the clothing does not become an obstruction while presenting the pistol. To present the pistol from the concealed pistol holster, the following steps are performed in one, continuous motion when the target appears:

- If the Marine is wearing a short jacket/sweater (waist-level) that is buttoned or zipped—
  - Move the left hand across the body and grasp the jacket or sweater just above the holster, at the same time place the right hand on the holster below the jacket or sweater. See figure 6-8 on page 6-6.
  - Pull up on the jacket or sweater with the left hand. At the same time, slide the right hand up the holster until the fingers come in contact with the pistol grip. Keep the thumb above the pistol to ensure the jacket or sweater clears the pistol. See figure 6-9 on page 6-6.

Note: Leaning slightly forward at the waist may assist in clearing the jacket or sweater from the pistol and in removing the pistol from the holster.

- If the Marine is wearing a long jacket (below waist-level), or a short jacket or sweater that is unfastened—
  - Place the heel of the left hand at the center of the torso to hold the left side of the jacket down with the fingers extended. At the same time, place the right hand on the right side of the jacket, with the thumb underneath the right side of the jacket below the left hand. See figure 6-10 on page 6-7.
  - Keep thumb of the right hand against the body and throw the jacket back and away from the holster. See figure 6-11 on page 6-7.

Note: Keeping a weighted object (e.g., keys, extra magazine) in the right-hand pocket of the jacket assists in throwing the jacket back and away from the holster.

  - Place the right hand on the holster and slide the hand up the holster until the fingers come in contact with the pistol grip.
- Grasp the pistol grip with the fingers and draw the pistol straight up. Continue withdrawing the pistol while moving the thumb to a position on the safety.

Note: Hand placement should allow the firing grip to be established once the thumb disengages the safety. Any adjustments made to the firing grip after the safety is disengaged should be minor.

- Rotate the muzzle forward, once it clears the holster, while sweeping the safety.
- Establish a two-handed grip on the pistol by joining the left hand with the right hand in front of the torso. At the same time, start to punch the pistol out toward the target.

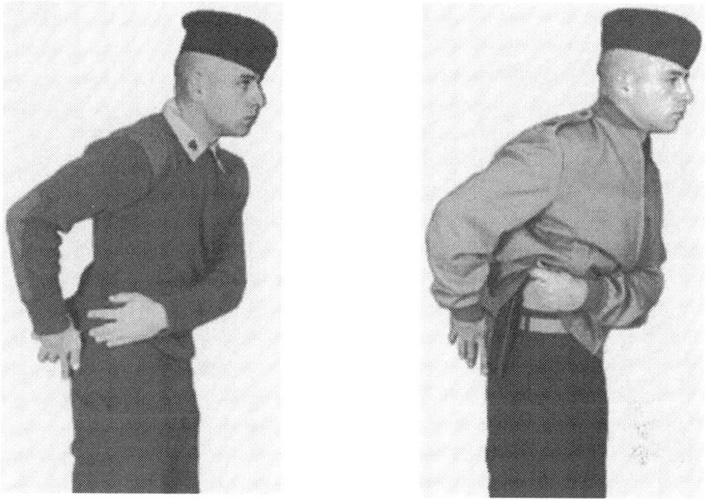

Figure 6-8. Presentation From the Concealed Pistol Holster—Step 1.

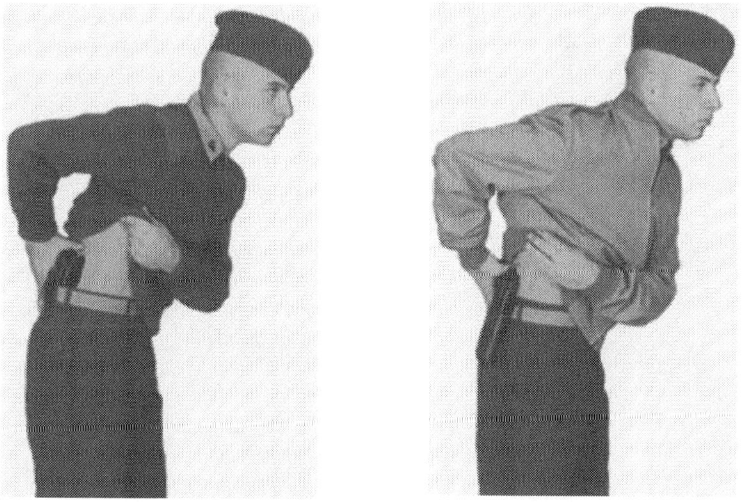

Figure 6-9. Presentation From the Concealed Pistol Holster—Step 2.

Figure 6-10. Presentation From the Concealed Pistol Holster with the Long Jacket—Step 1.

Figure 6-11. Presentation From the Concealed Pistol Holster with the Long Jacket—Step 2.

- Continue punching the pistol out and, at the same time, place the trigger finger on the trigger, acquire sight alignment and sight picture within the aiming area, and apply trigger pressure until the shot is fired.

Presentation While Assuming the Kneeling Position

In combat, the Marine must be able to quickly assume the kneeling position while presenting the pistol to the target. This usually takes advantage of cover and provides a more stable base for shooting. The following procedures are based on the Isosceles and Weaver variations of the standing to the kneeling shooting positions.

From the Ready

The Marine drops to the kneeling position while raising the arms to bring the pistol up to the target.

Note: The Weaver variation maintains the body's position in a 40 to 60 degree oblique to the target.

At the same time, the Marine sweeps the safety with the thumb of the right hand, places the trigger finger on the trigger, acquires sight alignment and sight picture within the aiming area, and applies trigger pressure until the shot is fired.

From the Alert (Arms Straight)

The Marine drops to the kneeling position while raising the arms to bring the pistol up to the target.

Note: The Weaver variation maintains the body's position in a 40 to 60 degree oblique to the target.

At the same time, the Marine sweeps the safety with the thumb of the right hand, places the trigger finger on the trigger, acquires sight alignment and sight picture within the aiming area, and applies trigger pressure until the shot is fired.

From the Alert (Close Quarters, Elbows Bent)

The Marine drops to the kneeling position while pushing the arms out toward the target.

> *Note:* The Weaver variation maintains the body's position in a 40 to 60 degree oblique to the target.

At the same time, the Marine sweeps the safety with the thumb of the right hand, places the trigger finger on the trigger, acquires sight alignment and sight picture within the aiming area, and applies trigger pressure until the shot is fired.

From the Holster Transport

The Marine withdraws the pistol from the holster while dropping to the kneeling position. The pistol should be rotated to the target by the time the knee hits the deck. The Marine sweeps the safety with the thumb of the right hand, establishes a two-handed firing grip and—

- In the Weaver variation, raises the arms to bring the pistol up toward the target.
- In the Isosceles variation, pushes the pistol out toward the target.

At the same time, the Marine places the trigger finger on the trigger, acquires sight alignment and sight picture within the aiming area, and applies trigger pressure until the shot is fired.

Presentation While Assuming the Prone Position

The prone position is generally assumed to take advantage of cover or to provide additional stability for shooting. The following procedures are executed from the standing holster transports.

Isosceles Prone

The Marine withdraws the pistol from holster. At the same time, the Marine eliminates body contact with the deck by either the squat method or the drop method. The pistol is rotated to the target as the left hand is placed on the deck. The Marine sweeps the safety off.

> *Note:* If the Marine wishes to thumbcock the pistol for a single action shot, it is done once the Marine is on the deck.

The Marine establishes a two-handed firing grip while spreading the legs a comfortable distance apart for stability. He places the trigger finger on the trigger, acquires sight alignment and sight picture within the aiming area, and applies trigger pressure until the shot is fired.

Weaver Prone

The Marine withdraws the pistol from the holster. At the same time, the Marine eliminates body contact with the deck by either the squat method or the drop method. The Marine maintains the 40 to 60 degree oblique to the target. The pistol should be rotated to the target as the left hand is placed on the deck. The Marine sweeps the safety off.

> *Note:* If the Marine wishes to thumbcock the pistol for a single action shot, it is done once the Marine is on the deck.

The Marine brings the left knee up, establishes a two-handed firing grip, and places the left elbow on the deck for stability. He places the trigger finger on the trigger, acquires sight alignment and sight picture within the aiming area, and applies trigger pressure until the shot is fired.

Search and Assess

To be successful in combat, the Marine must have the ability to assess the situation and take appropriate action following engagement. Once the Marine fires one or two rounds to engage a target, he must quickly assess the situation and the effectiveness of his engagement. This split-second assessment allows the Marine to determine the best course of action and prevents him from expending a number of rounds without assessing the results and his ability to accurately engage the target.

To search and assess, perform the following steps in sequence:

- Place the trigger finger straight along the receiver immediately after a target is engaged. Do not place the pistol on safe.
- Lower the arms just enough to look over the pistol sights and provide a clear field of view. Do not rotate the wrists in order to angle the pistol downward, this breaks the firing grip and changes the tension in the arms. Lowering the arms maintains the firing grip, keeps the sights level, and allows sight alignment and sight picture to be quickly re-established should follow-on shots need to be fired.
- Search the area by moving the head, eyes, and pistol left and right (approximately 45 degrees from center). Wherever the head moves, the muzzle moves (eyes, muzzle, target). Keeping both eyes open increases the field of view.

If the target has not been eliminated, the Marine must determine whether to re-engage the target:

- If the Marine decides to re-engage the target, the target's size, time and distance to the target, and Marine's capabilities dictate the technique used. Re-engagement techniques include re-engaging the target with a well-aimed precision shot, two shots, or offset aiming.
- If the target's distance exceeds the Marine's engagement capabilities or if the target is partially exposed and too small for the Marine to accurately engage, the Marine should not attempt to re-engage the target. The Marine may seek cover, seek out a better opportunity for engagement, or use support to better stabilize the pistol.

When it is determined that the area is clear of all enemy threat, place the pistol on safe without breaking the grip and assume a pistol carry or pistol transport.

CHAPTER 7
PISTOL ENGAGEMENT TECHNIQUES

The Marine must be able to quickly detect targets and employ the proper pistol engagement and firing techniques. The Marine must also be able to re-engage a target if the initial engagement is not successful.

> *Note:* The procedures in this chapter are written for right-handed Marines; left-handed Marines should reverse directions as needed.

Target Detection

Target indicators reveal an enemy's position. Most combat targets are detected by smoke, flash, dust, noise, or movement and are usually only momentarily visible. Target indicators are grouped into three general areas: movement, sound, and improper camouflage.

Movement

The Marine does not need to look directly at an object to notice movement. The eye is attracted to any movement, especially sudden movement. A slowly moving target is harder to detect than one with quick, jerky movements. Therefore, the ability to locate a moving target depends primarily on the speed of the object's movement.

Sound

Sound (e.g., movement, rattling equipment, talking) can be used to detect an enemy position. Sound provides only a general location, making it difficult to pinpoint a target by sound alone. However, sound can alert the Marine to the presence of a target and increases the probability of locating the target through other indicators.

Improper Camouflage

The improper use of camouflage creates three target indicators: shine, outline, and contrast with the background. The closeness of a typical pistol engagement somewhat diminishes these effects, but can still affect target detection. The Marine's ability to recognize target indicators aids in the detection of targets.

Shine

Shine is created from reflective objects such as metal, glass, pools of water, and the skin's natural oils. Shine acts as a beacon in locating a target's position.

Outline

The outline of camouflaged objects such as the body, head and shoulders, weapons, and gear can be recognizable.

Contrast With the Background

Objects contrast with a background because of differences in color, surface, and shape. The following are examples of objects contrasting with their background:

- A target wearing a dark uniform would be clearly visible in an area of snow or sand.
- Symmetrical shapes, such as helmets or rifle barrels, can be detected in a wooded area.
- Fresh soil around a fighting hole contrasts with the otherwise unbroken ground surface.

While observing an area, the Marine notes anything that looks out of place or unusual and studies it in more detail in order to increase the chances of spotting a hidden threat.

Techniques of Fire

To successfully engage a combat target with the M9 service pistol, the Marine must employ effective techniques of fire. The Marine's performance of these skills and the proper application of the fundamentals of marksmanship are critical to success in a combat situation.

Double and Single Action Firing

When the M9 service pistol is taken off safe, it is capable of firing in both a double and a single action mode.

Double Action Mode

The design of the M9 service pistol causes the first shot fired to be a double action shot.

In double action firing, two actions occur as the trigger is moved to the rear; the hammer moves to the rear, cocking the pistol, and then the hammer moves forward, firing the pistol. More pressure is required on the trigger to fire a double action shot due to the distance between the trigger and hammer and the weight of the trigger. A double action shot requires approximately 9 to 16 pounds of pressure to move the trigger rearward.

Maintaining sight alignment and sight picture are harder when firing a double action shot. Therefore, the sights will more than likely move outside the aiming area when applying trigger pressure.

Single Action Mode

In single action firing, the pistol is already cocked because the cycle of operation in the first double action shot leaves the hammer cocked to the rear. Therefore, the only action taking place as the trigger is moved to the rear is the hammer moving forward, firing the pistol. A single action shot requires approximately 4 to 6 pounds of pressure to move the trigger rearward. Therefore, the application of trigger control is easier when firing a single action shot.

To enable the first shot to be fired single action, the pistol's hammer can be manually cocked with the thumb:

> *Note:* The pistol must be taken off safe before it can be thumbcocked.

- Use the left thumb to pull back on the hammer to cock it. This ensures that the firing grip of the right hand does not have to be broken.
- Ensure that the hammer moves all the way to the rear.
- Ensure the trigger finger remains straight along the receiver until the pistol is fully cocked.
- Re-establish the firing grip with both hands once the pistol is cocked.

Single Action or Double Action Firing Factors

The decision to fire in a single or double action mode is made rapidly. Ultimately, the decision is based on the Marine's abilities, but the decision is also based on time and accuracy. The pistol is fired in the double action mode when trigger control, sight picture, and stability of hold are not as critical for accuracy (i.e., close range, large targets). The pistol is fired in the single action mode when the fundamentals are more critical to accuracy (i.e., long range, small targets).

Time

The Marine sacrifices time to fire a single action, precision shot; but what he sacrifices in time, he gains in accuracy.

For quick engagements at close range, there may not be time to thumbcock the pistol for a single action shot. Therefore, firing the first shot in the double action mode is preferred because shots are needed on target quickly and stability of hold and sight picture are not as critical to accuracy.

When time permits and for targets at longer ranges, the pistol may be thumbcocked to place it in the single action mode to reduce the weight of the trigger and the distance the trigger must travel rearward to fire the first shot.

Distance and Size of the Target

The smaller the target, the more critical the application of the fundamentals in order to engage the target accurately. To accurately engage a small target (e.g., head shot, long-range target) it is better to thumbcock the pistol for a single action shot. Engagement of a smaller target requires additional precision because sight alignment and sight picture are more critical to accuracy.

Two-Shot Technique

During combat, the preferred pistol engagement technique is to rapidly fire more than one shot on a target to eliminate it as a threat. Two shots fired in rapid succession increase the trauma (i.e., shock, blood loss) on the target and increase the Marine's chances of quickly eliminating the threat. Therefore, two shots are most often fired in rapid succession on a target at close range.

After the pistol is fired, the muzzle climbs with the recoil of the pistol. To fire two shots, the Marine must quickly recover the sights to the same area on the target while re-acquiring sight alignment and sight picture. The proper recovery automatically brings the sights back on target following recoil. A quick recovery allows more time for the Marine to align the sights and apply trigger control to fire the next shot. Recovery begins immediately after the application of the fundamentals to bring the pistol sights into alignment with the target in preparation for firing the next shot.

The amount the muzzle climbs during recoil depends on the amount of controlled muscular tension in the grip and wrists used to stabilize the pistol and create consistency in resistance to recoil. Controlled muscular tension allows the pistol sights to recover consistently back on target within a minimum amount of time. The speed of delivery of multiple shots depends on how fast the Marine can re-acquire sight alignment. Sight recovery is determined by the following:

- The key to proper recovery is a stable firing position and proper grip.
- If the Marine's firing position is not stable, recoil forces him out of his firing position, requiring him to re-establish his position before he takes his next shot.
- An improper grip or lack of controlled muscular tension causes the pistol to move in the Marine's hand after the shot is fired, disrupting sight alignment and requiring the Marine to re-establish his grip.

Slow Fire Technique

Sight alignment becomes more critical the smaller the target and the greater the distance to the target. In these situations, the Marine does not engage the target with two rapidly fired shots because he has to slow down his application of the fundamentals in order to fire a precision shot(s). To engage small targets (i.e., partially exposed) and targets at longer ranges where precision is required, the Marine must employ the following slow fire technique:

- Thumbcock the pistol for a single action shot.
- Slow down application of the fundamentals.
- Fire one well-aimed, precision shot on target.

Re-engagement Techniques

Once the Marine has quickly assessed the situation and determined that the threat still exists, he may make the decision to re-engage the target to eliminate it as a threat. Re-engagement techniques include engaging the target with a precision shot through slow fire, two shots, or offset aiming. The time, size, and distance to the target as well as the Marine's marksmanship abilities dictate the technique used to re-engage the target.

Slow Fire Technique

If the Marine has engaged a target and the target still poses a threat, the Marine may choose to slow down his application of the fundamentals and fire a slow fire, precision shot. The placement of one well-aimed, precision shot on a designated area of the target increases the chances of eliminating the target as a threat. However, the time the Marine has to engage the target determines if he can slow down his application of the fundamentals to fire a precision shot. Firing a precision shot takes time, so the Marine will sacrifice time for accuracy.

The key to successful shot placement is the Marine's ability to slow down his performance and focus on the application of the fundamentals of marksmanship. The Marine must determine where to place shot for maximum effectiveness. In order to make this decision, the Marine must consider distance and size of the target and time. For targets at close range and within the Marine's marksmanship abilities, a precision shot is placed in the head to immediately eliminate the target as a threat. Sometimes, even at close ranges, the target may only be partially exposed; therefore, the Marine is presented with a small target and aims his sights on the portion of the target that is exposed. For targets at long ranges, the target is smaller and requires a precision shot. A precision shot may be placed in the body to add trauma to the target and increase the chances of eliminating the target as a threat.

Two-Shot Technique

If a target is accurately engaged, but does not go down, the Marine may re-engage the target with additional shots. Two additional, rapidly delivered shots increases the trauma and increases the Marine's chances of eliminating the threat.

The size and distance to the target affect how quickly two shots can be delivered on the target. The speed at which two shots are fired also depends on the Marine's marksmanship abilities and how fast he can re-acquire his front sight. However, the Marine must not compromise accuracy for speed; the key to successful target engagement is to fire only as quickly as the Marine can fire effectively.

Offset Aiming

When the Marine assesses the situation and determines his shots are not successful and not striking the target in the designated aiming area, he may employ offset aiming. Since the pistol's sights cannot be adjusted, offset aiming is applied to adjust the aiming area and to cause rounds to strike center mass. This technique should only be applied when the Marine determines that he is applying the fundamentals of marksmanship correctly. Sight picture will change as the aiming area is adjusted.

The known strike of the round offset aiming technique requires shifting the point of aim to compensate for rounds striking off target center. To effectively engage a target using this technique, the Marine must be able to see where the rounds are striking and then aim an equal distance from the center of the aiming area opposite the observed strike of the round. For example, if the rounds are striking the target high and left of center mass, aim an equal and opposite distance low and right.

Multiple Targets

If engaging more than one target at a time, the Marine adheres to the fundamentals of marksmanship and employs the techniques of multiple target engagement. The introduction of multiple targets in a combat scenario requires additional skills that must be learned and practiced if a Marine is to be successful. To be effective in combat, the Marine must be able to detect targets through identification of target indicators (identified on page 7-1), prioritize the targets, and employ multiple target engagement techniques.

Combat Mindset

Successful engagement of multiple targets requires a somewhat different mindset from single target engagement. For example, following engagement of a single target, the Marine assesses the situation. During multiple target engagements, after the first target is engaged, he must immediately engage the next target and continue until all targets have been eliminated. Because split-second decisions must be made, the development of a combat mindset is important to success on the battlefield. A combat mindset allows the Marine to control the pace of the battle rather than react to the threat.

Mental preparedness is essential to successful engagement of multiple targets, and the required mindset must be developed until it becomes second nature to the Marine. When multiple targets appear, the Marine must prioritize the targets to establish an engagement sequence. To prioritize targets and establish an engagement sequence, the Marine must be aware of the surroundings, not focus on one target, and continuously search the terrain for additional targets.

Prioritizing Targets

Once multiple targets have been identified, they must be prioritized in terms of the threat each target presents. While the fundamentals of marksmanship must still be applied, prioritizing targets and planning the engagement are just as essential to successful multiple target engagement.

Target priority is based on factors such as proximity, threat, and opportunity. It also encompasses the Marine's proficiency level since a Marine should not attempt to engage a target beyond his proficiency level or the pistol's capability (e.g., a target 100 yards away). The principal method of prioritizing targets is to determine the level of threat for each target so that all targets may be engaged in succession, starting with the most threatening and ending with the least threatening.

Prioritizing targets is an ongoing process. As the engagement proceeds, new targets may appear or previously identified targets may take cover, delaying their engagement. Therefore, the Marine must remain constantly alert to changes in target threat, proximity, and the opportunity for engagement and revise target priorities accordingly.

Engaging Multiple Targets

Once targets have been prioritized, the Marine must quickly eliminate them. It is crucial to understand and practice the physical techniques for engaging multiple targets until they become second nature. When these techniques occur as automatic responses, the Marine maintains an awareness of the complete battlefield and is able to concentrate on the mental aspects of multiple target engagement.

Engagement Methods

The fundamentals of marksmanship are critical to the development of skills to support multiple target engagement. While responsiveness is important when engaging multiple targets, the Marine's primary concern is to place accurate rounds on target. A balance must be struck between placing shots quickly on targets at close range and slowing down to place precision shots on targets at long range because all targets pose a threat and must be accurately engaged. To engage multiple targets—

- Engage every target, move from target to target.
- After all of the targets have been initially engaged, assess the effectiveness of the engagements and, if necessary, re-engage targets that still pose a threat.
- If all of the targets are eliminated, search the area for new targets.

Considerations for Field Firing Positions

The selection and use of field firing positions are critical to engagement of multiple targets. As in any firing situation, if the situation permits, the Marine should make a quick mental review of the terrain to select a firing position that provides stability, mobility, and observation of the enemy. The firing position should also provide flexibility for engaging multiple targets. The more dispersed the multiple targets, the greater the lateral movement the position must afford to engage them.

The prone position provides the maximum stability for firing well-placed shots, but engaging multiple targets from this position may require adjustment in the position from shot to shot depending on the distance between the targets and their location. Because the arms are fully extended on the ground, the prone position can be restrictive, increasing the time it will take to recover the sights onto subsequent targets. If the Marine must make an adjustment in the prone position to engage multiple targets, he must ensure that he maintains proper arm placement. Improper arm placement may affect his stability of hold and his ability to acquire sight picture. The following identifies position adjustments:

- For targets that are not widely dispersed, the Marine adjusts the position of the lower part of the body to orient the upper body toward the target without disturbing the placement of the arms. This allows the Marine to maintain stability of hold and quickly recover on subsequent targets. Moving the legs to the left orients the upper body to the right; moving the legs to the right orients the upper body to the left.
- For targets that are widely dispersed, the Marine may need to adjust his entire position to orient his body in the direction of subsequent targets. Using the left hand, push the upper body off the deck just high enough to move the body in the direction of the target. Extend the right arm toward the target and keep the muzzle pointed in the direction of the target.

The kneeling and standing positions provide an increased field of view and allow maximum lateral adjustment to engage dispersed targets. The type of adjustment the Marine must make in the kneeling or standing position in order to orient his body in the direction of each target is related to the distance between the targets. It is also related to the Marine's ability to maintain stability of hold and recover from recoil. Once the first target is engaged—

- Maintain a proper shooting position, rotate the body in the direction of the target, keeping the feet in place. In the standing position, the Marine may bend slightly at the knees while rotating the body. This allows the Marine to distribute his weight forward to provide additional stability to the position and to better control the effects of recoil.
- Adjust the entire position by either adjusting the placement of the feet or knee(s) in order to face in the direction of the new target if the targets are some distance apart.
- Ensure that as the eyes move to a new target, the head and pistol muzzle should follow.

Moving Targets

The majority of combat targets will be moving; the enemy will move quickly from cover to cover, exposing himself for the shortest time possible. Therefore, a moving target must be engaged before it disappears. The Marine must engage a moving target with the same accuracy and precision used to engage a stationary target. Practice in the proper application of leads and the fundamentals of marksmanship enable the successful engagement of moving targets.

Types of Moving Targets

Moving Target

This type of target moves in a consistent manner (i.e., walking or running) and is in continuous sight as it moves across the Marine's field of vision. However, unless the enemy is completely unaware of the Marine's presence, this type of target is not likely to present itself.

Stop and Go Target

This type of target appears and disappears during its movement due to intermittent cover. It will present itself for only a short period of time before re-establishing cover. This type of target is most vulnerable to fire at the beginning and end of its movement because it begins slow and gains momentum as it leaves cover and then slows to avoid overrunning the new cover.

Leads for Moving Targets

Once the moving target has been identified it must be engaged. To engage a moving target, the Marine must aim at a point in front of the target, which is known as a lead. Lead is the distance in advance of the target that is required to strike the target when it is moving. To achieve success, determining the amount of lead to engage a moving target must be as precise as possible. When a shot is fired at a moving target, the target continues to move during the time of the bullet's flight. For this reason, the aim must be in front of the target; otherwise, the shot will fall behind the target.

Factors Affecting Lead

Factors that affect the amount of lead are the target's range, speed, and angle of movement.

Range. There is a time lag from the time a round is fired until the round strikes at the point of aim. This time of flight could allow a target to move out of the bullet's path if the round were fired directly at the target. Time of flight increases as range to the target increases. Therefore, the lead must be increased in proportion to the distance to the target.

Speed. A greater lead is required to hit a running man than a walking man because the running man moves a greater distance during the flight of the bullet.

Angle of Movement. The angle of target movement also affects the amount of lead required for target engagement. The angle of movement across the Marine's line of sight relative to the flight of the bullet determines the type (amount) of lead.

Types of Leads

Full Lead. The target moves straight across the Marine's line of sight with only one arm and half the body visible. This target requires a full lead because it moves the greatest distance across the Marine's line of sight during the bullet's flight.

Half Lead. The target moves obliquely across the Marine's line of sight (at about a 45-degree angle). One arm and over half of the back or chest are visible. This target requires half of a full lead because it moves half as far as a target moving directly across the Marine's line of sight during the bullet's flight.

No Lead. A target moving directly toward or away from the Marine presents a full view of both arms and the entire back or chest. No lead is required. This target is engaged in the same manner as a stationary target because it is not moving across the Marine's line of sight.

Point of Aim Technique

To engage a moving target, a Marine establishes a lead using a point of aim offset aiming technique. This technique uses predetermined points of aim to sector a man-sized target vertically, halfway between center mass and the leading edge of the target (both one point of aim and two points of aim) (see fig. 7-1 on page 7-8). The same units of measure can be applied off the target for holds of additional points of aim. To use the point of aim technique to establish a lead on a man-sized moving target at various ranges, speeds, and angles of movement, the following guidelines apply. See figure 7-2 on page 7-9.

- For a target moving at a distance of 15 yards away or less, no lead is required.
- For a target walking directly across the Marines line of sight (full lead) between 16 and 25 yards away, hold 1 point of aim in the direction the target is moving.
- For a target running directly across the Marine's line of sight (full lead) between 16 and 25 yards away, hold 2 points of aim in the direction the target is moving.

Methods of Engagement

Moving targets, although difficult, can be engaged by the tracking or ambush method or a combination of the two. See figure 7-2 on page 7-9.

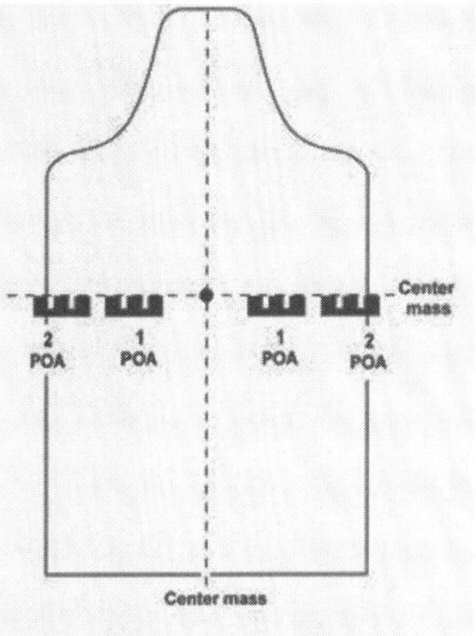

Figure 7-1. Points of Aim.

Tracking Method

In this method, the Marine "tracks" or follows the target with his front sight while maintaining sight alignment and a point of aim on or ahead of (leading) the target until the shot is fired. Sight picture is the aiming point in relation to the target while maintaining sight alignment (when a lead is established in moving target engagement, the sights are not entered on the target). To execute the tracking method—

- Thumbcock the pistol while presenting it to the target.
- Track the muzzle of the pistol through the target to the desired point of aim (lead). The point of aim may be on the target or some point in front of the target depending upon the target's range, speed, and angle of movement.
- Track and maintain focus on the front sight while applying trigger pressure and acquiring sight alignment.
- Continue tracking and applying trigger pressure while acquiring sight picture. When sight picture is established, engage the target while maintaining the proper point of aim (lead).
- Follow through so the lead is maintained as the bullet exits the muzzle. Continuing to track also enables a second shot to be fired on target.

Ambush Method

The ambush method is generally used to engage a stop and go target and when it is difficult to track the target with the pistol, such as in the prone position. With this method, the pistol is aimed at a predetermined engagement point ahead of the target and along its path, allowed to remain stationary, and fired when the target reaches the predetermined engagement point. The engagement point is based on the required point of aim (lead) to engage the target effectively. When the sights are settled, the target moves into the predetermined engagement point creating the desired sight picture. Once sight picture is established, the remaining pressure is applied on the trigger until the shot is fired. To execute the ambush method—

- Look for a pattern of exposure.
- Thumbcock the pistol while presenting it to a selected point of aim ahead of the target.
- Apply trigger pressure while obtaining sight alignment in the aiming area.
- Continue applying trigger pressure and hold sight alignment until the target moves into the predetermined engagement point and the desired sight picture is established.
- Engage the target once the sight picture is acquired.
- Follow through so the pistol sights are not disturbed as the bullet exits the muzzle.

Applying the Fundamentals of Marksmanship

The engagement of moving targets is a perishable skill that must be practiced frequently if it is to be maintained. The Marine must practice to develop the skill to calculate the point of aim (lead) and

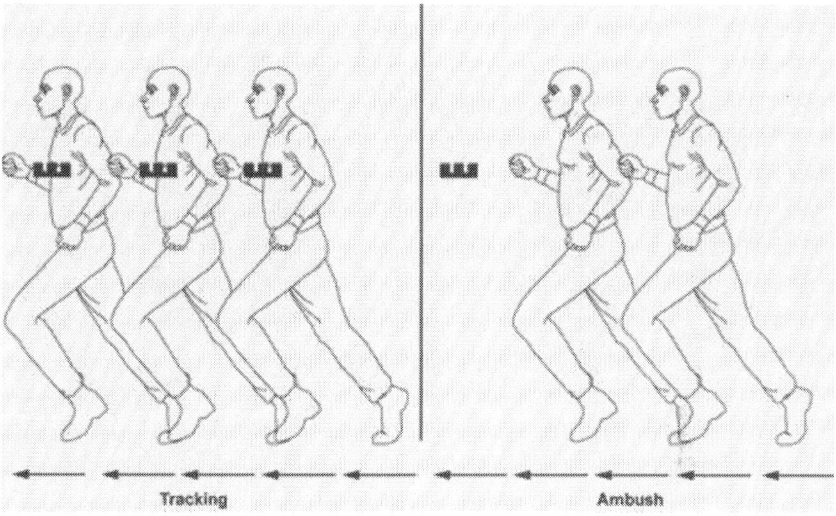

Figure 7-2. Moving Target Engagement Methods.

fire the shot while maintaining the proper point of aim (lead).

To engage moving targets using the tracking method, the Marine moves the pistol smoothly and steadily as the target moves. A stable position and firm grip are necessary to steady the pistol sights while tracking.

When using the tracking method, continue tracking the target while following through with the shot process so the point of aim (lead) is maintained as the bullet exits the muzzle. Continuing to track also enables a second shot to be fired on target if necessary. Concentration should be on continuing to track while applying the fundamentals. The fundamentals must be instinctively applied, allowing concentration on tracking the target and applying point of aim (lead).

Aiming

Sight alignment remains unchanged for accurate engagement of a moving target. The most common error when engaging moving targets is to focus on the target rather than focusing on the front sight. Sight picture is the point of aim in relation to the target while maintaining sight alignment. For both the ambush and tracking methods, sight picture is changed from the normal center mass picture, based on range, speed, and angle of movement of the target. It takes practice in moving target engagement to quickly establish the desired sight picture for a given point of aim (lead).

Breath Control

There is no difference in breath control when engaging moving targets; the breath is held to fire the shot.

Trigger Control

Trigger control is critical to firing shots while establishing and maintaining sight alignment and sight picture. Because the single action mode is the preferred method of engaging a moving target, the pistol is thumbcocked before trigger control is applied. The following guidelines apply:

- Do not use interrupted trigger control when engaging moving targets because the point of aim (lead) is lost or has to be adjusted if the trigger is held in order to re-assume the proper sight picture.
- Do not stop tracking as trigger control is applied. This causes the shot to impact behind the moving target. Trigger control should be uninterrupted while maintaining the point of aim so the trigger is pulled in one continuous, smooth motion to the rear.
- Do not interrupt trigger control when the target is masked from view. This is particularly true in the ambush method for targets that appear to stop and go when moving to and from cover. Trigger control should be uninterrupted.

CHAPTER 8
ONE-HANDED TECHNIQUES

In combat, a situation may arise where the Marine must engage a target by using only one hand. A Marine fires the pistol one-handed when—

- The Marine's other hand is occupied.
- A target presents itself so quickly and at a close range that the Marine does not have time to establish a two-handed grip and position.
- One of the Marine's hands is injured.

Note: The procedures in this chapter are written for right-handed Marines; left-handed Marines should reverse directions as needed.

Presentation and Shooting

Adjustment of the Standing Position

To fire accurately, the Marine must maintain the same stability and control of the pistol, manage recoil, and recover on target as effectively with one hand as with two. This is accomplished by adjusting the standing position as the pistol is presented to the target (see fig. 8-1).

The Marine's angle to the target determines what adjustments are made to the standing position. This adjustment is made while the pistol is being presented to the target by moving or sliding the feet to orient the body to the target. For stability and balance, the right foot may be slightly forward of the left foot. The further to the right of the Marine the target is the farther forward the right foot will be. To adjust the standing position when firing one-handed—

- Increase the muscular tension in the right arm and bend the elbow slightly to better manage the pistol.

Note: For some Marines, muscular tension in the firing arm and the position of the body in relation to the target causes them to cant the pistol inboard slightly. A slight inboard cant of the pistol and firing arm allows a natural bend in the elbow and allows a better management of recoil and recovery of the sights back on target. Any cant should be natural and not caused by excessive muscular tension. A slight cant does not affect shooting performance and should not alter the application of the fundamentals of marksmanship.

- Apply controlled muscular tension throughout the body to better manage the effects of recoil.
- Lean forward aggressively, roll the shoulders forward, and tuck the head into the shoulders.

Figure 8-1. One-Handed Presentation.

Note: The following step is done only during training to demonstrate how to stabilize the position and to simulate the Marine's other hand being occupied.

- Place the left hand in a fist on the center of the torso to increase stability. If the left arm is swinging freely it causes a corresponding movement in the pistol, reducing stability and control. Placing the left hand on the torso also ensures it is not covered by the muzzle of the pistol as the pistol is presented.

Searching and Assessing After Firing

Following one-handed target engagement, the Marine must quickly decide if it is to his advantage and if it is possible to free his occupied hand in order to place both hands on the pistol to search and assess. Whenever possible, the Marine places both hands on the pistol to increase stability and to be ready for re-engagement if necessary.

Reloading

Dry Reload

When only one hand is available, the dry reload can be performed as follows:

- Seek cover, if the situation permits.
- Press the magazine release button and allow the magazine to fall to the deck. If using the right hand, press the magazine release button with the thumb. If using the left hand, press the magazine release button with the trigger finger.
- Point the muzzle in a safe direction and position the pistol to facilitate loading the magazine. Depending on the firing position, this is accomplished using one of the following methods:

  Two-Knee Kneeling Position

  - Rotate the pistol so the magazine well faces up and the muzzle of the pistol is pointed away from the body.
  - Place the pistol between the thighs or knees and apply pressure to hold the pistol in place.
  - Withdraw a filled magazine from the ammunition pocket and insert it into the magazine well, seating it with the heel of the hand. See figure 8-2.

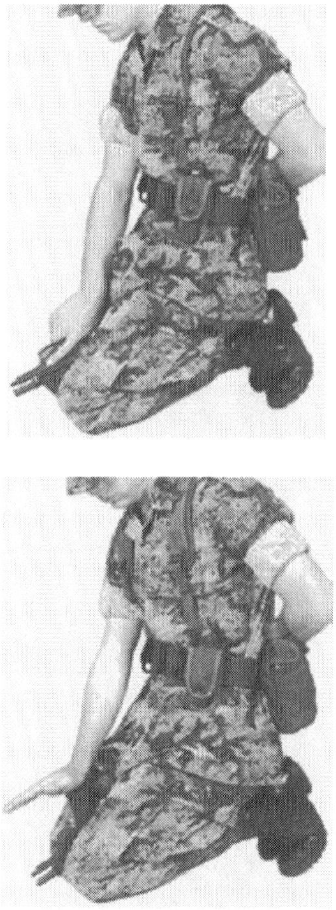

Figure 8-2. One-Handed Dry Reload: Two-Knee Kneeling Position.

Standing Position

- Rotate the pistol so the magazine well faces up and the muzzle of the pistol is pointed away from the body.
- Place the pistol between the thighs or knees and apply pressure to hold the pistol in place. Bend slightly at the knees to secure the pistol between the thighs.
- Withdraw a filled magazine from the ammunition pocket and insert it into the magazine well, seating it with the heel of the hand. See figure 8-3.

Kneeling Position (High, Medium, Low)

- Rotate the pistol so the magazine well faces outboard and the muzzle is pointed away from the body.
- Place the rear portion of the slide in the bend of the right knee and apply pressure with the leg to hold the pistol in place. For additional control, the Marine may drop to a two-knee kneeling position and place the pistol between his knees.
- Withdraw a filled magazine from the ammunition pocket and insert it into the magazine well, seating it with the heel of the hand. See figure 8-4 on page 8-4.

Prone Position

- The Marine places the pistol on the deck in front of him and against his body with the muzzle pointed in a safe direction. Alternately, the Marine may roll on his side and place the pistol between his knees, with the magazine well facing out.
- Withdraw a filled magazine from the ammunition pocket and insert it into the magazine well, seating it with the heel of the hand. See figure 8-5 on page 8-4.

- Grasp the pistol grip and remove the pistol from its secured location or from the deck.
- Press the slide release to allow the slide to move forward and chamber a round. If using the right hand, press the slide release with the thumb. If using the left hand, press the slide release with the index finger.

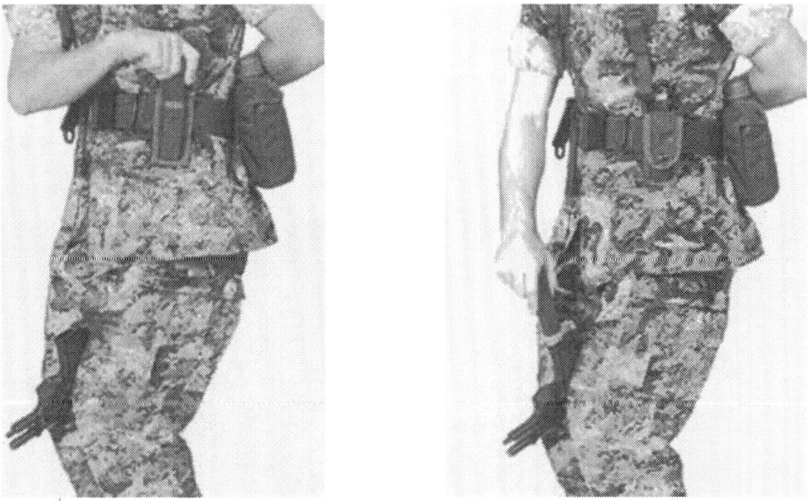

Figure 8-3. One-Handed Dry Reload: Standing Position.

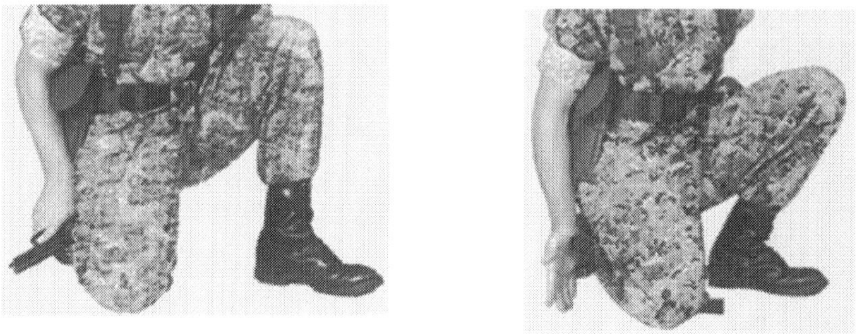

Figure 8-4. One-Handed Dry Reload: Kneeling Position.

Condition 1 Reload

When only one hand is available, a Condition 1 reload can be performed as follows:

- Point the muzzle in a safe direction and position the pistol to allow for removal of the magazine (e.g., between the thighs, in the bend of the knee, in the holster, tucked in the cartridge belt, on the deck against the body). Any position is correct if it allows the magazine well to be exposed.
- Withdraw a filled magazine from the ammunition pocket.
- Press the magazine release button and remove the magazine from the pistol. If using the right hand, press the magazine release button with the thumb. If using the left hand, press the magazine release button with the index finger.
- Insert the filled magazine into the magazine well, seating it with the heel of the hand.
- Stow the partially-filled magazine (e.g., inside the cartridge belt, in the cargo pocket).
- Grasp the pistol grip and remove the pistol from its secured location, keeping the trigger finger straight and off the trigger until ready to fire.

Remedial Action

Remedial action requires investigating the cause of the stoppage, clearing the stoppage, and returning the pistol to operation. When performing remedial action, seek cover if the tactical situation permits. Once a pistol ceases to fire, the Marine must visually or physically observe the pistol to identify the problem before it can be

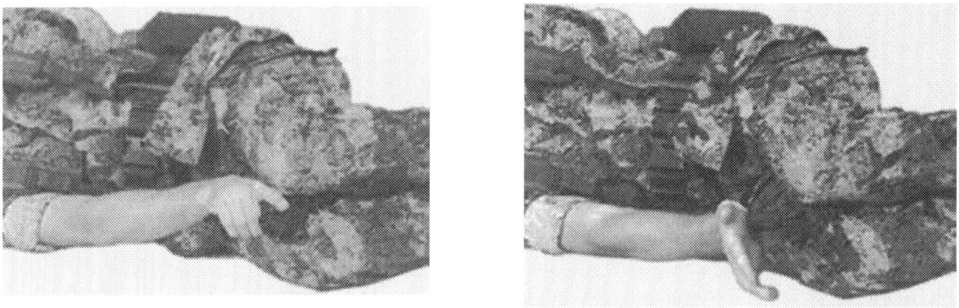

Figure 8-5. One-Handed Dry Reload: Prone Position.

cleared. The steps taken to clear the pistol are based on what is observed:

- Remove the finger from the trigger and place it straight along the receiver.
- Lock the slide to the rear. To pull and lock the slide to the rear, push up on the slide stop and maintain pressure with the right thumb. (Push up on the slide stop with the left index finger if remedial action is being performed with the weak hand.)
- Secure the rear sight on the top of the cartridge belt or any other surface that provides the resistance needed to stabilize the pistol. See figure 8-6. While applying pressure on the pistol to keep the rear sight secured, push downward on the pistol in one continuous motion to lock the slide to the rear.

  Note: Ensure the pistol does not move to safe when locking the slide to the rear.

- Place the pistol in a position to observe the chamber.
- Correct the stoppage.

If there is a round in the magazine but not in the chamber, release the slide and observe a round being chambered. If you do not observe a round being chambered, tap, rack, and bang to perform remedial action:

<u>Tap</u>

- Keep muzzle pointed in a safe direction and place trigger finger straight along the receiver.
- Strike the bottom of the magazine against a hard surface to ensure it is seated. In the kneeling position, strike the magazine against the thigh (see fig. 8-7 on page 8-6). In the standing position, bring the right knee up while striking magazine against the right thigh (see fig. 8-8 on page 8-6). In the prone position, strike the magazine against the deck.

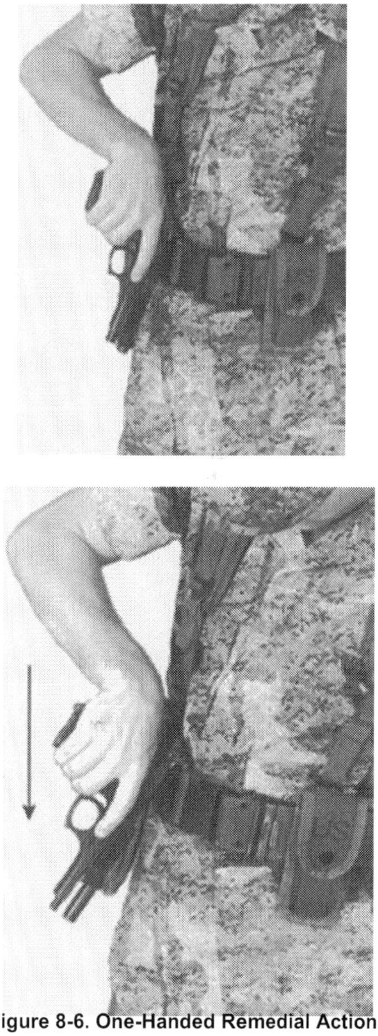

Figure 8-6. One-Handed Remedial Action.

Figure 8-7. Tap the Magazine Against the Knee: Kneeling Position.

Figure 8-8. Tap the Magazine Against the Knee: Standing Position.

Rack

- Rotate the pistol so the rear sight can be hooked on the top of the cartridge belt or any other surface (e.g., the edge of a table, wall, heel of the boot) that provides the resistance needed to rack the slide in a smooth, uninterrupted motion. Do not let the muzzle of the pistol cover the body. See figures 8-9 and 8-10.

Figure 8-9. Rack Slide Against Cartridge Belt.

Figure 8-10. Rack Slide Against Heel of Boot.

- Apply pressure on the pistol to keep the rear sight secured while pushing downward on the pistol to move the slide fully to the rear.
- Release pressure on the pistol to chamber the round.

Bang

- Recover the pistol on target, re-establish sight alignment/sight picture, and attempt to fire.

Presentation From the Holster With the Weak Hand

During combat, the Marine must be prepared to engage targets at any time. Therefore, the Marine must have the ability to present the pistol from the holster with his weak (left) hand if circumstances prevent him from using his strong (right) hand. He must apply the same smooth, controlled movement to present the pistol to the target. There are three methods for withdrawing the pistol from the holster with the weak hand. The Marine should practice each method to determine which works best for him.

Method One: Pistol Rotation

- Unfasten and release the D-ring with the left hand. See figure 8-11.

Figure 8-11. Method One: Release D-Ring.

- Wrap the fingers and thumb around the pistol grip so the index finger and thumb are around the base of the pistol grip. See figure 8-12.

Figure 8-12. Method One: Grasp the Pistol.

- Draw the pistol straight up until trigger guard clears the top of the holster. See figure 8-13.

Figure 8-13. Method One: Clearing Trigger Guard.

- Rotate the pistol so the magazine well faces the target and the trigger guard rests on the top of the holster. Ensure the trigger guard rests securely against the top of the holster. See figure 8-14 on page 8-8.

Figure 8-14. Method One: Rotating the Pistol.

- Maintain downward pressure on the pistol and slide the left hand around the pistol grip to establish a firing grip. See figure 8-15.

Figure 8-15. Method One: Establishing the Firing Grip.

- Grasp the pistol and continue to withdraw the pistol until the muzzle clears the holster and rotate the muzzle toward the target.
- Sweep the safety off with the thumb of the left hand while moving the pistol out toward the target. At the same time, place the trigger finger on the trigger and establish sight alignment and sight picture within the aiming area.
- Continue trigger pressure until the shot is fired.

Method Two: Hand Rotation

- Unfasten and release the D-ring with the left hand.
- Push the holster flap up and out of the way with the back of the left hand.
- Push the right hip out to facilitate grasping the pistol grip. Grasp the pistol grip between the holster flap and the pistol grip, the thumb is wrapped around the other side of the pistol grip against the backstrap. See figure 8-16.

Figure 8-16. Method Two: Grasping the Pistol.

Note: The cartridge belt may be pulled with the left hand to bring the holster closer to the body's center to facilitate grasping the pistol.

- Draw the pistol straight up while rotating the pistol so the magazine well faces inboard. Continue to remove the pistol until the trigger guard rests on the holster. See figure 8-17.
- Establish a firing grip and rotate the thumb to a position to operate the safety. See figure 8-18.
- Continue to withdraw the pistol until the muzzle clears the holster and rotate the muzzle to the target. See figure 8-19.
- Sweep the safety with the thumb of the left hand while moving the pistol out toward the

Figure 8-17. Method Two: Trigger Guard Rests on Holster.

Figure 8-18. Method Two: Establish Firing Grip.

target. At the same time, place the trigger finger on the trigger and establish sight alignment and sight picture within the aiming area.

- Continue trigger pressure until the shot is fired.

Method Three: Knee Placement

This method is particularly effective when firing from the two-knee kneeling position and the standing position.

Figure 8-19. Method Two: Rotating the Muzzle.

- Unfasten and release the D-ring with the left hand.
- Wrap the fingers and thumb around the pistol grip so the index finger and thumb are around the base of the pistol grip.
- Draw the pistol straight up until the muzzle clears the holster and rotate the muzzle forward. The magazine well should face upward.
- Place the pistol between the thighs, applying enough tension to keep the pistol in place. The Marine must ensure that he does not disengage the safety while performing this motion and that the muzzle of the pistol is clear of the body. In the standing position, the Marine bends slightly at the knees to facilitate securing the pistol between the thighs.
- Remove the left hand from the pistol grip and rotate the hand around the pistol grip to facilitate establishing a firing grip.
- Grasp the pistol grip and remove the pistol from the thighs, rotating the pistol until the muzzle is pointed toward the target.
- Sweep the safety with the thumb of the left hand while starting to punch the pistol out toward the target.
- Establish a firing grip and continue to punch the pistol out toward the target. At the same time, place the trigger finger on the trigger and

establish sight alignment and sight picture within the aiming area.

- Continue trigger pressure until the shot is fired.

Transferring the Pistol From One Hand to the Other

Situations may arise in combat where the Marine must transfer the pistol to the left hand to engage a target because the right hand or arm is injured. To safely transfer the pistol from one hand to the other, perform the following steps:

Note: This procedure is written to transfer the pistol from the right hand to the left hand.

- Point the muzzle in the direction of the target and place the trigger finger straight along the receiver. See figure 8-20.

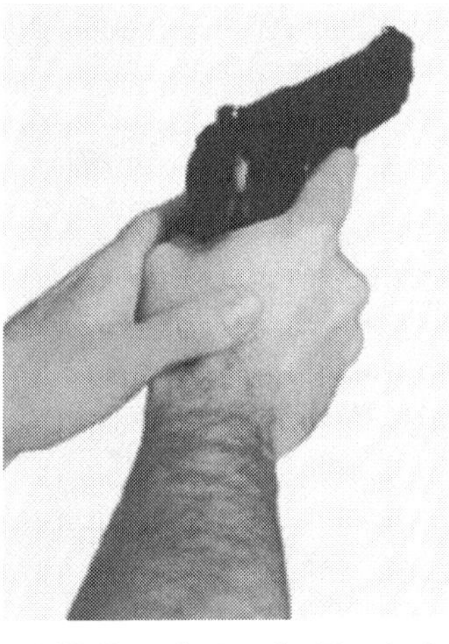

Figure 8-20. Transferring the Pistol—Step 1.

- Place the web of the left index finger and thumb under the backstrap of the pistol. See figure 8-21.
- Wrap the fingers of the left hand around the pistol grip while releasing the right hand grip.

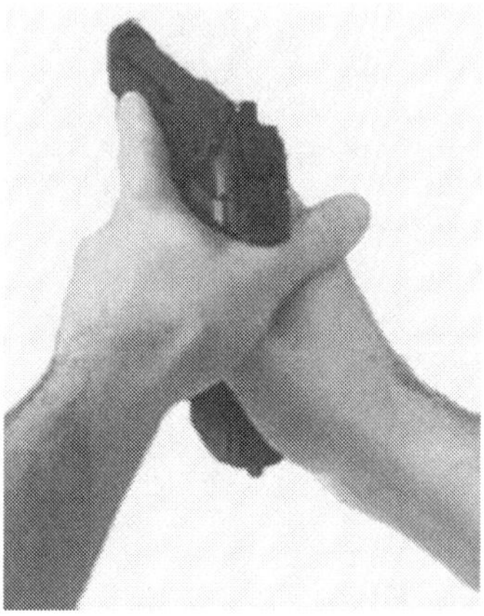

Figure 8-21. Transferring the Pistol—Step 2.

- Establish a firing grip with the left hand on the pistol. See figure 8-22.

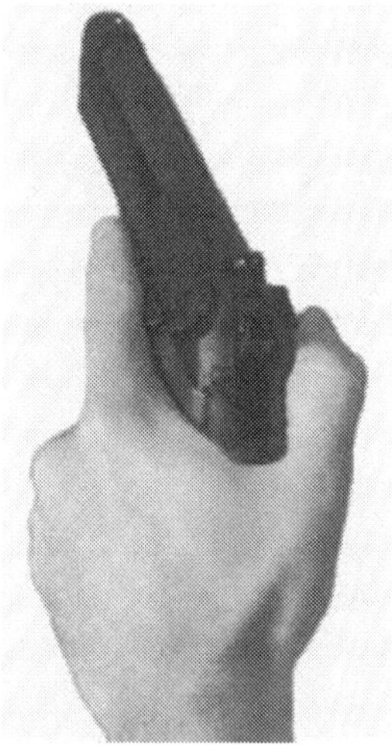

Figure 8-22. Transferring the Pistol—Step 3.

CHAPTER 9
ADVANCED TECHNIQUES

Some tactical situations are quite complex and require Marines to apply advanced pistol engagement techniques in order to effectively engage the enemy and survive. The ability of the Marine to rapidly shoot on the move and to turn and fire will be critical to a successful engagement.

Note: The procedures in this chapter are written for right-handed Marines; left-handed Marines should reverse directions as needed.

Shooting on the Move

When moving from one area of cover to another, the Marine may engage a target that presents an immediate threat. In this situation, the Marine relies on accurate fire to serve as his cover.

Moving With the Pistol

The likelihood of encountering a threat dictates how the pistol is carried while the Marine is moving. But regardless of the method of carry, the pistol's muzzle is pointed in the same direction that the head and eyes are looking—the eyes and muzzle move as one.

When carrying the pistol in the Alert position, the pistol is tucked in close to the body, finger off the trigger. This position allows the Marine freedom of movement and still allows for a quick presentation of the pistol.

If there is a high probability of encountering a threat, the Marine carries the pistol in the Ready position while moving.

Moving: The Glide Technique

To shoot accurately while moving, the Marine cannot run or walk fast. A normal running movement produces too much bounce in the Marine's body and makes it extremely difficult to use the sights of the pistol or to achieve the needed level of stability for accurate shooting. Therefore, the proper movement technique is similar to a glide and allows for accurate target engagement:

- A lower center of gravity is achieved by keeping the knees bent and the upper body erect.
- The bent knees also serve to absorb the shock generated by movement.
- The feet are not lifted as high as in a normal walk or run, allowing obstructions to be kicked out of the way.
- The feet and knees point in the direction of travel. If there is a need to engage a target off to the side, the Marine rotates at the waist to move the upper body in the desired direction. The upper body shooting position does not change.

The glide is not a technique that can be used for a long period of time or for long distances. Rather, the Marine can move in a normal manner until it becomes necessary to engage a target, he then slows and assumes the glide.

Engaging Targets: Using the Pistol Sights

When the Marine fires the pistol, he must concentrate on the pistol's front sight. The adherence to the fundamentals of marksmanship becomes even more vital when both the Marine and the enemy may be moving.

Continuing to Move

A Marine continues to move no matter what happens. Continuing to move makes it difficult for the threat to engage the Marine and can be one of the Marine's best defensive assets. When moving to cover or to a different position, the Marine should not stop to engage a target.

Reloading and Stoppages

If a stoppage occurs or a reload is required while moving, the Marine's primary consideration is to keep moving and seek cover. Ideally, reloads and clearing stoppages are performed behind cover.

Turn and Fire

The techniques for turn and fire allow a Marine to engage a target that is not directly in front of him. The key to turn and fire is smoothness and quickness of pivoting and presentation of the pistol to engage the threat. Turn and fire is based on the principle that wherever the head goes, the body follows.

Engaging Targets 90 Degrees to the Right and Left

The Marine turns the head toward the threat and identifies the target. Once the target is identified, the Marine maintains focus on the target for the rest of the presentation.

If engaging a target 90 degrees to the right, the Marine turns toward the target by raising the left foot while pivoting on the ball of the right foot. The Marine plants the left foot once he is facing the target squarely. The stance is natural. See figure 9-1. The Marine then engages the target.

Figure 9-1. Engaging Targets 90 Degrees to the Right.

If engaging a target 90 degrees to the left, the Marine turns toward the target by raising the right foot while pivoting on the ball of the left foot. The Marine plants the right foot once he is facing the target squarely. The stance is natural. See figure 9-2. The Marine then engages the target.

Engaging Targets 180 Degrees to the Rear

The Marine turns the head toward the target, looking over either the right or left shoulder, and identifies the target. Once the target is identified, the Marine maintains focus on the target for the rest of the presentation. See figures 9-3 on page 9-4.

If the Marine pivots to the right to engage the target, pick up the left foot, move it across the right foot, while pivoting 180 degrees on the ball of the right foot, plant the left foot, and squarely face the target in a natural stance. See figure 9-4 on page 9-4.

If the Marine pivots to the left to engage the target, pick up the right foot, move it across the left foot, while pivoting 180 degrees on the ball of the left foot, plant the left foot, and squarely face the target in a natural shooting stance. See figure 9-5 on page 9-4.

The Marine then engages the target.

Figure 9-2. Engaging Targets 90 Degrees to the Left.

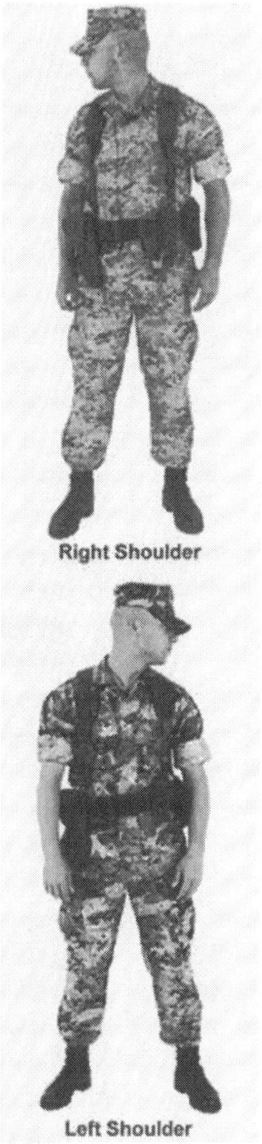

Figure 9-3. Engaging
Targets 180 Degrees to the Rear.

Figure 9-4. Engaging Targets
180 Degrees to the Rear (Right Shoulder).

Figure 9-5. Engaging Targets
180 Degrees to the Rear (Left Shoulder).

Chapter 10
Low Light and Darkness Techniques

An effective combat marksman must be prepared to detect and engage targets under a variety of conditions. Factors such as terrain and opportunity often dictate that the Marine engage combat targets at night or under low-light conditions.

The fundamentals of marksmanship are employed for engagement of targets in low light and darkness just as they are in daylight. However, the principles of night vision and target detection when engaging targets at night must be applied.

Combat Mindset

In the stress of the combat environment, Marines must eliminate any hesitation, fear, or uncertainty of action and focus on the actions to fire well-aimed shots. This is important during low light and darkness when attention is more easily diverted because the sense of vision is reduced. This may create a shock of awareness as the Marine relies more heavily on his other senses. Every noise, movement, and muzzle blast appears intensified at night. The physical acts of shooting must become second nature to the Marine, so his focus will not be diverted from firing well-aimed shots. This is accomplished through mental preparedness and training until shooting actions become instinctive.

Night Vision Adaptation and Maintenance

Night Vision Adaptation

There are two methods for acquiring night vision.

The first method is to remain in an area of darkness for about 30 minutes. This area can be indoors or outdoors. The major disadvantage of this approach is that a Marine is not able to perform any tasks while acquiring night vision in total darkness.

The second method is to remain in a darkened area under a low intensity red light (similar to the light in a photographer's darkroom) for about 20 minutes, followed by about 10 minutes in darkness without the red light. This method produces almost complete night vision adaptation while permitting the performance of some tasks during the adjustment period.

Night Vision Maintenance

Because the eyes take a long time to adjust to darkness, it is important to protect night vision once it is acquired. To maintain night vision, the Marine avoids looking at any bright light and shields the eyes from parachute flares, spotlights, or headlights. When using a flashlight to read a map or other material—

- Put one hand over the glass to limit the area illuminated and the intensity of the light.
- Keep one eye shut to reduce the amount of night vision lost.
- Cover the light with a red filter to help reduce the loss of night vision.

Target Detection Techniques

Once the Marine has acquired night vision, he is prepared to locate targets. Some of the daylight observation techniques, such as searching for target indicators, also apply at night or in low light. But night observation techniques must allow for the limitations of night vision and the need to protect it.

Off-Center Vision

Because of the placement of the cones in the center of the retina and the rods around the edges, the angle at which the Marine observes an object at night affects how well it is seen. Off-center vision is the technique of keeping the attention focused on an object without looking directly at it. To maximize the use of the eye's rods that provide night vision—

- Never look directly at the object being observed.
- Look slightly to the left, right, above, or below the object.

Each Marine must experiment and practice this technique to find the best off-center angle. Typically, the best off-center angle is 6 to 10 degrees away from the object or about a fist's width at arm's length. See figure 10-1.

Scanning/Figure Eight Scan

Scanning is the use of off-center vision to observe an area or object and involves slowly moving the eyes in a series of separate movements across the objective area. See figure 10-2. To execute a figure eight scan, move the eyes in a figure eight pattern in short, abrupt, irregular movements over and around the area. Once a target indicator has been detected, focus should be concentrated in that area, but not directly at it. It is more effective to scan from a prone position or a position closer to the ground than the object being observed, creating a silhouetted view of the object. When scanning an area, look and listen for the same target indicators as in daylight: movement, sound (which seems louder at night than during daylight), and improper camouflage. Also, objects in bright moonlight/starlight cast shadows just as in sunlight.

Effects of Illumination

Both ambient (natural) light and artificial illumination can affect a Marine's perception of the target's distance and size and his night vision. The following situations produce less of an effect at the close ranges that pistol engagements occur, but they still affect target detection and engagement:

- Light behind or between the Marine and the target illuminates the front of the target and makes it appear closer than it actually is.
- Light beyond the target displays the target in silhouette and makes it appear farther away than it actually is.
- The introduction of light requires the eyes to make a sudden, drastic adjustment to the amount of light received, causing a temporary blinding effect that occurs when night vision is interrupted abruptly. Ambient light can also have the

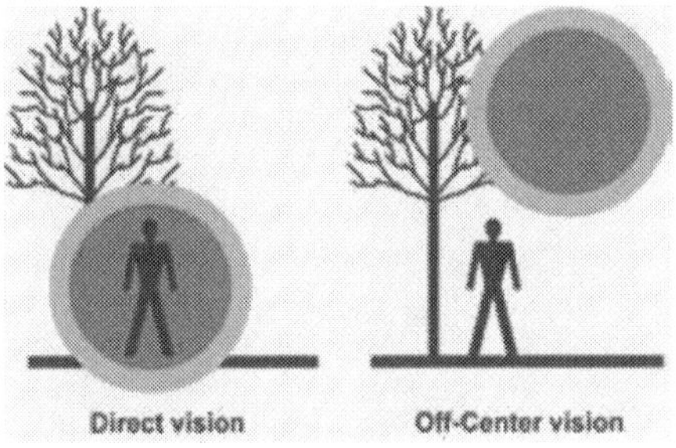

Figure 10-1. Off-Center Vision.

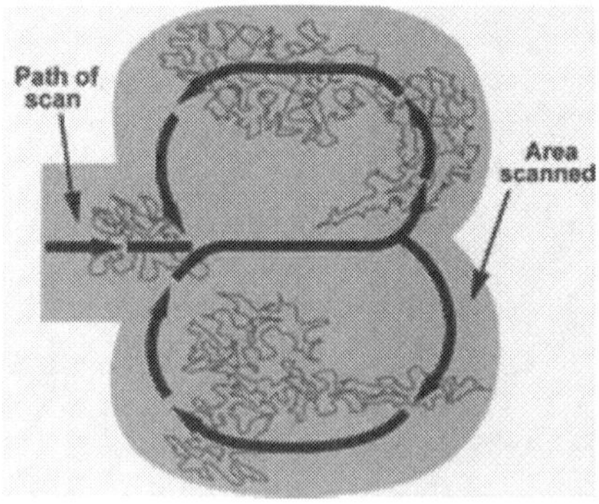

Figure 10-2. Figure Eight Scan.

same blinding effect; e.g., when a bright moon suddenly appears from behind clouds.

Acquiring Targets at Night

The Marine must keep both eyes open to get maximum visual coverage of the target area. Keeping both eyes open also improves depth perception and peripheral vision.

Sight Alignment/Sight Picture

Sight alignment and sight picture are obtained the same way in darkness as in daylight. There is normally enough ambient light to perceive objects as far away as 50 meters, especially if they are moving. However, when pistol sights are placed on a dark background, such as a camouflaged target, the Marine may not be able to acquire and align the sights clearly. Instead, the Marine may have to rely almost entirely on his presentation, firing position, and grip to get the pistol on target. To check for sight alignment or acquire the sights:

- Move the pistol just off the edge of the target to an area that provides a good contrast.
- Acquire sight alignment.
- Bring the sights back on line with the target while applying pressure to the trigger so the shot breaks once the sights are on target.

Flashlights

A flashlight can be used in low light and darkness to identify and illuminate targets. It is also used to acquire sight picture.

Types of Flashlights

There are a variety of flashlights used throughout the Marine Corps. They come in many different shapes and sizes; however, they are either straight/tubular or L-shaped.

Straight/Tubular

The straight/tubular flashlight is the most common. The body of the flashlight is a straight/tubular shape with the lens at the head of the flashlight. The on/off mechanism is located along the body or at the base of the flashlight.

L-Shaped

The L-shaped flashlight has a straight/tubular body, with the head/lens of the flashlight positioned at a 90-degree angle to the body of the flashlight. The on/off mechanism is located along the body or at the base of the flashlight.

Activation Devices

Depending on the type of flashlight, there are several ways that it can be activated.

On/Off Button

On/off buttons are located either along the body of the flashlight or at the base. On/off buttons are operated in a number of ways, depending on the flashlight.

On/Off Switch

On/off switches are mounted on the body of the flashlight. This switch has three settings that allow the flashlight to be either on, off, or in manual mode. In the manual mode, the flashlight is turned on by pressing and holding the button located next to the on/off switch.

Rotating Head

Rotating the head of the flashlight, either clockwise or counterclockwise, turns the flashlight on. Continuing this motion allows the width and intensity of the beam to be adjusted.

Types of Lenses

Clear Lenses

Clear lenses are used primarily for target detection and to illuminate the pistol sights. They are the most commonly used lens.

Colored Lenses

Colored lenses include red, amber, and blue. These lenses are used primarily for map reading and signaling, but can be used to illuminate the pistol's sights.

Target Detection

When a target is illuminated, the front sight may become silhouetted against the target, providing the sight picture needed to engage the target. During target detection, the Marine's focus should be twofold: to scan the area to identify possible targets and to assess the area to formulate a plan for engagement or cover (e.g., identifying the quickest route to cover, determining the best method for engagement based on terrain).

Searches

The Marine executes a hasty search immediately upon entering a new area in order to identify a threat that poses immediate danger. To execute a hasty search—

- Quickly scan the area with the flashlight taking note of obvious points throughout the area that could cover or conceal the enemy. Wherever the eyes move, the flashlight should move.
- Aim the flashlight beam on the ground about 8 to 10 feet in front of the Marine's location, this allows the eyes to follow the beam and quickly establishes a reference point for the light. Aiming the beam at the outer edge of the search area strains the eyes to find the beam and decreases the field of view.

The Marine executes a detailed search after the hasty search. A detailed search focuses on target indicators identified during the hasty search. To conduct the detailed search—

- Focus the most direct or intense portion of the flashlight beam on the target indicators.
- Move the beam of light slowly across the target indicator from right to left or left to right. Wherever the eyes move, the flashlight should move.

Considerations

During night operations, the Marine should use a flashlight whenever possible. A flashlight not only helps the Marine locate targets at night, but the beam can be aimed directly at the target once it is detected and temporarily blinds the target. This gives the Marine the advantage to react before the target does. However, the Marine must consider the following:

- Anytime a flashlight is on, the location may be revealed to the enemy. The Marine should keep the flashlight pointed out in front of his body to help avoid illuminating himself and revealing his position.
- Light shined directly from in front of the Marine at the target obscures the Marine.
- Light shined from the side of the Marine at the target illuminates the Marine.
- Light can bounce off surfaces and reflect back onto larger areas that the light is not focused on. Therefore, whenever a Marine shines a flashlight onto a surface, some of the light may reflect back and illuminate the Marine.
- When positioned at the outside corner of a room, building, or other cover, the Marine should avoid pointing the flashlight beam directly at the corner because the beam reflects off the corner and illuminates the Marine. To prevent this, the head of the flashlight should extend just beyond the corner.
- Many flashlights have adjustable beams that aid in target detection. The beam's intensity must be adjusted to provide the best illumination of the area depending on the distance between the Marine and the area of observation.
- Diffused light from a wide beam creates a softer light and illuminates a greater area, but the beam does not travel a great distance. A wide beam is best for observing larger areas at close range.
- Concentrated light from a narrow beam illuminates a smaller field of view, but the beam travels a greater distance. A concentrated beam is effective for observing a specific area or an area that is further away. Concentrated light

blinds the enemy and prevents him from focusing on the Marine or determining his location.

- When searching an area, the Marine can alter the position where he is holding his flashlight so that the beam is perceived as coming from various locations. For example, the Marine can turn the flashlight on from a standing position, quickly search for targets, turn the flashlight off, assume a kneeling position, and search again. This keeps the enemy from obtaining an exact location of the Marine's position.
- Keep both eyes open to get maximum visual coverage of the target area. Keeping both eyes open also improves depth perception and peripheral vision.

Target Engagement

Once a target is detected using a flashlight, the flashlight can be used to acquire sight picture and facilitate engagement of the target. The fundamentals of marksmanship are employed for engagement of targets in darkness just as they are in daylight. However, the Marine must apply the principles of night vision and target detection and must be able to employ a flashlight properly when engaging targets at night.

Grip

To engage a target accurately while using a flashlight, the flashlight must be held to maintain the pistol's stability, control, and recoil during firing. Normally, the left hand (two-handed grip) provides the pistol's stability and ability to manage recoil. However, when firing with a flashlight, the left hand holds and operates the flashlight. Therefore, some stability, control, and management of recoil is lost.

Securing the Flashlight

Most flashlights come equipped with a retaining loop located at the base for ease of transportation. To position the cord properly:

- Feed one end of the cord through the retaining loop of the flashlight and tie the ends of the cord together.
- Slip the left hand through the cord loop. The loop should be just big enough so that, if the flashlight needs to be dropped, it can be retained on the wrist.
- Rotate the flashlight until the cord is twisted to the desired tension and length necessary to best stabilize the rear end of the flashlight in the hand.
- Grasp the flashlight with the left hand in a position that allows it to be operated easily.

If the flashlight has a retaining loop, attaching a cord to the loop helps support and stabilize the flashlight when the Marine must fire the pistol with the flashlight in his left hand. The cord also helps retain the flashlight.

Acquiring Sight Alignment/Sight Picture

When holding a flashlight on a target, the tendency is to look at the target rather than the sights. Sight alignment is still necessary for effective target engagement. Using a flashlight to illuminate a target allows the Marine to—

- Acquire sight picture by silhouetting the sights against the target.
  - The Marine establishes sight picture by focusing the sights in the soft, diffused light area of the target.
  - At close ranges, colored lenses produce enough light on the target to silhouette the sights; at long ranges, colored lenses will not provide enough light off the target to illuminate the pistol sights. The Marine must be able to see the target by some other means, (e.g., contrast, moonlight).
- Acquire sight picture by illuminating the sights.
  - A colored lens can be held directly over the sights to shine light on the top of the pistol to illuminate the sights. This allows sight alignment to be established without revealing the Marine to the enemy. This method can also be used with a clear lens to make a precision shot at long ranges but the Marine risks being illuminated by the light.
  - A colored lens can be shined from directly behind the pistol sights to illuminate them. This

allows sight alignment to be established, but reveals the Marine's position to the enemy.

Target Engagement Techniques

Two-Handed Grip Technique. This technique is typically used with a straight/tubular flashlight. The pistol and flashlight must be side by side and level so the Marine can engage a target without making adjustments to the pistol or the flashlight. This technique is also effective for firing multiple shots because the flashlight and pistol recoil as a unit.

Whenever possible, the flashlight is just in front of the muzzle of the pistol so that it does not illuminate the pistol. Placement of the flashlight alongside the pistol may need to be adjusted depending on the size of the flashlight or the location of the on/off mechanism.

This technique is most often used with an Isosceles position (see fig. 10-3):

- Grasp the flashlight with the left hand, lens pointing down range.
- Wrap the thumb and index finger of the left hand around the body of the flashlight, thumb resting on the on/off switch to allow easy operation of the on/off switch without disrupting the grip on the flashlight.
- Extend both arms toward the target and bring the flashlight alongside the pistol so the fingers of the left and right hands touch.
- Wrap the bottom three fingers of the left hand around the fingers of the right hand, incorporating the flashlight into a two-handed grip.
- Apply isometric pressure against both hands to aid in stabilizing the pistol.

This technique can also be incorporated into a Weaver position; the more the left hand wraps around the right, the more the body can be angled and the left elbow bent. Apply push-pull pressure on the grip to stabilize the pistol. See figure 10-4.

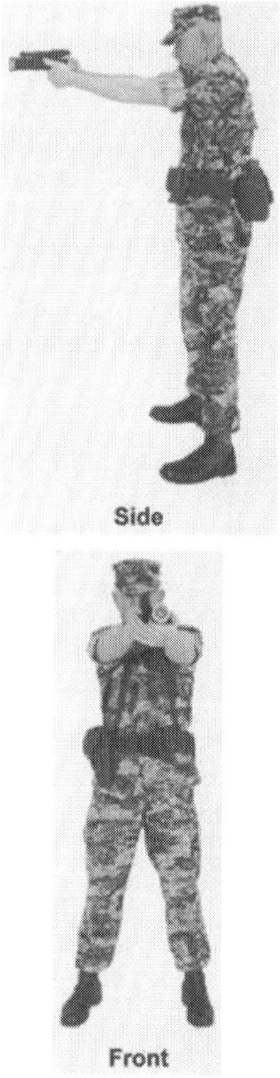

Figure 10-3. Two-Handed Grip Technique With Flashlight (Side and Front View).

Cross-Hand Technique. The cross-hand technique is used primarily with a Weaver position and can be used with either a straight/tubular flashlight or an L-shaped flashlight (see fig. 10-5):

- Grasp the flashlight with the left hand, fingers wrapped around the top of the flashlight and thumb wrapped around the bottom.

Figure 10-4. Two-Handed Grip Technique With Flashlight (Weaver Position).

Figure 10-5. Flashlight Cross-Hand Technique.

- Place the thumb on the on/off button while maintaining a firm grip on the flashlight.
- Extend both arms toward the target and bring the left hand under the pistol so that the back of the left hand is resting firmly against the back of the right hand.
- Apply isometric pressure against both hands to aid in stabilizing the pistol.
- Extended the right arm fully and bend the left arm at the elbow.

Considerations for the Carry/Transport

When searching an area for targets, the Marine moves with the pistol at the carry or transport dictated by the threat level.

The Ready. If enemy contact is expected (contact imminent), the Marine carries the pistol at the Ready and carries the flashlight in his left hand and incorporates it into his firing grip. In the Ready, the Marine can search for targets and readily present the pistol and flashlight to the target for engagement without making adjustments to either the grip or the flashlight. See figure 10-6 on page 10-8.

The Alert. If enemy contact is likely, the Marine carries the pistol at the Alert and the flashlight in his left hand and incorporates it into his firing grip. The Marine must ensure the flashlight is not pointed at the deck because it illuminates the Marine. If necessary, the head of the flashlight may have to be tilted up to elevate the beam to increase the field of view. If the flashlight has been tilted, it has to be lowered so it is level with the pistol muzzle and the light shines directly on the target in order to establish sight picture. See figure 10-7 on page 10-8.

Holster Transport. If there is no immediate threat, the pistol is holstered and the Marine carries the flashlight with his left hand to search for targets. Should a target present itself, the Marine presents the pistol from the holster and engages the target. Time and distance to the target dictate whether the Marine incorporates the flashlight and his left hand into the firing grip.

Figure 10-6. Flashlight Ready Carry (Cross-Hand and Two-Hand Grip).

Figure 10-7. Flashlight Alert Carry (Two-Handed Grip and Cross-Hand).